THE BLACK BOX
OF BIOLOGY

THE BLACK BOX OF BIOLOGY

A History of the Molecular Revolution

Michel Morange

Translated by
Matthew Cobb

Harvard University Press

Cambridge, Massachusetts, and London, England   2020

Library of Congress Cataloging-in-Publication Data

Names: Morange, Michel, author. | Cobb, Matthew, translator.
Title: The black box of biology : a history of the molecular revolution / Michel Morange, translated by Matthew Cobb.
Other titles: Histoire de la biologie moléculaire. English
Description: Cambridge, Massachusetts : Harvard University Press, 2020. | This volume is revised and expanded from the first English-language edition, published as A History of Molecular Biology by Harvard University Press. Originally published (in French) as Histoire de la biologie moléculaire—Title page verso. | Includes bibliographical references and index.
Identifiers: LCCN 2019040562 | ISBN 9780674281363 (hardcover)
Subjects: LCSH: Molecular biology—History.
Classification: LCC QH506 .M7313 2020 | DDC 572.8—dc23
LC record available at https://lccn.loc.gov/2019040562

Contents

Introduction

TWO DECADES AGO, when I published *A History of Molecular Biology,* the predecessor of this book, barely a day went by without the media highlighting another new development in biology—gene therapy, the human genome project, the creation of new varieties of animals and plants by genetic engineering, and even the possibility of cloning a human being. Naturally enough, the public was fascinated. Back then, everyone knew that these developments were the products of molecular biology, which had appeared in the middle of the twentieth century.

The endless cycle of new discoveries still continues, but the public is less fascinated and more anxious, and it is not always clear that these developments are the results of the rise of the new science in the middle of the twentieth century. All these advances—such as those enabled by the genome editing system CRISPR—reignite hopes and debates that first emerged decades ago, while all the progress in gene and cell therapies, and new treatments for cancer, are still the more or less direct consequences of the development of molecular biology.

Faced with the immense discoveries of the last two decades, and their promised—or threatened—future, I felt it was time to produce a new book. Although this is based on my original *History*—in particular the first three parts—this is much more than a new edition. The entirely new fourth section deals with developments that were only dreamed of in the late twentieth century, and I have taken the opportunity to extensively revise and update the analysis in the earlier sections.

Molecular biology is not the description of biology in terms of molecules—if this were the case, it would include not only biochemistry, but also all those nineteenth century studies in chemistry and physiology that led to the characterization of biological molecules. With such a broad definition, even Pasteur would have been a molecular biologist![1] Rather, molecular biology consists of those techniques and discoveries that make it possible to carry out molecular analyses of the most fundamental biological processes—those involved in the stability, survival, and reproduction of organisms. It is not only a level of observation and explanation of living phenomena, it also provides a method for intervening into those systems in order to understand and manipulate them. Although it is not the only possible level at which biological research should be carried out—sometimes it is not even the best one—it clearly represents an important approach to our understanding of the living world.

Molecular biology is a result of the meeting of genetics and biochemistry, two branches of biology that developed at the beginning of the twentieth century, each of which had clearly defined research objects: the gene for genetics, and proteins and enzymes for biochemistry. Molecular biology emerged when the relation between these two objects became clearer; scientists identified the gene as a macromolecule (DNA), determined its structure, and described its role in protein synthesis.

Strictly speaking, molecular biology is not a new discipline but rather a new way of looking at organisms as reservoirs and transmitters of information.[2] This new vision opened up possibilities of action and intervention that were first revealed during the early years of genetic engineering.

The techniques necessary for the study of macromolecules were developed between 1920 and 1940, and the new conceptual tools for analyzing biological phenomena were forged between 1940 and 1965. The consequent operational control was acquired between 1972 and 1980 with the development of genetic engineering. In the subsequent decades, molecular biology has completely transformed our understanding of biological phenomena. This book covers the molecular revolution in its entirety, from its earliest days right up to tomorrow.

Molecular biology and genetic engineering are too intimately linked for their histories to be separated; genetic engineering cannot be understood without molecular biology, but it was genetic engineering that highlighted the importance of the conceptual changes that molecular biology had produced. The discovery in 1983 of a technique for amplifying DNA, called the polymerase chain reaction (PCR), had its origins in the theoretical framework developed in the 1950s and in the experimental tools devised in the 1970s. Better than any other example, it shows the effectiveness of the theoretical and practical tools forged by biologists in the second half of the twentieth century.

Molecular biology was born when geneticists, no longer satisfied with an abstract view of the role of genes, focused their attention on the nature of genes and their mechanism of action. It was also a result of biochemists trying to understand how proteins and enzymes—essential agents of organic specificity—are synthesized and how genes intervene in this process.

The end of this history is more difficult to discern. At various points over recent decades, a number of biologists, historians, and philosophers have announced the death of molecular biology. In the last part of this book I briefly describe the key changes that have occurred in the last forty years, exploring how deeply they have affected the framework created by molecular biologists. My conclusion is that we still live within the molecular paradigm,[3] and that contemporary biologists use the same conceptual framework that was established more than fifty years ago. Despite the repeated announcements of its demise, molecular biology is still alive and kicking.

Parts 1, 2, and 3 contain very detailed descriptions of the experiments that underpinned the development and consolidation of molecular biology. That approach could not be repeated in Part 4 because each new development would have required a book in itself. The chapters in the final part therefore address this recent history in a more synthetic way. This also allows a more direct focus on the major issue that lies at the core of the whole book: whether the "molecularization" of so many biological subdisciplines has transformed molecular biology itself, up to the point where it has disappeared as a discrete subject and been replaced by the emergence of a new form of biology.

These final chapters also deal with the key participants in a rather different way. Although the construction of scientific knowledge is always the result of a collective effort, in the first three parts I tried to outline major individual contributions. This was not possible in the fourth part of the book—thousands of researchers would have deserved to be mentioned. The individuals who are named are the authors of articles or experiments that can be seen as landmarks of the transformation of all branches of biology that occurred over the last forty years.

A major problem in writing the history of the molecular revolution in the life sciences is the sheer mass of documentation available. Many of those who made this revolution have written their own accounts; there are also a large number of studies by scientists, historians, and philosophers.[4] Four books in particular, with very different approaches, have made a major contribution to the history of molecular biology. The British historian Robert Olby wrote a detailed account of the path that led to the discovery of the double helix structure of DNA.[5] Horace F. Judson interviewed more than one hundred of the most important participants in the history of molecular biology and reconstituted the technical and conceptual debates that surrounded its birth.[6] Lily E. Kay showed how the molecular vision was established, emphasizing the role of techniques and funding organizations.[7] More recently Matthew Cobb has reexplored the discovery of life's greatest secret and added new and original insights to a story that many thought was settled.[8]

The aim of this book is not to repeat what has already been done by these authors. Unlike their works, my history of molecular biology describes the development of genetic engineering (this term now sounds rather old-fashioned, but it has not been improved upon), a story that has been told only in a fragmented manner and has generally tended to focus on the development of biotechnology.[9] Exploring the history of genetic engineering makes it easier to appreciate the originality of the new view of life that constitutes molecular biology. On the other hand, Olby's, Judson's, and Kay's books have been complemented or criticized in a series of historical studies that try to explain how and why key discoveries took place. These articles have been published in specialist journals, but their insights have yet to be gathered together in a general history of molecular biology. Many of these ideas are included in this book, an important part of which deals with the question of the origin of molecular biology.

Studying contemporary science poses a number of problems for the historian. First, there are many sources—in addition to the traditional material (books, articles, laboratory notebooks), there are the scientists' own accounts of their discoveries, which can take the form of autobiographical books or articles, interviews, and oral accounts. This wealth of material does not always make for clear history. Although oral history can be vivid and fascinating, as shown by Judson's work, Dominique Pestre has pointed out that extreme care is needed in dealing with oral accounts, which are generally partial reconstructions rather than trustworthy historical documents.[10] This has been demonstrated by Frederic Holmes in his painstaking analysis of the Meselson-Stahl experiment on DNA replication.[11] In the case of molecular biology, the authors of autobiographical accounts have understandably tended to justify their own role, or the role of their discipline—most have been produced while their authors still occupied important positions and were playing major roles in the shaping of scientific policy. Consciously or unconsciously, such accounts are often marked more by strategic motivations than by a concern for historical truth.

Another difficulty in arriving at a historical analysis of recent scientific developments is that historians tend to interpret the past through the eyes of the present. This danger, intrinsic to any historical analysis, is even greater when the words and techniques are (or appear to be) the same as those used today. The historian runs the risk of projecting a contemporary mind-set onto past experiments. This problem is particularly significant when the history is written by a scientist who believes that a discipline's past is merely the beginning of its present.

Although I recognize that many of my readers will be actual or budding scientists, my aim has been to write a book that could be read by the general public. Too many articles and books require the reader to fully understand the object of the history before reading the first word. There is a bit of that here, I admit, but I have devoted a substantial amount of space to explaining discoveries and as far as possible to describing the techniques involved. Because general readers may find it difficult to understand some of the terminology, I have added an appendix that summarizes the key results of molecular biology.

I also wanted to write a history that is as complete as possible. Many previous accounts have emphasized a particular research school or approach. I have tried to provide a balanced presentation of the disciplines that contributed to the development of molecular biology, in particular by emphasizing the role of biochemistry.[12]

Finally, this book contains biographical sketches of some of the scientists who played major roles in the birth of molecular biology. This reflects the rich biographical and autobiographical material that is available and has an additional advantage: it enables me to step outside the framework of a purely internal history and to outline the role of external factors, such as the cultural context in which molecular biology was born. The occasionally chaotic path followed by the founders of the new science, shifting between disciplines or between countries, all contributed to the cultural mix that produced the new biology.[13]

That is a lot to ask of one book! Inevitably, I have not been able to devote as much space as I would have liked to some aspects, in particular in the fourth part which covers more recent developments. Each chapter is devoted to a different theme—a discovery, a research group, a line of research, or a particular historical question. Although this thematic presentation helps the reader by making each chapter independent, it has the disadvantage of not precisely following the chronological order of events, and it can also lead to some repetition.

It is normal practice in the history of science for the author to explain their strategy at the outset. Although I will not try to situate myself in the complex world of different academic schools and intellectual approaches, it is worth making a brief statement of aims, given that the history presented here raises the fundamental question of whether there has in fact been a molecular revolution, or instead a slow evolution of biology toward the study of biological phenomena at the molecular level. These two apparently contradictory interpretations—revolution or evolution—are in fact both possible and complementary.

The French historian Fernand Braudel has shown that history follows different rhythms and tempos that are superposed on each other.[14] With the partial exception of the work of Frederick Holmes,[15] it has rarely been applied to the history of science; however, the history of molecular biology becomes clearer when it is seen as a result of three parallel but different histories.

The longest time frame is that of reduction—the reduction of biology to physics and mechanics, which began in the seventeenth century, and the reduction of biology to chemistry, which began at the end of the eighteenth century. The history of this reductionist current is complex, with steps forward and backward, but it formed a powerful wave that, in the middle of the twentieth century, swept biology to the feet of the structural chemists.

The shorter histories of the various biological disciplines are superimposed on this long time frame. The key events here are the birth of biochemistry and, more significantly, the appearance of genetics at the beginning of the twentieth century. Seen from this point of view, molecular biology is the fruit of the convergence of these two disciplines and also the beginning of the reconciliation of heredity and development, of genetics and embryology. Microbiology, including bacteriology and virology, was the key place at which this meeting of genetics and biochemistry took place.

Finally, these slow transformations form the backdrop to the history of events, the history that involved experiments and theories. The events that influenced the birth of molecular biology do not belong only to the history of science. Other factors, such as the migration of many scientists to Great Britain and the United States before World War II, growing global communication needs, and a wartime focus on code-breaking, as well as the linked birth of computing, all helped give molecular biology its current form.

Viewing molecular biology as the result of three different histories frees us from the sterile counterposition of evolution and revolution: what seems to be revolution at one level may be revealed as evolution when history is seen on a longer scale.

Whereas the reductionist approach was intimately linked with the birth of modern science in the sixteenth and seventeenth centuries, the chemical form it later adopted was the product of development of chemistry, both fundamental and applied, in the nineteenth century. Studying the history of disciplines, which exists at the intermediate level, is the most appropriate way of analyzing the articulation between conceptual history and social history. For example, the history of genetics is the history of the discovery of genes, but it is also the history of the social recognition of the importance of heredity. Similar insights can be gained by using this approach to look at biochemistry and the other disciplines that were present at the birth of molecular biology, such as virology and bacteriology. Finally, event-based history can also be understood as the consequence of the confrontation between internal developments in biology (for example, the fact that Avery, who was not a geneticist, discovered that genes are made of DNA) and the external influences that led physicists to be the midwives of the new discipline.

Riding the crest of these three waves of history throughout the second half of the twentieth century and beyond, molecular biology has transformed the life sciences.

Even in a field like molecular biology, whose history has been so well studied, large areas remain barely explored or are even untouched. For example, whereas the history of the discovery of the double helix has been described over and over again, the discovery of DNA polymerase, the enzyme that duplicates DNA, is barely mentioned in even the most thorough historical studies. Like many other disciplines, the history of science cannot avoid intellectual fashion,[16] and some historians tend to focus on the most spectacular and highly publicized aspects of science. Refusing to take the usual path in studying the history of science—reading scientific publications and understanding the possibilities and limits of the techniques employed— they prefer to interview the "stars" and to pore over their correspondence. This attitude, frequently justified by long methodological explanations,[17] often goes against the proclaimed objective: it reinforces an intellectual and elitist vision of science, which is presented as being made by "great men" (and it usually is men). This danger is probably even greater today, given the dominant position recently acquired by biology over other scientific disciplines, and its central place in the media.

The fact that some historians are too distant from the material also explains their biased interpretation of David Bloor's principle of symmetry, according to which historians should give equal weight to success and failure.[18] Bloor argues that the same analytical framework should be employed to account for a theory that succeeded and for a theory that failed. This is an excellent principle, because the natural tendency of scientists is to consider that the best theory was necessarily the one that triumphed, the one that history has retained. But this approach has sometimes been used to relativize the objectivity of science, to suggest that all theories are equal and those theories that "lost" were eliminated for political or social reasons rather than because they were less accurate.[19]

This position does not take into account how science works and the fundamental role played by experimental constraints.[20] Scientists can conjure up all the models and theories they like, but they cannot force their experiments to work. Even a brief visit to a biology laboratory will show that biological material is difficult to work with, and that experiments that fail have an important place in everyday laboratory life. Scientific knowledge is indeed constructed, and scientists are free to define their strategy and to elaborate their models, but only within the narrow limits allowed to them by the experimental systems they use.

Above all, I have tried to be precise in historical terms: to describe, as faithfully as possible, the known and unknown aspects of the history of molecular biology. I have paid particular attention to what Thomas Kuhn described as "normal science." Whatever the value of the interpretations put forward here, this book and the historical information it contains will enable others to take us further in understanding the molecular revolution in biology and to acknowledge the importance it still has in research today and tomorrow.

I AM UNABLE TO acknowledge all those who contributed to this book—my colleagues and my students, in both biology and in the history and philosophy of biology. But I want to thank Janice Audet, executive editor at Harvard University Press, who supported me in what I initially thought of as a new edition of *A History of Molecular Biology* and then encouraged me to convert the text into a new book. I am also grateful to the three reviewers who made very helpful and positive comments. Above all, I want to acknowledge a journal and its editors, and a friend. The journal is the *Journal of Biosciences* published by the Indian Academy of Sciences. Some years ago, the then-editor, Vidyanand Nanjundiah, suggested I write a short historical contribution for every issue of the journal, in a series that he called "What History Tells Us." This was a challenge that turned out to be an extraordinary opportunity for me to explore many episodes of the history of molecular biology, acquiring insights that now form part of this new book. The friend is Matthew Cobb, and of course we remain friends despite the decision of his countrymen to vote for Brexit. He was the translator of *A History of Molecular Biology* more than two decades ago, and he agreed to thoroughly revise this version, translating the new sections with Barbara Mellor. He tells me that working on the original book encouraged him to turn his hand to writing history, including a recent book on the history of the genetic code. Through his insight and his style he has enriched this new book as much as he did my previous work, over twenty years ago.

PART ONE

THE BIRTH OF
MOLECULAR BIOLOGY

1

The Roots of the
New Science

AT THE BEGINNING of the twentieth century, a new science emerged as biochemistry replaced a series of research areas that were roughly grouped under the title physiological chemistry.[1] Unlike its predecessor, biochemistry provided medicine with scientific methods of diagnosis; as a fundamental science, it attempted to understand the ways in which molecules are transformed within organisms.

The first truly biochemical experiment took place in 1897, when Eduard Buchner, a German chemist, succeeded in obtaining sugar fermentation in vitro, using a cell-free yeast extract. This was particularly significant because forty years earlier Pasteur had argued that fermentation represented the "sign" or "signature" of life.[2]

Biochemistry developed in two directions. On the one hand, it studied the transformation of molecules (in particular, sugars) within organisms. On the other, it characterized proteins (including enzymes), which were not only the essential constituents of life but also the agents of the molecular transformations that interested biochemists.

The first half of the twentieth century was an important period for biochemistry, marked by the decoding of the major metabolic pathways and cycles—the glycolysis pathway, the urea cycle, the Krebs cycle, and so on—and by a large number of studies on cell respiration. At the same time, advances in physical chemistry led to the creation of systems that made it possible to study enzyme activity in vitro.[3] A quantitative scale of acidity (pH) was developed, as were "buffer solutions" that reproduced the properties of the intracellular medium. These developments laid the basis for the study of the fundamental principles of the kinetic activity of enzymes: enzymology. They also made it possible to stabilize the activity of enzymes and therefore to purify them. In turn, the crystallization of enzymes made it possible to study their structure.

In biochemistry, the first two decades of the twentieth century were dominated by what was known as colloid theory.[4] At the interface of chemistry and biology, this theory explored the existence of what was said to be a new state of matter—the colloidal state—with physical and chemical properties that were characteristic of life but which could nevertheless be studied by physics and chemistry. The boundaries of this theory were vague and ill-defined, and a number of studies marked by its approach are now considered classics in physical chemistry. Although colloid theory has long sunk into oblivion, it was extremely important; indeed, several Nobel Prizes were awarded for work on colloids.

We now know that this theory was completely wrong on a number of points. One of its key postulates was that colloids were formed when low-molecular-weight elementary molecules were grouped together. Supporters of colloid theory thought that large molecules could not exist and that there could only be aggregates of smaller molecules.

One important battle between supporters and opponents of colloid theory took place over the measurement of molecular mass, following the development of appropriate techniques. The isolation of crystallized proteins and enzymes, and the development of X-ray diffraction images of these crystals, showed that the components of organisms have well-defined structures, something that was incompatible with colloid theory.[5] As a result of these debates and discoveries, colloid theory was gradually replaced by macromolecular theory. The term macromolecule was introduced in 1922 by the German chemist Hermann Staudinger to describe high-mass molecules in which atoms are held together by strong bonds.

Another key concept in the history of biochemistry and the subsequent development of molecular biology is "specificity." Specificity initially referred to the ability of enzymes to recognize the chemical structure of the molecules on which they act (their substrate). This concept, which was thought to be characteristic of biological molecules, was omnipresent in the biology of the first half of the twentieth century but is much less common today. The first clear reference to this idea was made in 1890 by the German chemist Emil Fischer, who had undertaken an extensive study of proteins. To illustrate specificity, Fischer used the metaphor of a lock and key—the substrate interacts with its enzyme like a key with its lock.

The idea that all biological molecules have a chemical specificity found its most striking development in the study of the immune system, where immunologists soon made an analogy between the substrate–enzyme interaction and the interaction between antigens and antibodies.

After the insights provided by the German biologist Paul Ehrlich and his model of the immune response, the chemical study of antibodies—immunochemistry—developed in the first half of the twentieth century under the leadership of Karl Landsteiner.[6] Landsteiner, an Austrian immunologist who worked at the Rockefeller Institute in New York, injected animals with various molecules and studied how antibodies were formed against them. The results were impressive: whatever the chemical nature of the molecules, the organism was capable of producing antibodies against them, provided they were coupled with a carrier macromolecule.

Landsteiner made small molecular modifications to the injected substances and showed that the animal's body could recognize these variations and synthesize antibodies that reacted specifically with the new molecules, showing that specificity of recognition is an intrinsic property of the animal. Immunochemistry, a discipline in its own right, proved to be a tool for studying the composition of an organism—a subject that had previously been dominated by physiology and anatomy. If an animal is injected with proteins extracted from other organisms, antibodies directed specifically against a given protein from a given organism are produced. The specificity of immunological recognition can thus reveal the specificity of the components of an organism.

Between 1936 and 1940 the concept of specificity changed substantially, evolving from a biological concept into what is called a stereochemical one. This rapid development, which was essential for organic properties to be explained at the physicochemical level, was the result of Landsteiner's meeting with the American chemist Linus Pauling.[7] Landsteiner wanted to find a chemical explanation for the specificity shown by the antibodies that were produced against the molecules he injected. For Pauling—who wanted to study biological molecules—Landsteiner's results were excellent material for characterizing the chemical bonds responsible for the interaction between antigens and antibodies.

Pauling was already well known for having applied quantum mechanics to the study of molecules.[8] The theoretical work of the Austrian physicist Erwin Schrödinger, together with Walter Heitler and Fritz London's use of quantum mechanics to understand the hydrogen molecule, had shown that the formation of chemical bonds could be explained by quantum mechanics and could be predicted from the structure of the atoms involved in the bond. However, the calculations involved were so difficult that the new quantum theory could not be applied to complex molecules. Pauling simplified these calculations and showed, using many examples, that quantum mechanics could explain the existence and characteristics of the chemical bond. He subsequently extended this work to a number of other molecules where the calculations gave results for bond length and strength that did not match the observed values. Pauling suggested that this was due to the molecules "resonating" between several different structures. The observed difference was the direct translation of this equilibrium between two forms, the result of the resonant energy that was thus created. Pauling used this insight to reinterpret previous experimental results, in particular those from crystallography. His semi-empirical approach, which continually shifted between structural studies and simple theoretical rules derived from the principles of quantum mechanics, gave chemistry a new form. Pauling's personality, his charisma, and his gift for teaching also played an important part in this transformation.

This approach had previously allowed Pauling to distinguish strong covalent bonds from weak bonds. Strong bonds are classically called chemical bonds—they arise when two atoms share electrons. Weak bonds—for instance, hydrogen or ionic bonds—are produced by the partial sharing of electrons; despite their name, they play an important role in biology.

Landsteiner's data were a striking demonstration of the importance of weak bonds in molecular interactions. Pauling explained the specificity of the antibody–antigen interaction by the formation of a certain number of weak bonds and in particular by hydrogen bonds between the antigen and the antibody. Weak bonds are formed only between atoms situated close to each other; the existence of a large number of weak bonds showed that antigens and antibodies have complementary structures and fit into one another, thus confirming the model proposed half a century earlier by

Emil Fischer. With Pauling's work, the concept of specificity acquired real chemical credentials and became understood in terms of stereospecificity, structural complementarity, and collections of weak bonds.[9]

Applying the concept of weak bonds to protein structure, Pauling studied the denaturation of proteins by heat, acids, bases, or chemicals such as urea, arriving at a correct interpretation of this process in 1936.[10] Denaturation does not involve the breaking of a covalent bond in the molecule nor the separation of a colloidal aggregate, but instead is based on the loss of hydrogen bonds that stabilize the protein's three-dimensional structure. Although correct, Pauling's answer was incomplete because it omitted other weak bonds, in particular the interactions between apolar amino acids responsible for the formation of a hydrophobic core within proteins.

Pauling thus played an essential role in showing that specificity could be understood in terms of ("reduced to") physics and chemistry. The notions of weak bonds and structural complementarity are still the basis of our understanding of macromolecular interactions, and these principles govern the structure and functioning of all forms of life.

THE HISTORY OF genetics is just as compelling. In 1866 the Moravian monk Gregor Mendel first formulated the laws of the segregation of characters present in a hybrid, now generally wrongly known as his "laws of heredity," which in 1900 were rediscovered by Hugo de Vries, Erich von Tschermak, and Carl Correns.[11] Between these two dates, August Weismann and other biologists developed complex theoretical models of inheritance. After the rediscovery of Mendel's laws in 1900, it was several years before they assumed the form in which they are known today. For example, it was only in 1909 that Wilhelm Johannsen introduced the fundamental distinction between genotype and phenotype. *Genotype* refers to the factors that are transmitted down the generations, one of which in many organisms is contributed by the father, the other by the mother. These factors are called *genes. Phenotype* refers to the totality of adult characters; many characters have several forms that can be easily distinguished.

The expansion of genetics was closely linked to the choice of the vinegar fly *(Drosophila)* as a model organism. The American biologist Thomas H. Morgan and his collaborators at Columbia University chose *Drosophila* because of its rapid reproduction rate, which made studying changes in its characters particularly easy.[12] In 1910 Morgan discovered that some genes were transmitted differently depending on the sex of the fly that carried them, in the same manner as the sex chromosomes. At a stroke, this confirmed both the role of the sex chromosomes in sex determination and the suggestions made a few years earlier by both Theodor Boveri and Walter S. Sutton that genes are carried on chromosomes.

The recombination of forms of genes (alleles) carried by a pair of chromosomes was explained by the exchange of chromosome fragments that takes place during the formation of germ line cells. The closer together the genes, the rarer the recombination. Recombination allowed Morgan's group to order genes on the chromosome and thus to establish chromosome maps.

The choice of *Drosophila* proved to be particularly favorable. The salivary glands of the fruit fly contain giant chromosomes that can easily be seen under the microscope, where they appear as a series of alternating light and dark bands. By the 1930s, changes observed in these bands when genes were altered were used to draw up a physical map of the genome, which could be superimposed on the genetic map established on the basis of recombination frequencies. This major discovery strongly suggested that genes were physical entities and that they were localized on chromosomes.

Geneticists also described mutations and chromosome alterations and analyzed their effects on morphological and physiological characters, and they showed that some types of physical or chemical treatments (such as X-rays) could increase the mutation rate. The genetic analysis of *Drosophila* became a model for the genetic analysis of other organisms and confirmed that genetics provided a new and accurate view of how reproduction works. Above all, it helped to establish genetics as a science.

THE ESSENTIAL ROLE OF genetics in the birth of molecular biology can be explained as follows. Genetics became a leading biological science because of its powerful and elegant discoveries and its influence on agronomy. In both the United States and Great Britain, farmers and seed merchants contributed to the development of research and the creation of genetics institutes, aided by public funding.[13] Genetics developed much faster in the United States than in France or Germany; the flexibility of American universities, unlike the French or German systems, enabled them to respond to the challenge of the new science.[14] Genetics rapidly developed into a separate discipline that had little contact with other branches of biology. Far from being a handicap, this institutionalized isolation helped the new science, particularly in the United States. In the countries where genetics was not recognized as an independent discipline (for example, in Germany), genetic research was carried out in departments of physiology or biology and did not develop to the same extent.[15] Finally, genetics also appeared to be closer to physics than any other biological discipline. In particular, it used mathematics. Genes were considered as the atoms of biology, and their mutations were comparable to the transmutation of elements that was begun by physicists at around the same time (as will be discussed later).

To understand why the isolation of genetics was necessary for its development as a discipline, we have to go back to Mendel's discovery of the laws of heredity. Mendel's results had to wait forty years before they were understood and accepted. During this period, heredity became separated from embryology, a process that was essential for the birth of genetics. The study of sexual reproduction explores how two organisms give rise to a third, similar, organism. It therefore involves two intimately linked topics: the development of the organism from egg to adult, and the similarities between the organism and its parents. The success of genetics shows that the separation of these topics—the separation of heredity and embryology—was necessary, at least for a time.[16]

Because of its intellectual isolation, many biologists considered genetics to be an abstract science that was somewhat removed from reality. Geneticists studied genes without apparently being interested in either the way they worked or their chemical nature. Hermann Muller was the only one of Morgan's students to show an immediate interest in these questions. For Muller, genes were the "basis of life," the place where the "secrets" of life could be found. In 1927 Muller discovered the mutagenic effect of X-rays; just as Ernest Rutherford's 1919 transmutation of chemical elements had opened the road to an understanding of the atomic nucleus, so Muller was convinced that his experiment would help unravel the mysteries of genes.[17] Muller was particularly interested in the work of the German physicist-turned-geneticist Max Delbrück, who was trying to define the properties of genes by studying variations in mutation rate as a function of radiation energy (see Chapter 3).

Nevertheless, there are two reasons that geneticists took only a limited interest in such research; one was experimental, the other theoretical. In the 1930s the direct chemical study of chromosomes had shown that they were composed of nucleic acid (deoxyribonucleic acid, or DNA) and protein. Although biologists rapidly accepted this finding, it did not imply that the two components were equally important.[18] Genes were thought to be composed of proteins, while DNA was either a material support or an energy reserve. The chemical nature of genes was not considered important for the simple reason that most geneticists thought the problem had at least partly been solved.

The second reason geneticists were not immediately interested in studying how genes worked is that they already had an idea of the role of genes in the cell. In a 1945 lecture Muller described genes as "guides . . . relatively invariable . . . that serve as a frame of reference." Taken together, the genes in an organism were thought to form "a relatively stable controlling structure to which the rest is attached."[19] Geneticists may have considered that genes were the basis of all organisms, but this did not mean that they thought that genes were chemically distinct from other components of the organism.[20] This can be seen in the erroneous model of the structure of chromosomes proposed in 1934 by the Soviet geneticist Nikolai Koltsov, according to which chromosomes were long chains of amino acids linked to different molecules, such as hormones, that were active within the organism.[21]

Geneticists did, however, believe that genes had a fundamental and unique property: self-replication. This capacity was inherently linked to the nature of genes but could be studied independently of their functioning. Geneticists therefore placed much more emphasis on the self-replicating function of the gene than on its action in the cell. Self-replication was particularly intriguing and gave rise to many models—some scientists related it to the pairing of identical chromosomes that could be seen under the microscope, arguing that like attracted like in both cases.[22]

The German theoretical physicist Pascual Jordan proposed a model of gene replication based on the principles of quantum mechanics (of which he was one of the founders), according to which genes replicated by attracting their constituent elements from the surrounding medium, the interaction between two identical elements being favored by the existence of a long-range resonant energy.[23] In 1940, Pauling and Delbrück refuted Jordan's model by showing that it was incompatible with the results of quantum mechanics. They suggested that the force of attraction between identical molecules stemmed from the existence of a structural complementarity at the submolecular level.[24]

Pauling and Delbrück's paper may seem to anticipate the complementary double helix structure of DNA and its replication, but such an interpretation would be profoundly anachronistic and would give the paper more weight than even its authors attributed to it. Pauling himself did not refer to the paper when he tried to determine the structure of DNA in the 1950s. This paradox indicates that the theoretical study of the self-replicating properties of genes was not a functional research program at this time.

Although genetics was separated from other fields of biology such as embryology or biochemistry, its relations with evolutionary biology were complex. Genetic models were initially received with little enthusiasm; many evolutionary biologists (both Darwinians and neo-Lamarckians) had learned to reason in terms of gradual evolution and continuous variations that were accumulated over generations. The mutations discovered by the geneticists produced far greater effects than the microvariations that, ever since Darwin, many evolutionary theorists thought were the basis of evolution. Furthermore, there were close links between the Darwinians and the biometricians, who measured the variability of hereditary characters and their transmission from generation to generation. Although the biometricians apparently studied the same objects as the geneticists, they had a completely different theoretical approach. The traits they studied (such as height or weight) are now known to be polygenic characters (produced by the action of a number of genes), which makes it particularly difficult to study how they are transmitted.

The reconciliation of genetics and Darwin's theory of evolution was a slow process, in which the mathematical geneticists Ronald A. Fisher, J. B. S. Haldane, and Sewall Wright all played an important part. Population genetics emerged in the 1930s as a result of their debates and research, and was followed by what is called the evolutionary synthesis, which took place under the influence of Theodosius Dobzhansky, George Simpson, Julian Huxley, and Ernst Mayr.[25] The Modern Synthesis, as it was termed by Julian Huxley, explained evolution by a combination of two phenomena: the existence of spontaneous small-scale genetic variations (mutations) and the selection of individuals with a higher rate of reproduction.

Reinforced by the prestige of genetics, the new interpretation of evolution prevailed.[26] Like genetics, however, the Modern Synthesis faced a number of major problems. One of its key postulates was that variations (mutations) were small in scale and could have either positive or negative effects. This was problematic given that nothing was known about the nature of mutations, genes, or gene products. From this point of view, it could be argued that the evolutionary synthesis was just as abstract and divorced from the real world as the formal genetics from which it emerged.

2

The One Gene–One Enzyme Hypothesis

ACCORDING TO MANY ACCOUNTS, in 1941 George Beadle and Ed Tatum showed that genes control enzyme biosynthesis and that there is a different gene for each enzyme.[1] This is often considered to be the first step toward the unification of biochemistry and genetics and also the first major discovery of molecular biology. To use the term coined by Joshua Lederberg and Harriet Zuckerman, Beadle and Tatum's discovery was postmature; given the spectacular developments in biochemistry and genetics that had taken place at the beginning of the century, it is surprising that the discovery was not made earlier.[2]

A number of studies, dating back to the origins of genetics, had suggested that there was a connection between genes and the chemical reactions that take place within organisms.[3] The most easily detectable genetic differences are color differences produced by the presence or absence of pigments. By the end of the nineteenth century, many pigments had been at least partly characterized chemically.

The first precise relation between genes and metabolism was described in 1902 by Archibald Garrod, a physician at Saint Bartholomew's Hospital in London.[4] In 1898, one of Garrod's patients was a young boy suffering from alkaptonuria, a condition in which the patient's urine darkens on contact with air. This disorder was well known, and the chemical responsible for the dark color had been characterized in 1859. It had subsequently been shown to be produced by the transformation of an amino acid—tyrosine—that is present in food.

In 1901, a fifth child suffering from alkaptonuria was born into the same family. By studying his patient and the new sibling, Garrod was able to show that the symptoms were due to a metabolic disorder, the chemical equivalent of an anatomical deformation. Garrod also noted that the children's parents were first cousins. On studying other examples of children with alkaptonuria, he discovered that in three out of four cases, the children were born to parents who were first cousins. Aware of the work of the English geneticist William Bateson, Garrod concluded that the problem was due to a rare Mendelian factor: "the mating of first cousins gives exactly the conditions most likely to enable a rare, and usually recessive character, to show itself."[5]

In his book *The Inborn Errors of Metabolism* (1909), Garrod described the exact chemical nature of the disorder found in patients with alkaptonuria—in such patients one of the first stages of the metabolism of tyrosine, the splitting of the benzene ring, does not occur. Garrod concluded, "We may further conceive that the splitting of the benzene ring in normal metabolism is the work of a special enzyme, [and] that in congenital alkaptonuria this enzyme is wanting."[6]

Garrod's work was favorably received by British geneticists and biologists such as J. B. S. Haldane, but it could not lead to a viable research program because the experimental material—human beings—could not be studied using genetics or biochemistry. Furthermore, in 1909 genetics was still in its infancy, and population genetics did not yet exist. Understanding the metabolic pathways involved in the disease also proved problematic; it was not until around 1950 that they were finally unraveled through the use of chromatography and radioactive markers. Garrod worked in physiological chemistry, a field that aimed to characterize the chemicals present in organisms. This discipline gradually lost ground to biochemistry, which was concerned with fundamental metabolic reactions. It could also be argued that there was little medical pressure to understand this phenomenon. After all, alkaptonuria was not a disease, and the patients lived normally without any ill effects apart from the color of their urine.

Three other sets of data, collected between the 1920s and the 1940s, also suggested a relation between genes and enzymes: the study of plant pigments, or anthocyanins (carried out first by Muriel Wheldale, then by Rose Scott-Moncrieff); Alfred Kühn's work on eye color in the butterfly *Ephestia kuhniella* and the work of Boris Ephrussi and George Beadle on eye color in *Drosophila,* first performed in Thomas Hunt Morgan's laboratory at Caltech and later at the Institut de Biologie Physico-Chimique in Paris.[7]

All these studies met the same obstacles that Garrod had encountered—the complexity of the metabolic pathways and the difficulty of characterizing the substances that were involved. The choice of the subject that was being studied turned out to be fundamental in determining whether these research projects were successful. Beadle and Tatum's merit was that they studied the genetic control of essential metabolic pathways rather than peripheral ones—pathways that were well known or at least easy to study experimentally. However, the separation of biochemistry and genetics had other causes that flowed directly from the research carried out by the dominant current in genetics—Morgan's school.

Morgan and his students collected a remarkable amount of data on *Drosophila.* They localized genes to specific positions on chromosomes and studied the frequency of different forms (alleles) of genes.[8] Such a concentration of effort is not uncommon in science. When a technology—in this case gene mapping—is highly productive and provides more and more results, it is always difficult for scientists to abandon it, even if the novelty value of the results soon declines. For example, during the Renaissance wonderful anatomical descriptions were published, but despite their beauty and their accuracy, there was no immediate concomitant breakthrough in our understanding of physiology or medicine. Most geneticists were simply not interested in two problems that now seem to be both obvious and essential: the nature of genes and their mechanism of action. There were some exceptions—in the 1940s, attempts to solve these questions led to the formation of an informal research network (the "phage group") that was to be particularly important in the history of molecular biology (see Chapter 4). One of Morgan's students, Jack Schultz, went to work with the Swedish physical chemist Torbjörn Caspersson (see Chapter 13). Richard Goldschmidt, a German geneticist and a sharp critic of formal genetics, laughed at those who drew up chromosome maps based on the frequency with which characters were transmitted, but deliberately ignored how genes control characters.[9]

Apart from a few rebels such as Conrad H. Waddington and Goldschmidt, the geneticists were equally uninterested in the role of genes in development.[10] This was even more surprising given that several of the founders of genetics, including Morgan, had initially been embryologists.[11] Stranger still, Morgan's late "conversion" to genetics was linked to the discovery of the role of chromosomes in a key aspect of development—sex determination.[12]

Some historians have interpreted the geneticists' apparent lack of interest in the very biological problems that had initially oriented them toward genetics as the result of a strategy that was both cognitive and institutional.[13] As noted earlier, genetics was able to develop only on condition that it separated, albeit arbitrarily, heredity and gene transmission from development and gene action. From an institutional point of view, this enabled the geneticists to acquire a certain degree of autonomy, becoming increasingly isolated from embryologists and other biologists.

Other reasons, less political and more scientific, led Morgan to turn to genetic research. He was convinced that in the future scientists would indeed investigate the mechanism of gene action and the role of genes in development. However, even as late as the 1930s he still felt that the study of such links was premature because of the complexity of the relation between gene and phenotype.[14] Given the problems Garrod encountered in interpreting the chemical defect associated with alkaptonuria, and the partial failure of genetic studies of anthocyanins and insect eye pigments, Morgan's caution seems justified. In retrospect, it seems even more legitimate, given that molecular biology began to make inroads in the field of developmental genetics only in the 1980s.

As described earlier, George Beadle had worked with Boris Ephrussi on *Drosophila* eye pigments.[15] This experience convinced Beadle that this system would not reveal the chemical nature of the substances involved in gene action. He gradually realized that the problem had to be looked at the other way around, by starting off from well-known metabolic compounds and studying the genetic control of their production or function.

This required an organism that could be studied genetically and that also presented the possibility of simple biochemical investigation. Beadle chose a fungus, *Neurospora,* that had recently been analyzed genetically by Carl C. Lindegren in Morgan's group. Beadle's question was straightforward. In order to grow, *Neurospora* did not need vitamins (with the exception of biotin) because it could generate these compounds itself. Beadle decided to isolate mutants that had lost this ability and therefore needed to be supplied with a given vitamin.

Neurospora has a complicated reproductive cycle: the organism has a haploid form (that is, it contains only one set of chromosomes instead of two) in which each gene is present in only one copy. Thus, any modification of the gene is immediately expressed because it is not masked by the other copy. Two haploids can be crossed to get a diploid form, and the haploid products of this cross can in turn be isolated. While it might seem complicated, this reproductive cycle made it easy and quick to isolate mutations.

Beadle and Tatum irradiated *Neurospora* haploid spores in order to produce a large number of mutations. After crossing these spores, they added them to either a normal medium or to one containing vitamin B1 and vitamin B6. This procedure enabled Beadle and Tatum to isolate several dozen mutant strains that required either vitamin to survive. By crossing the strains, they showed that each nutritionally deficient strain was the result of a mutation in a single gene. They then found mutant strains that could not synthesize a given amino acid—tryptophan—and therefore required this substance in the medium. They characterized several mutants and genes that enabled them to describe the tryptophan biosynthetic pathway and to show that each step was controlled by a different gene.

Beadle and Tatum's results were enthusiastically received. In 1944 Beadle was elected to the National Academy of Sciences, and in 1958, together with Tatum and Lederberg, he was awarded the Nobel Prize.

The success of Beadle and Tatum was based on a marked shift in focus with regard to the link between genes and phenotype. When Beadle began his work with Ephrussi years earlier, they were not trying to understand the relation between genes and enzymes (proteins). Instead, they wanted to characterize the role of genes in the formation of the *Drosophila* eye during embryogenesis, and in particular the development of its color. Beadle's new approach represented a smart reorientation of efforts that had previously been fruitless. As it turned out, a year before Beadle and Tatum's work was published, the substance that was essential for *Drosophila* eye pigment formation, which was absent in certain mutants and the synthesis of which was therefore controlled by the action of genes, had been identified by Adolf Butenandt in Germany in 1940. Had Beadle stuck with *Drosophila,* another approach to revealing the relation between genes and enzymes might have been possible.

Historians have highlighted a number of anomalies in the traditional, textbook version of Beadle and Tatum's discovery. For example, the first article describing their work on the metabolism of *Neurospora* did not in fact proclaim the establishment of the famous relation between genes and enzymes (indeed, in its canonical "one gene–one enzyme" form, it was not said by them at all). What was initially important for Beadle and Tatum was that they had designed a new experimental system that made it possible to explore the link between genes and the products of gene action in a relatively simple way.

Perhaps the most surprising aspect of Beadle and Tatum's articles from 1935–1940, written in the run-up to their breakthrough, was the recurrent use of the word "hormones" to describe the substances produced by gene action. (This term also occurred in the articles by Boris Ephrussi from the same period.) It is extremely unlikely that they were not aware of the precise meaning of this term—at the time hormones were being purified and studied by biologists, work that was rewarded by successive Nobel Prizes.

The inevitable conclusion is that Beadle and Tatum expected to find that hormones would be both the products of gene action, and the agents of this action. This was not far-fetched—the role of hormones (thyroid hormone, steroid hormones, growth hormone) in the development of animals and in the differentiation of their tissues was well known. The same was true for plants—the role of growth hormones was intensely studied at the time. But this link between genes and hormones was tacit, made explicit only by some researchers such as Richard Kuhn, who carried out similar work to that of Beadle, Tatum, and Ephrussi on the flour moth *Ephestia*. Key to Beadle and Tatum's success was their decision to abandon a model of gene action that was limited to embryology, replacing it by a general model of gene action that was valid for all organisms. In other words, they adopted an approach that was more compatible with the future spirit of molecular biology.[16]

The historian Lily E. Kay uncovered a hidden—or at least forgotten—applied aspect of Beadle's work that allowed Beadle and Tatum to pursue their research during World War II.[17] Beadle's research had two potential consequences: it could lead to a better understanding of metabolism and perhaps to the discovery of essential metabolic factors, such as vitamins or amino acids. But above all, Beadle and Tatum's mutants provided an indirect but elegant method of measuring the nutrients present in different types of food—the growth of mutant strains of *Neurospora* depended on the supply of amino acids or vitamins. Beadle and Tatum were well aware that their research could have significant applications. Tatum had trained at the University of Wisconsin at Madison, where the Department of Biochemistry had close ties with the dairy, agricultural, and food industries. These links had partly been established by Tatum's father, Arthur Tatum, who was a well-known university biochemist.

Furthermore, Beadle and Tatum received grants from various pharmaceutical and agricultural companies and from the Nutrition Foundation. Alongside this funding stream, Beadle continued to receive major financial support from the Rockefeller Foundation, which had backed his project from the outset and had been behind Tatum's recruitment. Most important, the possible applications enabled Beadle to keep his students doing research at a time when the war effort channeled research money to war-related topics and conscripted young scientists into the army.

It is not easy to measure the impact of Beadle and Tatum's data on the development of molecular biology, although it was clearly well received. In 1945, Beadle gave the prestigious Harvey Lecture in New York, in which he stated that his work had led to the unification of biochemistry and genetics, which had previously been separated by "human limitations and the inflexible organization of our institutions of higher learning."[18]

The genetic approach developed by Beadle turned out to be a remarkable tool for the study of metabolic pathways. In the space of a few years, biochemists were able to establish the complex map of organic metabolic pathways, using radioactively labeled molecules and the bacterium *Escherichia coli (E. coli)*.

Beadle and Tatum's work led to the experimental association of biochemistry and genetics; but despite what the textbooks tell us, it is not at all clear that their work truly clarified the link between genes and enzymes. All of Beadle and Tatum's results, and those obtained by other groups using the same approach, showed that each elementary chemical step in a metabolic pathway is controlled by a single gene. Because each step is under the control of an enzyme, the one gene–one enzyme hypothesis followed logically. But a retrospective, anachronistic reading of this hypothesis is potentially misleading. Although we now know that the structure of each protein and thus of each enzyme is "coded" by a separate gene, the notion of coding, of the genetic determination of the detailed structure of proteins and enzymes, was completely absent from Beadle's thinking for the simple reason that no one had yet thought of it.

Two points will help to scale down the importance of this discovery, or at least to place it in its proper scientific context. The fact that a single gene seemed to control each metabolic enzyme did not mean that one gene was sufficient for the synthesis of that enzyme. Enzymes are proteins; because proteins are formed of many amino acids, it was thought that for a given enzyme to be synthesized, several enzymes and thus several genes would be necessary. The gene revealed by Beadle and Tatum's experiments was the one that conferred the final structure, form, and specificity of the newly synthesized enzyme. Max Delbrück, the ever-skeptical leader of the informal phage group and one of the acknowledged founders of molecular biology, doubted the significance of Beadle and Tatum's results.[19] Like most biochemists, the physicist Delbrück thought that several genes controlled the synthesis of a protein or enzyme. From this standpoint, some of these genes, including those discovered by Beadle and Tatum, were specific to a single enzyme; others controlled the synthesis of several enzymes and therefore had not been selected in Beadle and Tatum's procedure. This implied that the one gene–one enzyme hypothesis was the consequence of an experimental artifact.

On the other hand, Beadle and Tatum's results seemed to bring genes and enzymes closer, even to the extent of identifying them with each other in a way that now seems strange. Many biologists had an enzymatic view of genes, which went as follows: the astonishing catalytic properties of enzymes were responsible for organic specificity, genes control both the organism and enzyme synthesis, thus the simplest hypothesis was that genes were self-synthesizing, self-replicating enzymes.[20] This protein or enzymatic theory of genes was based on results from experiments on bacteriophage reproduction carried out by John Northrop's group at the Rockefeller Institute, and Wendell Stanley's work on the tobacco mosaic virus, or TMV (see Chapters 4 and 6). The key point is that the immediate interpretation of Beadle and Tatum's experiments was not the one we now use, the one that has been retained by history. This will be particularly important when it comes to understanding the context in which Oswald Avery described the chemical nature of what he called the transforming principle in 1944 (see Chapter 3).

In the 1930s, the German geneticist Franz Moewus was one of the first biologists to attempt a genetic study of a microorganism—the alga *Chlamydomonas*.[21] Together with the German biochemist Richard Kuhn, Moewus showed that the various stages of the *Chlamydomonas* reproductive cycle were controlled by carotenoid hormones that were synthesized and converted through the action of enzymes, each of which was controlled by a different gene that Moewus localized to the alga's chromosomes. Moewus's results were initially well received, but things soon began to unravel. Some of the data were statistically too good to be true, scientists could not get hold of his strains, and eventually his experiments could not be replicated, even in his presence.

Moewus was the first to have appreciated the potential of microorganisms for forging links between biochemistry and genetics, but his work did not represent a break with earlier studies carried out on other systems—it was analogous to Beadle and Ephrussi's research on eye color in *Drosophila*. Moewus's results were a perfect proof of the model of gene action through the production of hormones that was initially adopted by Beadle and Ephrussi, and by many other biologists. This is not surprising—fraudulent results often fulfill scientists' expectations. Unfortunately for the fraudsters, in the long run these expectations are rarely fulfilled! Unlike Beadle and Tatum, Moewus did not approach the problem from the other end—studying well-known biochemical reactions by genetic means. Beadle and Tatum's great merit was not only to have chosen an appropriate system, but also to have obtained enough observations for the one gene–one enzyme hypothesis to take on a quantitative and not merely a qualitative meaning.

3

The Chemical Nature of the Gene

THE FIRST EXPERIMENT that showed that genes are made of DNA was carried out by the American microbiologist Oswald Avery and his associates, and was published at the beginning of 1944 in the *Journal of Experimental Medicine*.[1] Strikingly, the earliest accounts of the origins and development of molecular biology did not mention Avery's work but instead focused on another, less conclusive experiment, which was performed eight years later.[2] Indeed, in textbooks, this second experiment is still often presented together with Avery's work, as though the two were of equal significance. This has given rise to a substantial literature according to which Avery was an unrecognized genius, molecular biology's equivalent of Mendel.[3]

The truth is both simpler and more complex. It is simpler because Avery was not at all unknown; it is more complex because, although his discovery was understood and its importance was recognized at the time by many scientists, a large number of researchers rejected it for various reasons, above all their conviction that genes must be made of proteins, not DNA. In the words of Vivian Wyatt, everybody knew about Avery's discovery, but that information did not "become knowledge."[4]

Avery worked in the prestigious Rockefeller Institute in New York, where he spent most of his scientific career studying pneumococci (pneumonia-causing bacteria, now included in streptococci). Before the discovery of sulfonamides and antibiotics, pneumonia was a severe, often lethal disease:

during the influenza pandemic of 1918–1919, secondary bacterial pneumonia was probably a major cause of death.[5] Avery's main scientific tool was immunology: after infection, the organism reacts by producing antibodies against the pathogenic agent. This enabled Avery to characterize various kinds of pneumococci—antibodies react with the bacterial envelope, or capsule, which is made up of sugars. In collaboration with Michael Heidelberger, Avery had made an important contribution to the characterization of the structure of this capsule in different types of pneumococci. He had acquired a reputation for the high quality of his work, to the extent that in the 1930s and 1940s he was repeatedly nominated for a Nobel Prize for this research.

Pneumococcal infection can be produced in mice. Because this bacterium caused such serious diseases, the pneumococcus-mouse system became a model for the study of human infectious diseases. Avery's work thus fell squarely within the scope of the Rockefeller Institute's research interests—understanding pathological phenomena at a fundamental level using the most up-to-date techniques from physics and chemistry.

During infection, pneumococci spontaneously shed their sugar capsules, and their colonies take on a rough appearance. These rough (R) pneumococci are not infectious, unlike smooth (S) pneumococci, in which the capsule is intact. When antibodies attach themselves to the capsule, they encourage phagocytosis by white blood cells and thus help eliminate the pathogen: although R pneumococci are not infectious, they escape the defenses of the host and survive in the organism.

In 1928, the British physician Fred Griffith discovered the strange phenomenon of transformation. When a mouse was simultaneously injected with R pneumococci (derived, for example, from type I bacteria) and with type III S pneumococci that had been killed by heat, the animal rapidly succumbed to infection and died. Griffith was able to isolate colonies of living virulent type III S from the blood of the dead mouse. The only possible explanation was that the dead type III pneumococci had transformed the living, nonpathogenic type I bacteria.[6]

In 1931, Martin Dawson and Richard Sia reproduced this result in vitro in Avery's laboratory, and J. Lionel Alloway showed that it was possible to transform in vitro the nonvirulent pneumococci of one type with extracts of killed virulent pneumococci of another type. Alloway's experiment opened the way to the purification of what became known as the transforming principle—whatever it was in the extracts of dead pneumococci that was responsible for the transformation of pneumococcal type.

Avery began his attempts to purify the transforming principle. First with Colin MacLeod and then Maclyn McCarty, Avery continued his work for nearly a decade with few interruptions, apart from a period of severe ill health.[7] The 1944 article Avery published in the *Journal of Experimental Medicine* represented a substantial amount of work. Several pages were devoted to explaining well-defined and easily reproducible conditions for carrying out transformation in vitro, and the purification of the transforming factor (or principle) was described in some detail.[8] But the key part of the paper was the characterization of the purified transforming factor via a series of physical and chemical methods.

All these techniques pointed in the same direction: the transforming factor was not a protein but a deoxyribonucleic acid. The factor was able to withstand temperatures that would denature a protein; colorimetric tests showed that the purified transforming principle contained only DNA, with no traces of protein or of ribonucleic acid (RNA). Simple chemical analysis confirmed that proteins made up less than 1 percent of the purified material, and enzymatic tests showed that the transforming factor was not affected by enzymes that cleaved proteins, nor by phosphatases that degraded RNA, but that it could be destroyed by unheated serum. This last point was important—serum was known to contain an enzyme capable of degrading DNA.

Using the state-of-the-art techniques from physics and chemistry that were the hallmark of the Rockefeller Institute (see Chapter 9), Avery was able to show that the purified transforming factor had a high molecular weight and was probably a nucleic acid.

Finally, a significant part of the paper was devoted to an immunological analysis showing that the purified material did not react with antibodies directed against the capsule. This seemed to suggest that although the transforming factor was responsible for the formation of the capsule, it did not itself contain any sugars; it was therefore chemically distinct from the chemical structure it transformed. The article concluded with a discussion of the nature and mechanism of action of the transforming factor.

The usual explanation for the fact that Avery's discovery did not have a significant impact is that his paper was not read, or at least was not read by those who would really have understood it. In particular, it was supposedly not read by geneticists. The *Journal of Experimental Medicine* was aimed more at physiologists and pathologists than at protein biochemists or geneticists, and the title and the abstract of the article, which were essentially centered on transformation, did not emphasize the significance of the key result. Even the discussion was cautious—Avery did indeed put forward the hypothesis that the transforming agent might be related to or of the same nature as genes or viruses (especially the tumor viruses discovered some years before by Peyton Rous), but he merely outlined a possible relation without going any further. This caution was symptomatic of his personality. He was sixty-seven years old at the time and was really interested only in bench work; he never sought an important position and rarely attended conferences.[9]

In reality, none of these issues prevented his paper from being read and discussed, as Vivian Wyatt has clearly shown.[10] It should be remembered that the Rockefeller Institute was one of the most prestigious research institutes in the United States, that Avery was well known and his work widely appreciated, and that the journal in which he published his article had a large circulation and a solid reputation.

Avery's work did not go unnoticed; his discovery, popularized by journals like *American Scientist,* was brought to the attention of the key biochemists and geneticists and the main practitioners of the nascent science of molecular biology. The result seemed to be very important, but it raised such profound problems that many researchers found it was difficult to appreciate its true significance.

Max Delbrück, who got wind of Avery's result even before it was published, recalled, "You really did not know what to do with it."[11] This comment reveals both the state of knowledge at the time and the initial lack of interest shown by Delbrück's phage group in discovering the physical nature of the gene. They do not appear to have thought that this was an important question, or to have realized that insights would be gained by studying the link between form and function.

Transformation was a new and unusual phenomenon that initially appeared to be limited to pneumococci. Even its extension to the bacterium *E. coli* by André Boivin in France did not support Avery's view. Boivin's results could not be reproduced by other laboratories, which meant that transformation became even more marginalized as an air of doubt accumulated around it.[12] Furthermore, the pneumococcus was poorly understood in terms of its biochemical makeup; before Avery's work, the only nucleic acid that had been characterized in this bacterium was RNA. The existence of genes in bacteria was not universally accepted.[13] Some scientists such as the British biochemist Sir Cyril Hinshelwood thought that the properties of bacteria, including adaptation and mutation, could be explained by changes in biochemical equilibria (see Chapter 5).

Another problem was that nucleic acids were poorly understood at the time. The chemical structure of nucleotides and the nature of the bonds that bind them in polymers were still disputed. The dominant view was the tetranucleotide model proposed by Phoebus A. Levene in 1933, according to which DNA was formed by the repetition of an elementary unit containing each one of the four bases (adenine, thymine, guanine, and cytosine), implying that the DNA molecule was monotonous ("boring" was a term that was later used). Its function was poorly understood, although detailed cytological experiments, mainly carried out by Torbjörn Caspersson in Sweden, had shown that it was associated with chromosomes. According to the nucleoprotein chromosome model, DNA was the material support that carried the proteins thought to be responsible for genetic specificity.

A few years earlier, the American biochemist Fritz Lipmann had discovered that the nucleotide adenosine triphosphate (ATP) acted as an intracellular energy reserve. This had suggested an energy-related dimension to the nucleoprotein component of chromosomes; DNA might play a role in providing the energy needed for gene replication. Avery's discovery thus took place in a context that was extremely unfavorable to the idea that DNA played a specific role in heredity. Avery himself was well aware of this problem; one of the questions discussed in detail in his article was the fact that the DNA molecule appeared to be nonspecific.

In addition to these difficulties, which were caused by ignorance of the structure and properties of nucleic acids, another problem with this study was that it came up against the widely held view (which Avery initially agreed with) that proteins were the most likely candidates for genetic specificity (see Chapter 1).[14] The idea that proteins were the essential constituent of genes was not new: it was accepted by both geneticists like Muller and the young molecular biologists of the phage group, whose work will be described in the next chapter.[15] Wendell M. Stanley gave support to this idea when in 1935 he identified the tobacco mosaic virus (TMV) as a protein (see Chapter 6), for which he won the Nobel Prize two years after the publication of Avery's breakthrough discovery. This protein model of the gene agreed with all the experimental data that had been accumulated over the previous half century, which seemed to show that proteins were the carriers of organic specificity, as seen by the wide range of cellular reactions controlled by proteins. John Northrop and John B. Sumner, two of Avery's colleagues at the Rockefeller Institute, had purified and crystallized several enzymes and had confirmed that they were proteins.

Specificity of protein form and function can be studied by making antibodies against these molecules. Immunological reactions were originally used to reveal structural differences between protein molecules that no physicochemical method could detect. Immunological criteria later became "the" criteria defining specificity—a molecule was considered specific if it could induce specific antibodies. Up until this point, DNA had failed to pass the test.

Other arguments supported the hypothesis that proteins were the main constituents of genes—they were one of the two key components of chromosomes, and Beadle and Tatum had shown that enzymes responsible for various metabolic reactions were closely controlled by genes. Beadle and Tatum's work tended to link genes and enzymes and, as far as many biologists were concerned, had reinforced the more or less explicit identification of genes with enzymes and proteins. Furthermore, Avery had obtained a relatively large quantity of the transforming factor: as a result, the trace levels of contaminating proteins that were found in it could explain the observed data. Or so the partisans of the protein theory of genes, such as Alfred E. Mirsky of the Rockefeller Institute, argued.

In the years that followed, Avery's colleagues improved the purification of the transforming principle. Contamination by amino acids—constituents of proteins—was reduced to less than two parts in 10,000. Ironically, much later it turned out that even these trace amino acids were not due to protein contamination but to the degradation of a nucleotide (adenine). The identification of DNA as the transforming agent was further confirmed by its direct inactivation by deoxyribonuclease, an enzyme recently isolated at the Rockefeller Institute. Nevertheless, these improvements of Avery's experiment did not shake the convictions of those who thought that proteins played an essential role in heredity.

The main problem preventing the importance of Avery's discovery from being accepted by all scientists was not so much the difficulties posed by

DNA or the objections raised by the supporters of the protein theory of genes, as the impossibility of imagining how DNA could produce a protein, which is what genes do. Avery's letters to his brother and the testimony of his colleagues show that despite the cautious conclusion to his 1944 paper, he was in fact profoundly convinced that the isolated material—the transforming factor—was a pure gene.[16]

Although pneumococcal transformation had been an ideal system for discovering the chemical nature of genes, it was most certainly not suited to the extension and generalization of this result, and even less so to its interpretation.[17] Transformation was a highly particular phenomenon, mainly of medical interest, and was initially limited to pneumococci. In subsequent years, Avery's colleagues showed that transformation could also apply to other characters apart from the capsule. But reliably extending the phenomenon to other bacteria (for example, to *E. coli*) turned out to be hard. Furthermore, the observation of transformation in pneumococci is particularly difficult; indeed, Avery's paper emphasizes the care needed in carrying out experimental observations of transformation.

The most intriguing aspect of the result was coming up with a hypothesis as to how a nucleic acid could alter the structure of a capsule made of sugars. The problem was well presented in Avery's 1943 report to the Rockefeller Institute: "The transforming principle—a nucleic acid—and the end type of synthesis it evokes—the type III polysaccharide—are each chemically distinct. . . . The former has been linked to a gene, the latter to a gene product, the accession of which is mediated through enzymatic synthesis."[18]

Avery's model required a stage involving the presence of an enzyme or a protein. He had to find out how the nucleic acid—which was thought to be a nonspecific compound—could control protein activity, or more precisely, control the enzymes responsible for capsule synthesis. The supporters of the protein theory of genes did not have this problem. It was easier for them to explain how proteins (genes) could control the activity of other proteins than it was to imagine how a nucleic acid could control enzymes. Or at least, so it seemed.

Two obstacles had to be overcome before it could seriously be considered that genes were formed of DNA. Contrary to what might be expected, the idea of a genetic code was not a precondition. This idea was in fact introduced gradually, much later on, after Jim Watson and Francis Crick's discovery of the double helix structure of DNA in 1953. In fact, it was only completely accepted in 1961, when the genetic code was first cracked. The true precondition to accepting the role of DNA in heredity was the separation of the dual problems of the nature of the gene and its mechanism of action.

Unlike biochemists, geneticists had been prepared for this kind of approach. From the late 1940s some physicists began to think of genetics in terms of information transfer, and the gene as a carrier of information. This insight in fact resulted from their lack of understanding and their despair when faced with the seemingly irrational complexity of biochemistry. The physicists were able to approach the problem from another angle and abstract the notion of specificity from its biochemical and protein connotations. They separated the gene from its effects within the cell and made it possible to conceive of relations between genes and proteins apart from a direct protein–protein interaction.

This alternative way of looking at genes could prevail only if it was associated with a new experimental system in which the study of the transfer and replication of genes was more important than the characterization of their functions such as a virus. This system could have been the study of plant or animal viruses; instead, it turned out to be the bacteriophage, an experimental system that will be described in Chapter 4. This alternative conception of the role of genes gradually led to the abandonment of the idea that genes control organic functions from a distance and to its replacement by the current model of molecular biology, in which genes determine the development and functioning of the organism down to the smallest chemical detail.

Avery's experiment was difficult for many of his contemporaries to accept, and even more difficult for them to interpret. As a result, although the experiment was well known, it did not provoke the immediate conceptual revolution that might have been expected.

Avery was perhaps not as single-minded in his research as he could have been. Between 1937 and 1940, he turned away from transformation for a while.[19] It should not be thought, however, that Avery's results had no influence on the subsequent development of molecular biology.[20] The fact that his work was incompatible with Levene's tetranucleotide model intrigued Erwin Chargaff, an Austrian-born American biochemist. Chargaff repeated the experimental measurements on which Levene's model was based and measured the percentages of the four nucleotides in the DNA of several different species.[21]

Chargaff's data were significant in a number of ways. He showed that the proportions of the four nucleotides changed depending on the species being studied, thus removing a major obstacle in understanding the genetic role of DNA. Furthermore, Chargaff separated the nucleotides by a highly sensitive chromatographic technique, which was subsequently widely applied in molecular biology. Last but not least, he rather reluctantly reported a certain regularity in the proportions of the four nucleotides and the four bases that constituted them. According to modern versions of the story, he found that the proportions of adenine and guanine were equal to those of thymine and cytosine, respectively. In reality, because of poor measurement accuracy, the data reported by Chargaff were never so precise—hence his hesitation and his persistent reluctance to generalize his findings. Much later, this effect was called Chargaff's rule, and, as will be seen, was retrospectively considered to be strong evidence in favor of Watson and Crick's double helix model of DNA structure (Chapter 11).

Vivian Wyatt has shown that the phage group was well aware of Avery's result (Delbrück had seen a personal letter from Avery announcing the discovery in May 1943). But although some members of the group tried to reproduce it, for others it simply meant they had to consider the hypothesis that DNA might be the essential component of bacteriophage reproduction.[22] Strikingly, it was several years before they began to explore the physical nature of the gene. Avery's experiment was the first to undermine the conviction of many biochemists and geneticists that genes were proteins. Although it did not immediately convince them all that genes were made of DNA, it opened the door to that possibility, revealing a scientific trail that would be blazed in the years to come.

4

The Phage Group

THE TERM "phage group" refers to all those scientists who, between 1940 and 1960, used bacterial viruses (bacteriophages) as a model system to study how organisms reproduce. The group never had an organized structure, nor any official existence. Furthermore, the research carried out by its members was extremely diverse and evolved over the years. What gave the group its identity—apart from the viruses they studied—was a state of mind, a new approach to biological problems. This was largely the result of the influence of Max Delbrück, who is universally acknowledged as the founder and driving force behind the group.[1]

Delbrück was born in 1906 into an upper-class German family—his father was a professor of history at Berlin University. In 1930, Delbrück was awarded his doctorate in theoretical physics.[2] The following year, funded by a grant from the Rockefeller Foundation, he joined Niels Bohr's laboratory in Copenhagen. His meeting with Bohr, and especially Bohr's lecture on "Light and Life" (delivered in Copenhagen on August 15, 1932), had a decisive influence on Delbrück and led him to turn to biology.[3] (Bohr's lecture will be discussed in Chapter 7, together with his ideas about biology.) Bohr was not a very clear speaker, and the lesson that Delbrück took home from the lecture was probably not what Bohr intended. What Delbrück understood was that even if the study of biological molecules were to be taken as far as possible, it would still not lead to an understanding of life. Life could be understood only through an entirely new approach, which would complement existing methods. On his return to Berlin, Delbrück set up a small informal group of biologists and physicists, which met at his mother's house. His first piece of biological research was performed in collaboration with the Russian geneticist Nikolaï Timofeeff-Ressovsky and the German physicist Karl Zimmer.[4]

The idea behind this study was taken from nuclear physics. Physicists could not study the nucleus directly; they had to bombard it with particles. By varying the number and energy of these particles and relating that variation with the results of the bombardment, it was possible to deduce some of the properties of the target—the nucleus. Delbrück decided to use the same approach to study the gene. In 1927, Muller had shown that mutations could be induced by X-rays. Timofeeff-Ressovsky's role in the team was to determine the number of mutations induced in *Drosophila* by varying the radiation doses delivered at different temperatures. Zimmer converted radiation doses into the number of ion pairs formed (it was assumed that ion pairs were responsible for the mutagenic effect). On the basis of these experiments, Delbrück estimated the size of the gene to be a few thousandths of a millimeter. Variations in mutation rate as a function of temperature were interpreted in terms of a quantum model of the gene, according to which the gene had several stable energy states; a mutation was interpreted as a passage from one stable state to another.

Their article, known as the Three-Man Paper, was published in 1935 in German in a Göttingen-based journal with a small circulation.[5] Nevertheless, it came to the attention of Erwin Schrödinger, who devoted considerable space to it in a series of lectures, published in 1944 as a highly influential book entitled *What Is Life?*[6] Salvador Luria, a physicist working in Enrico Fermi's laboratory in Rome, and who later worked closely with Delbrück, also heard of the paper.[7] On the basis of this study, Delbrück and Timofeeff-Ressovsky subsequently published an article in *Nature* concluding that cosmic radiation had little effect on speciation.[8]

Several research groups had used, or would use, similar approaches to study gene structure and the nature of mutations, but in the absence of any data on the chemical nature of genes, the results were difficult to interpret. Although Delbrück's model of mutation by ionization turned out to be wrong, it was nonetheless a "successful failure."[9] These and similar experiments helped to show that genes could be studied using the tools of physics.

In 1937, funded by another grant from the Rockefeller Foundation, Delbrück visited the key U.S. genetics laboratories, including Thomas Hunt Morgan's laboratory at Caltech, in Pasadena, California. Delbrück was not impressed by *Drosophila* as a model system—he thought it was too complex to reveal the secret of life. Ever since his meeting with Bohr, he had been convinced that only the simplest biological system would do. The principles of quantum mechanics had been discovered only when matter had been studied at its most elementary level—that of the atom and its constituents.[10] The same principle should operate in biology, reasoned Delbrück. Contradictory experiments and paradoxes—which for a physicist are the starting point for new theories—appear only when simple systems are studied. For Delbrück, this elementary biological system was the bacteriophage, which he discovered during his visit to Morgan's laboratory. Emory Ellis, who had recently started studying phage, worked in the same Caltech department as Morgan.[11]

The bacteriophage, or bacterial virus, immediately struck Delbrück as being particularly suited to the study of the key characteristic of life: self-replication.[12] Bacteriophages infect bacteria and reproduce rapidly inside their hosts. They appeared to be elementary biological particles—they were less than one ten-thousandth of a millimeter in length, which meant that, according to Delbrück's earlier study, they were smaller than genes.

Bacteriophages had been widely studied for many years. Initially described by the British bacteriologist Frederick Twort in 1915, they were rediscovered in 1917 by the Canadian / French scientist Félix d'Hérelle. They were subsequently studied by several other groups, particularly at the Pasteur Institutes in Paris and Brussels, and at the Rockefeller Institute in New York.[13] Despite their small size, phages could be easily detected: a layer of bacteria was grown on the surface of a Petri dish; after infection with phage, light plaques appeared ("lysis plaques") where the bacteria had been killed and lysed by the bacteriophage.

Despite a great deal of work, the nature of the phage and its relation to bacteria remained unclear and were widely disputed.[14] Some scientists thought that the bacteriophage was a real infectious particle; others, such as Albert Krueger and John Northrop, saw phage multiplication merely as evidence of the autocatalytic transformation of an inactive protein precursor that was present in bacterial cells before the addition of phage. According to this widely accepted model, bacteriophage replication—which showed an s-shaped (sigmoid) kinetic—was analogous to the autocatalytic activation of enzymes such as trypsin.[15]

Delbrück began to work with Ellis, who was already studying the phage growth curve. The first step was to develop statistical tests that could validate their method of counting phages. They showed that bacteriophage multiplication was irregular, entailing a series of steps and plateaus (a phenomenon already observed by d'Hérelle). They interpreted the sudden increases in the number of phages that took place every thirty minutes at 37°C as the result of bacterial lysis and the liberation of large numbers of bacteriophages. The phages rapidly attached themselves to new hosts, which they infected, and thirty minutes later the sudden increase would be seen again.

This experiment, known as the one-step growth experiment, was published in the *Journal of General Physiology* and had a major impact because of its clarity and its elegant statistical methods.[16] But what was most important was that it was in complete opposition to the autocatalytic model proposed by Krueger and Northrop. There was no sign of any sigmoid curves in Ellis and Delbrück's results.

Delbrück's Rockefeller Foundation grant was renewed in 1938. Because of political developments in Germany, he wanted to remain in the United States. At the end of 1939 Vanderbilt University, in Nashville, Tennessee, offered him a job teaching physics. This permanent position enabled Delbrück to carry out his work on phages in the years that followed.

At this point, it is difficult to speak of a "phage group." Ellis, Delbrück's only collaborator, had to stop work on phages and return to research more directly centered on cancer. The phage group emerged as a result of Delbrück's meeting with Luria at the end of 1940 and their first joint research, carried out in the summer of 1941 at Cold Spring Harbor Laboratory on Long Island in New York.[17] This collaboration between Luria and Delbrück also gave rise to bacterial genetics. (Luria's scientific career and his first experiment with Delbrück are described in Chapter 5.) The third founding member of the group, Alfred Hershey, joined Luria and Delbrück in 1943.[18]

The members of the phage group performed a large number of experiments as they tried to understand bacteriophage replication. One of Luria's findings was that if the same bacterium is infected with two different types of phage, a kind of interference or mutual exclusion takes place between them. A series of other experiments was carried out to uncover what happened to the bacteriophage during the "eclipse phase," its period of replication.

An examination of the research output of the phage group between 1940 and 1953 (that is, until the discovery of the double helix structure of DNA) reveals relatively few major findings. Many biologists will be surprised by this—after all, the phage group was responsible for the birth of bacterial genetics, while an experiment conducted by Alfred Hershey and Martha Chase supposedly showed that DNA was the hereditary material. But the link between these two discoveries and the phage group is, in fact, quite complex. Luria and Delbrück's 1943 experiment, which showed that there were mutations in bacteria, was in reality a marginal by-product of their work on phage. Soon afterward, Delbrück turned away from bacterial genetics and concentrated on phage.[19] And as will be seen, Hershey and Chase's finding was in fact merely a new and ultimately inconclusive confirmation of the discovery made by Avery and his coworkers. It was mainly because of the phage group's subsequent prestige, amplified by the fact that the earliest historical accounts were written by its members, together with the near-simultaneous discovery of the double helix structure of DNA, that the Hershey–Chase experiment had such an impact on how history has been presented.

Results from the key focus of the phage group's work—bacteriophage replication—were relatively unimportant. In particular, Delbrück's initial dream of understanding replication as a simple phenomenon without opening the biochemical "black box" proved to be an illusion. In 1942, Thomas Anderson and Luria turned the electron microscope on the phage and found that it was not an elementary biological particle but a complex entity, highly organized and structured.[20] (Similar images had been obtained in 1941 by Ernst Ruska in Germany.[21]) After this discovery, experiments that disturbed bacteriophage replication through various kinds of physical treatment, in particular by radiation, appeared less appropriate as a way of characterizing the various stages of replication.

At this point, Luria's understanding of the problem was as follows.[22] The physicists' way of investigating the gene was like someone shooting at a tree with guns of different calibers to find out about the tree's fruit. Depending on the effectiveness of the different sizes of bullet, one could deduce whether the fruit was the size of a cherry or the size of an apple. But further progress would not be possible simply by shooting at the tree; rather, the fruit would have to be isolated and characterized.

The phage group's initial project was thus a failure.[23] Not only had no new physical principle been discovered and no paradoxical result been recorded, but research on bacteriophages was becoming increasingly influenced by biochemistry, which Delbrück found depressingly complex.[24]

The scarcity of the results obtained by the phage group seems to have been inversely proportional to its fame. There are several reasons for this apparent paradox. Membership of the group was not restricted to people who worked with Delbrück. Between 1940 and 1945, many laboratories began to work on the bacteriophage.[25] For example, Milislav Demerec, a Cold Spring Harbor *Drosophila* geneticist, started to study bacteriophages, and Leo Szilard, a physicist who had helped set up the Manhattan Project and had subsequently turned to biology, organized monthly discussion meetings on phages. A large number of scientists, trained in both biology and physics, were thus beginning to study the phage.

This rapid growth of the phage group was partly the result of Delbrück's personal influence.[26] His scientific approach was particularly elegant—using simple statistical tests, he was able to clarify confused questions. His approach also benefited from the prestige that surrounded any technique or concept that had its origin in physics. But most importantly, Delbrück showed that it was possible to develop a revolutionary approach to biology. The new research dealt with what was thought to be specific about life: self-replication. The credo of Delbrück and his colleagues, as set out later, was that the same principles should be able to explain the functioning and reproduction of all organisms, from the virus to humans.

The allure of these ideas was reinforced by Delbrück's exceptional charm. He had a brilliant mind and a charismatic personality. He was both kind and somewhat brutal; at the end of a seminar Delbrück would often say to the speaker, "This was the worst seminar I ever heard."[27]

On its own, however, Delbrück's charisma would not have made the phage group famous. Delbrück also introduced a new style into biological research. In his youth he had been profoundly influenced by the atmosphere in Bohr's group in Copenhagen, with its apparent absence of hierarchies, freedom of discussion, and close mixture of work and pleasure. (For example, Bohr's young physicists would often go on hiking trips in the mountains together.) Delbrück introduced this new research style into the phage group, first in his Nashville laboratory, then, beginning in 1947, at Caltech.

In biology more than in other disciplines, Delbrück's approach marked a significant change in the way laboratories functioned. He wanted his researchers to spend at least one day a week away from the bench, simply thinking about their experiments. Experiments and papers were discussed by the whole group and were subject to lively criticism. These discussions (which often took place during camping trips to the California desert) and the many parties that were hosted at the Delbrück home made a great impression on the participants. A visit to Delbrück's laboratory was a must for anyone starting out in the new science of molecular biology. Many modern molecular biology groups still function in the style of Delbrück, even though most will be unaware of the origins of this tradition.

In addition to this psychological and cultural influence, there was the more direct role of the annual Cold Spring Harbor Laboratory bacteriophage practical course, designed to initiate both students and established scientists into the arcane mysteries of phage. The first of these courses took place from July 23 to August 11, 1945. A certain level of mathematical ability was required because the course was initially aimed at physicists who wanted to do biology. Delbrück had become famous through Schrödinger's presentation of his work in *What Is Life?* and any physicist who wanted to turn to biology was more or less obliged to attend the Cold Spring Harbor course. The course created an informal network centered on Delbrück and the founders of the phage group, and all participants learned to use the same material and the same techniques, thereby reinforcing the influence of Delbrück's approach.[28]

THERE ARE AT LEAST TWO TYPES of scientific experiment, each with a different nature and function. Avery's experiment on transformation is an example of the first type: without any a priori expectations, a scientist discovers a new and unexpected phenomenon. The experiment carried out by Hershey and Chase in 1952 is an example of a second type: there are alternative hypotheses available, so the aim of the experiment is to distinguish between them as clearly as possible. In this case, the experiment was designed to reveal the roles of protein and DNA in phage reproduction (strikingly, none of Avery's papers were cited). The textbooks tell us it was a complete success, but the reality was rather different, as was recognized at the time.

The Hershey–Chase experiment was in fact made up of a series of experiments, all of which were later interpreted as showing that bacteriophage DNA is fundamental for phage replication.[29] History has focused on one of these experiments, the "mixer" or "Waring blender" experiment. This was one of the first experiments in molecular biology to use radioactively labeled molecules, the sign of its entry into the new atomic era. As Angela Creager has shown, the rapid and widespread adoption of the use of radiolabeled molecules was an indirect consequence of the Cold War.[30] The driver behind this policy was the need to show that the investments involved in nuclear research had a positive impact on health and knowledge. The results obtained by using radiolabeled molecules were often more clear-cut than those provided by traditional methods, enabling researchers to choose between alternative hypotheses, or so it appeared.

In the Hershey-Chase experiment the bacteriophages reproduced in bacteria that were grown on a medium labeled either with sulfur (^{35}S), which was ultimately incorporated into phage proteins, or with phosphorus (^{32}P), which was incorporated into phage DNA. The radioactive phages were then added to a culture containing nonradioactive bacteria, and after a suitable delay the mixture was placed in the famous blender. The vigorous agitation produced by the blender (in fact a piece of laboratory equipment) removed 80 percent of the phage proteins from the bacterial surface, but only 30 percent of the phage DNA. Furthermore, this treatment did not prevent the bacteria from being infected or the phage from growing in either of the conditions. (We now know that when the phage lands on the bacterial surface, it quickly injects its DNA into it.)

Other experiments described in the same article suggested that the phage was merely a protein shell containing and protecting the inner DNA; a drastic reduction in salt concentration separated the proteins from the DNA, which could then be degraded by enzymes. Furthermore, there was evidence DNA was released from the phage during its interaction with the bacterial membrane: if the phage was adsorbed onto killed bacteria or membrane fragments, DNA became available for enzymatic digestion. Finally, the offspring of radioactive phage contained less than 1 percent of the original protein (less than 1 percent of the radioactive sulfur), but at least 30 percent of the DNA (30 percent of the ^{32}P).

All the experiments described in the article thus seemed to suggest that the part of the bacteriophage responsible for replication was DNA. The protein shell that enclosed the phage DNA was simply there to protect it and inject it into the bacterium like a syringe. Given the iconic position of the article in many accounts of molecular biology, its conclusion is surprisingly hesitant: "This protein probably has no function in the growth of intracellular phage. The DNA has some function. Further chemical inferences should not be drawn from the experiments presented." At the June 1953 Cold Spring Harbor symposium, after the double helix structure of DNA had been described, Hershey presented his findings, which had been published the previous year. He then explained his conviction that DNA could *not* be the sole hereditary molecule:

> None of these, nor all together, forms a sufficient basis for scientific judgement concerning the genetic function of DNA. The evidence for this statement is that biologists (all of whom, being human, have an opinion) are about equally divided pro and con. My own guess is that DNA will not prove to be a unique determiner of genetic specificity, but that contributions to the question will be made in the future only by persons willing to entertain the contrary view.[31]

Despite this contemporary uncertainty, early histories of molecular biology attributed the discovery of the genetic role of DNA to Hershey and Chase.[32] It took protests from Avery's colleagues for his experiment to be restored to the pantheon of molecular biology. Even today, molecular biology textbooks often put the two experiments on an equal footing, ignoring the fact that they took place eight years apart and that their authors interpreted their findings with very different degrees of certainty.

Vivian Wyatt has shown that many books describing the Hershey–Chase experiments "blend" or distort different aspects in order to make them more convincing.[33] For example, the percentages of sulfur and phosphorus that remained associated with the bacteria after agitation are often massaged in the "right" direction. Or it is suggested that the two kinds of label—sulfur for the proteins and phosphorus for the DNA—were added in the same experiment. In still other accounts, the additional experiments described in the 1952 paper are not even mentioned. And nobody ever points out Hershey's own lack of conviction about the significance of his finding.

These omissions and alterations make it impossible to understand why Hershey and Chase's results were eventually so readily accepted. They were in fact the end point of a long series of convergent experiments. The first electron microscope images showing phages attached to the bacterial surface but not penetrating the cell implied that only a part of the phage was responsible for its reproduction. The shape of the phage—a kind of syringe with paws—suggested that it injected one of its constituents into the bacterium. The phage was chemically very simple, consisting only of DNA and proteins. Immunological experiments had shown that the proteins formed the phage envelope, whereas DNA seemed to be found within the protein shell.

Even if it was not completely accepted, Avery's discovery had gradually become an influential part of scientific thinking.[34] All the members of the phage group knew of it, and it had encouraged them to study the DNA found in the phage. Furthermore, Levene's tetranucleotide model had been undermined by the work of Chargaff and others, thus weakening some of the obstacles to the idea that DNA had a specific function in self-replication. Biochemical studies by André Boivin with Roger and Colette Vendrely, together with cytochemical data, had shown that DNA was present in the nucleus in constant amounts, except in sperm, where there was only half as much.[35] All these experiments helped to make it possible to conceive of DNA as having an essential role in genetic phenomena.

Presented for the first time in Paris in 1952, Hershey and Chase's results were publicly outlined for a second time in July 1953 at the Cold Spring Harbor Symposium, at the same time as the double helix structure of DNA was described by Jim Watson (see Chapter 11). In many people's minds, the two sets of results became associated and were considered additional proofs of the genetic role of DNA.

Nevertheless, it is impossible not to compare Hershey and Chase's experiment with Avery's. Alfred Mirsky attacked Avery's results—or rather, their interpretation—because there were traces of protein components (0.02 percent) in the DNA that could conceivably account for the transforming power of the pneumococci. But in the Hershey–Chase experiment at least 1 percent of bacteriophage radioactive proteins penetrated the bacteria, just as unlabeled proteins with no sulfur-containing amino acids might have done, all of which might have contributed to phage replication. These criticisms were not voiced at the time, and the results soon helped convince the biological community that the heredity role of DNA was at least a good working hypothesis. This disparity shows that a scientific experiment does not have an intrinsic value: it counts only to the extent that it forms part of a theoretical, experimental, and social framework.

Things had changed between 1944 and 1952. Even the skeptics now accepted that DNA might play some role in heredity. Avery had found an unexpected result, and had gone on to answer the fundamental question that nobody else was even asking: What is the nature of the hereditary material? In Hershey and Chase's hands, radioactive labeling gave a suggestive answer as to which of the two components—DNA or proteins—potentially played this role.

The two experiments were performed in very different conceptual environments. Furthermore, Avery was an isolated researcher who did little to draw attention to his data, whereas Hershey and Chase's results were spread by the informal but highly efficient phage group network. Hershey and Chase's experiment also benefited from the publicity given to Watson and Crick's virtually simultaneous discovery of the double helix structure of DNA.

5

The Birth of Bacterial Genetics

MICROBES WERE FIRST DISCOVERED by Anton Van Leeuwenhoek in the seventeenth century, but it was not until the work of Louis Pasteur, Robert Koch, and many other microbiologists in the middle of the nineteenth century that scientists began to understand how they reproduce.[1] It was finally accepted that these organisms are not the result of spontaneous generation but, like primitive plants, they reproduce by division.

In its rush for understanding, genetics had bypassed bacteria, focusing on sexually reproducing organisms with hereditary variations that were easy to observe. Bacteria, which appeared not to reproduce sexually, seemed to have very complex life cycles that enabled them to go through different states or forms. The transformation of smooth (S) pneumococci into rough (R) pneumococci was thus interpreted in the light of what were called cyclogenic theories.[2] This confused understanding of bacterial reproduction blocked research on hereditary variation in these organisms, or at least made it more difficult. In the first half of the twentieth century, the study of bacterial physiology made great advances. For instance, bacteria were found to be able to adapt to new sources of food, but it was not known whether this adaptation was genetic.[3] Models suggesting that bacteria contained no genes and were stable simply because the chemical reactions within them were in equilibrium had tended to separate microorganisms and the rest of the living world.[4] This isolation of bacteria was all the more surprising given that viruses, which are lower on the scale of organic complexity than are bacteria, were considered by some to be almost pure genes.

Between 1920 and 1940, there was a major unification of the life sciences, as it was widely accepted that the same basic mechanisms explained the functioning of all forms of life, and the evolutionary synthesis succeeded in unifying population genetics and the Darwinian theory of evolution.[5] Results from biochemistry played a key role in this unification of biology, in particular the work of Albert Kluyver. In the 1930s, André and Marguerite Lwoff also showed that all organisms use the same vitamins and coenzymes. The same metabolic pathways and the same transformations of biological molecules were found in bacteria and yeast, and in the muscles of birds and mammals. George Beadle and Edward L. Tatum's experiments showed that in fungi each of the enzymes that catalyzed these essential metabolic steps was precisely controlled by a single gene; it seemed very unlikely that in bacteria these steps could be independent from genes.

At the same time, a growing number of physicists turned to biological research, hoping to discover the elementary principles of the functioning of life. Breezily unaware of the complexity of the biological world, they were convinced that the same principles must be at work in a bacterium (and in a virus) as in an elephant, to paraphrase an adage that is generally attributed to Jacques Monod but which was first coined by Kluyver. For Max Delbrück, the mysterious replication of bacteriophages had to obey the same rules as the clearly genetic reproduction of animals and plants.

The first experiments performed by the physicists to uncover the secrets of life reinforced their unifying vision. X-rays, which Hermann Muller had shown could alter genes, also perturbed bacterial development.[6] The same turned out to be true for ultraviolet rays and various chemicals. Similar doses and wavelengths acted in the same way on all forms of life.

A number of experiments had already suggested that bacteria showed hereditary variations. In 1934, I. M. Lewis proposed a novel experimental approach that would prove whether metabolic variations in bacteria were due to individual changes or to the selection of favorable mutants. All that was required was to measure the fraction of bacteria that could grow on a new metabolite before it was added to the medium and to compare this value with that found after the metabolite had been added. If the two values were equal, metabolic variations must be hereditary. Lewis carried out two experiments that confirmed his selective model.[7]

In 1943, Salvador Luria and Delbrück carried out a key experiment, known as the fluctuation test experiment, that heralded the study of bacterial genetics. This took place in an extremely favorable context and fitted their views of how bacteriophages replicated within bacteria.[8] Luria and Delbrück had previously shown that when two different phages were added simultaneously to the bacteria, interference occurred: one of the phages blocked the development of the other.[9]

To conduct their experiment, Luria had to determine the nature of the phages that managed to grow in coinfected bacteria. This involved using bacteria that were resistant to one of the phages, obtained by pre-incubation with that phage. Most of these preincubated bacteria died, but a small minority survived because they were resistant to the phage. By picking out these colonies and allowing them to multiply, Luria found that he could grow bacterial cultures resistant to one or another type of phage. By adding phage of an unknown type to cultures with a known resistance profile, he could determine the nature of the added phage.

At this point, Luria was more or less obliged to investigate the origin of this bacterial resistance to bacteriophages, to be sure that this characteristic would remain stable in the conditions of the experiment. He needed to know whether the bacteria became resistant after their preincubation with the phage or some of them were already resistant before any contact and were selected only later, during the infection period.

Luria's story is an interesting one. He was born in Turin in 1912 into a middle-class Jewish family.[10] After studying medicine without any great enthusiasm, he was convinced by one of his friends that the new physics was the place to be, and he found his way to Enrico Fermi's laboratory in Rome. Luria quickly discovered that his mathematical training was not sufficient for conducting research in physics, but his stay in Fermi's laboratory was nevertheless a turning point in his life—Franco Rasetti, who lectured on spectroscopy, told him about Delbrück's recent article on the structure of the gene. Luria was fascinated by this article because he felt it addressed what he later called "the Holy Grail of biophysics." It was quite by chance that he heard about bacteriophages. One day he found himself in a stalled trolley bus seated next to the bacteriologist Geo Rita. They got to talking, and Rita invited Luria to visit the laboratory where he worked on bacteria and phage.[11]

Although he did not know it, Luria's conversion to phage took place at the same time as Delbrück's. In order to begin work, Luria applied to the Italian government for a grant to go to Berkeley in California to study radiation biology. This grant was approved shortly before Mussolini proclaimed Italy's racial laws, whereupon it was subsequently cancelled. Knowing what had happened in Germany on Kristallnacht, Luria quickly left Italy for Paris. There he was able to get a grant from the Fonds de la Recherche Nationale, the forerunner of the Centre National de la Recherche Scientifique (CNRS), to work on the effect of radiation on bacteriophage with the French physicist Fernand Holweck in the Institut du Radium, founded by Marie Curie. During his stay in the French capital, Luria also worked with Eugène Wollman, a phage specialist at the Pasteur Institute.[12] In June 1940, two days before German troops marched into the city, Luria left Paris on a bicycle; he managed to get to Marseille, where he obtained a visa for the United States.

With the help of the Rockefeller Foundation, and thanks to a scientific reputation based on his first publications, Luria was given a position at the College of Physicians and Surgeons at Columbia University. In late 1940, he met Delbrück at a conference of the American Physical Society at Philadelphia, and the two men struck up a partnership that was to last many years, and which was renewed each summer at Cold Spring Harbor. During his first two years in the United States, Luria helped Thomas Anderson take electron micrographs of the phage.[13] In 1942, he worked with Delbrück at Nashville for a year, then in 1943 he obtained a permanent position at Indiana University at Bloomington, where he soon performed the famous fluctuation test experiment.

According to Luria's own account, the idea for this experiment came to him when he was idly looking at a slot machine during a dance at Indiana University.[14] When a coin is inserted into a slot machine, the gambler generally wins nothing—but every now and then they hit the jackpot. Overall, no more coins are won than are inserted, but in terms of amounts paid out, the slot machine does not behave statistically; if it followed a Poisson distribution, the player would recover one, two, or no coins each time, rarely more. But in reality, it was possible to win the jackpot.

Luria applied this reasoning to the origin of phage-resistant bacteria. Take twenty independent bacterial cultures, prepared from the same number of bacteria, and add phage. If resistance to viruses is induced by encountering the phage ("acquired hereditary immunity") then there is a fixed chance of a given bacterium surviving, and although the number of resistant bacteria would vary from culture to culture, this variation would be relatively small, following a Poisson distribution. If, by contrast, resistance to phage is the product of randomly occurring spontaneous mutations, then the number of such bacteria will vary considerably from culture to culture. In particular, the number would be very large if a mutation that causes resistance appeared during the early stages of bacterial replication—that culture would have hit the jackpot (this term was used in the published article). By measuring the variation between cultures, it was possible to infer from this "fluctuation" which process explained the origin of resistance—acquired immunity or mutation.

Luria described the experiment to Delbrück in a letter and then set about conducting it without delay. Delbrück quickly replied, sending him a mathematical analysis of the experiment, which showed that it would also be possible to use the results to calculate the rate of mutation, an essential measure for any genetic analysis.[15] This was the first time that mutation rate had been measured with such precision. The Luria–Delbrück experiment was a complete success and proved that the resistant strains were produced by spontaneous mutation.

For many, this experiment marks the birth of bacterial genetics. In classical genetics, characters are defined and their transmission and recombination are followed over a number of generations, but the Luria–Delbrück experiment had none of these characteristics. The experiment was important because it brought bacteria into the general understanding of evolution that had been outlined a few years earlier in the Modern Synthesis: organisms show variation, and variant organisms are selected over time.

In retrospect, therefore, the Luria–Delbrück experiment represented the triumph of Darwinism, a victory in the heart of what for many years had been "the last stronghold of Lamarckism"—microbiology—linking Darwinism and molecular biology once and for all (see Chapter 23).[16] Published in *Genetics,* the journal of the American Society of Genetics, their experiment also announced to the wider scientific community the kind of approach to biology that Delbrück had in mind—a mixture of simple, clean experimentation and thorough mathematical analysis of the underlying theory. Strikingly, however, it made no mention of Darwin or natural selection, nor of the recent neo-Darwinian synthesis, and said nothing about the implication of the study for our understanding of evolution—Luria and Delbrück went no further than to say that in this particular case they had proof that resistance was due to spontaneous mutations.

Despite its success, the experiment did not immediately give geneticists access to the world of bacteria. Genes can be localized only when they recombine, reorganizing themselves into new groups, and the simplest phenomenon that leads to recombination is sexual reproduction. At the time, it was accepted that bacteria reproduced only by fission. A few experiments had been carried out to detect sexual recombination, but they had all produced negative results, though a positive result would probably have been found only if there had been a very high recombination rate.

In *The Bacterial Cell,* published in 1945, René Dubos of the Rockefeller Institute showed that there was no strong experimental argument in favor of sexual reproduction in bacteria.[17] But this absence of evidence could not be construed as evidence of absence—it might simply indicate that the phenomenon had not been studied sufficiently. The unification of biology around the Modern Synthesis and the important evolutionary role this view gave to sexual reproduction tended to suggest that this mode of reproduction must have originated in the simplest, most "primitive" organisms, such as bacteria.[18]

Nevertheless, looking for the existence of sexual reproduction in bacteria was a risky business. The whole field of bacterial genetics was full of contradictory results and pitfalls. Any scientist who realized that the key to success is to attack "soluble" problems would instinctively have avoided the subject.[19]

The only scientists who can allow themselves the luxury of studying such a subject are those for whom the marginal costs of such an attempt are low, to borrow terminology from economics. This would include very young scientists, who can allow themselves a moment of rashness, or very famous scientists, whose reputation would not be harmed by failure. Joshua Lederberg and Ed Tatum formed just such a pair.[20]

Lederberg, born in 1925, was a medical student at Columbia University, and at the age of eighteen began work on *Neurospora* in Francis Ryan's laboratory. Ryan had just returned from a postdoctoral visit to Stanford and had been inspired by George Beadle and Ed Tatum's experimental approach. As soon as Lederberg read Avery's article suggesting that DNA was the genetic material, he was enthused and tried to reproduce the transformation experiment in *Neurospora*. Although he failed, he had an idea for a new approach that could detect sexual recombination in bacteria. For this experiment, he required mutant bacteria, like the mutations obtained by Beadle and Tatum in *Neurospora,* which needed certain vitamins or amino acids to grow. Tatum had just begun work on bacteria and had obtained *Escherichia coli* mutants that were just right for Lederberg's experiment.

Toward the end of 1945, after Tatum had left Stanford for Yale, Lederberg wrote to him explaining his project and asking for help. Tatum was busy moving and, with his change of subject, did not have much time to spare, but he strongly encouraged Lederberg and invited him to visit his new laboratory in March 1946.

In less than six weeks, Lederberg showed that his genetic markers were relatively stable and, after crossing, obtained recombinant bacteria, thus proving that genetic material had been exchanged and that some form of sexual reproduction must be taking place in bacteria. The result was presented for the first time at the Cold Spring Harbor Symposium in July 1946.[21] It was well received with two notable exceptions: the perpetual skeptic Delbrück thought the phenomenon uninteresting because it did not immediately lend itself to a kinetic analysis, while the French scientist André Lwoff suspected that the complementation observed between the various deficient strains of *E. coli* was not genetic but was simply the result of an exchange of metabolites between bacteria. With Max Zelle's help, Lederberg isolated recombinant bacteria under the microscope and was able to answer Lwoff's criticism by proving that the lack of mutations did indeed show genetic complementation. At the age of twenty-one, Lederberg had brought bacteria into the world of sexual reproduction and had made his own entry into the prestigious world of the young molecular biologists.

"What if" is generally a sterile question in history, but knowing what we know today, Lederberg's result appears to be a near miracle. He chose a strain of bacteria (K12) that was not widely used but that we now know was one of the few strains in which conjugation (mating) could be detected. If Lederberg had not been lucky, he would probably have had to test at least twenty different strains before getting a single positive result. After several negative results, would he have bothered to continue? Even after he chose K12, Lederberg's luck still held. By chance, he decided to follow the transmission of genetic markers developed by Tatum that all turned out to be located on the same region of the *E. coli* chromosome. As François Jacob and Elie Wollman later showed, during conjugation chromosomal transmission begins at a precise physical point on the chromosome; the markers Lederberg used all happened to be close to this point and were thus exchanged with a relatively high frequency.

In fact, genetic exchange in bacteria is a rare event of little physiological importance in "ordinary" bacterial life. For biologists, however, its discovery confirmed the fundamental unity of the living world. Above all, it provided new tools for studying genes.

Over the next fifteen years, bacterial genetics made remarkable progress and was crowned by the Nobel Prize for Lederberg in 1958, at the remarkably young age of 33.[22] Phage genetics joined bacterial genetics when Luria showed that just as there were mutant bacteria capable of resisting phages, so too were there mutant phages that could grow on these resistant bacteria.[23] At the same time, Alfred Hershey isolated other phage mutants that produced lysis plaques with different shapes.[24] In 1946, Hershey and Delbrück independently proved that if bacteria were simultaneously infected by two phages carrying different mutations, "recombinant" phages containing either both mutations or no mutations at all could be obtained.[25]

The power of bacterial genetics derives from the fact that it is possible to select extremely rare recombination events in relatively short periods of time through the use of antibiotics, phage, or—in the case of phage genetics—mutant bacteria. Seymour Benzer and a number of other researchers used these possibilities to take genetic analyses to a level well below that of the gene, revealing its inner structure. Bacterial genetics thus helped bring the gene closer to the real world—and to molecules.[26]

The contribution of bacterial genetics to the growth of molecular biology was particularly complex. Many important discoveries in molecular biology, such as the deciphering of the genetic code, took place outside the field of bacterial genetics. And yet the tools developed by bacterial geneticists often made it possible to find an elegant verification or demonstration of results that had been obtained through biochemical studies. Bacterial genetics made its biggest contribution to the growth of molecular biology in the field of gene regulation, in particular in the work of the French school of molecular biology (see Chapter 14). From a longer-term perspective, bacterial genetics also provided the essential tools for the development of genetic engineering (see Chapter 16).

Bacterial genetics is at least as abstract and formal as *Drosophila* genetics, but it is also thoroughly practical, based as it is on a set of simple and
precise techniques for manipulating bacteria and phage. However, although
the techniques may have been simple, the data they generated were initially
hard to understand.

As soon as bacterial recombination had been discovered, Lederberg tried
to establish a genetic map of *E. coli*. Unfortunately, the data could not be
easily interpreted. In 1951, Lederberg and his colleagues finally proposed
that the bacterial chromosome was a branched structure with four arms (in
reality most bacteria have circular chromosomes).[27] In 1952, the Irish scientist William Hayes discovered that sexual reproduction in bacteria was
extremely unusual, and his results provided scientists with a new interpre-
tative framework. Hayes showed that the transfer of different genetic
markers during bacterial conjugation was neither simultaneous nor sym-
metric: one bacterium behaved like a donor (male), while the other acted
like a recipient (female).[28] In reality, the use of the terms male and female
to designate the two types of bacteria was more the result of the assumptions of male scientists than of scientific objectivity.

The true nature of bacterial sexuality was finally revealed by Elie
Wollman and François Jacob. Wollman was the son of Eugène and Elisabeth Wollman, who during the interwar years had worked on lysogeny at
the Pasteur Institute in Paris (see Chapter 14); he had been a student of
Lwoff and from 1948 to 1950 had worked in Delbrück's laboratory at Caltech
on phage T4 adsorption.[29] To study conjugation, Wollman and Jacob used
a bacterial strain called Hfr (high frequency of recombination), which
showed high levels of conjugation. Wollman had the idea of interrupting
mating by violently agitating the bacterial culture in a laboratory
blender—the same apparatus (and principle) used by Alfred Hershey and
Martha Chase, and by Thomas Anderson before them, to separate bacteria
from infecting phage. Wollman and Jacob showed that gene transfer took
place regularly over time.[30] They hypothesized that the order in which genes
were transferred was related to their position on the bacterial chromosome
and that this order could be directly converted into a genetic map.

After this discovery, bacterial conjugation became a powerful genetic
tool. Contrary to the initial hypotheses, it turned out to be very different
from sexual reproduction in other organisms. But at least in terms of the
use that geneticists could make of it, it was similar both to the traditional

techniques of genetics and to generalized transduction in bacteria, which had been discovered by Norton Zinder and Joshua Lederberg in 1950.[31] The initial objective of Zinder and Lederberg was to show the existence of conjugation in bacteria other than *E. coli*.[32] The bacterium they chose was *Salmonella*. Studies with antibodies had revealed that there were several antigens on the surface of this microorganism and that different strains of bacteria carried different combinations of these antigens. Lederberg felt that this complex antigen profile was the result of sexual recombination between strains.

Initial experiments suggested that there was indeed a transfer of genetic material between different strains of *Salmonella*, but that this did not require the two strains to be in physical contact, as in conjugation. Genetic transfer was carried out by an unknown agent that could pass through filters and was not affected by the action of the enzyme DNase. This new phenomenon of genetic transfer was thus different from Avery's transformation (see Chapter 3). The properties of this agent turned out to be identical to those of the bacteriophage, which passively transported a fragment of genetic material from one bacterium to another.[33]

When similar effects were discovered in *E. coli*, transduction became a powerful tool for studying the structure of genes and their chromosomal organization. Beadle and Tatum had shown that each step in the metabolic pathways of biosynthesis or degradation was catalyzed by a different enzyme, corresponding to a separate gene (see Chapter 2). Transduction experiments revealed that in bacteria the genes involved in a given metabolic pathway were found close together on the chromosome, often in the same order as they intervened in the pathway.[34] This spatial organization suggested that these genes might be coregulated (Chapter 14).

Transduction nevertheless remained mysterious. In particular, it was unclear how bacterial genetic material became associated with the phage. The answer came indirectly, from the study of what were called temperate phages, which produced an effect known as lysogeny. It was known that such phages could be integrated into the bacterial genome at specific positions; they could also transduce the host's genetic material, but the only genes that became associated with the phage were those situated near the site where the phage had integrated itself into the bacterial chromosome. Transduction thus corresponded to an imprecise "excision" of the temperate phage, which took parts of the surrounding genes with it when it left the

chromosome. The simplest hypothesis that could explain generalized transduction was that transducing phages could also integrate themselves into
the chromosome, but with no spatial "preferences."[35] Upon excision, they
would take with them surrounding bacterial genetic material from wherever on the chromosome they happened to be.

Conjugation and transduction made it possible to draw up a genetic map
of *E. coli* that had a higher resolution than those of organisms classically
studied by geneticists, such as *Drosophila*. These new techniques thus
helped genetics descend to the molecular level.

6

The Crystallization of the Tobacco Mosaic Virus

THE EARLY HISTORY of molecular biology shows that the virus was the most appropriate system for studying the chemical nature of the hereditary material. Viruses had been discovered at the end of the nineteenth century and characterized at the beginning of the twentieth century. Because of their minute size, they appeared to lie between the world of chemistry, with its molecules and macromolecules, and the world of biology, the simplest representatives of which were bacteria.

Like organisms and genes, viruses could reproduce and switch to a new stable state; that is, they could mutate. The fact that viruses might be genes, or primitive forms of genes, had been suggested by Hermann Muller in the 1920s.[1] Viruses appeared to be an ideal material both for studying genes and for understanding the secret of life. In addition to these scientific reasons, a number of practical factors made the study of viruses a key research topic: viruses are responsible for serious pathologies in humans, and also for diseases in agricultural animals and plants.[2] Large sums of money have always been available for research on viruses. In fact, President Franklin D. Roosevelt, who contracted poliomyelitis at age thirty-nine, was instrumental in securing funds for research on the polio virus and the development of a vaccine.

Viruses will not reproduce in synthetic media. This problem was overcome only by the development of the cell culture technique in the 1950s, which explains why, for many years, the structural study of animal viruses was extremely difficult. Although the study of bacteriophages was much easier, their viral nature was firmly established only in the 1940s. Throughout this period, plant viruses, and in particular the tobacco mosaic virus (TMV), were the model system for studying viral reproduction.[3] The leaves of the infected plant were the equivalent of cell culture dishes: viruses were placed on a leaf, and after a period of incubation their number could be directly estimated by measuring the area of dead tissue.

The purification and crystallization of TMV, carried out by Wendell Stanley in the summer of 1935, was a revolutionary event and was widely reported at the time.[4] Stanley was compared to Pasteur, and in 1946 he was awarded the Nobel Prize for chemistry for this work.

This success was typical of the physicochemical approach to the study of biological phenomena that was particularly well-rooted at the Rockefeller Institute. Stanley was trained as a chemist and had done his doctoral thesis on the synthesis of bio-organic compounds. In 1931, after a year-long stay in Munich, he was recruited by the director of the Rockefeller Institute, Simon Flexner, to work in the new laboratory of the Department of Animal and Vegetable Pathology, located in the institute's annex at Princeton.

Stanley found the atmosphere at Princeton extremely favorable for a number of reasons, one of which was financial: the Rockefeller Institute was one of the research bodies that was able to weather the budget restrictions after the Crash of 1929.[5] There were also scientific advantages: one of Stanley's Princeton colleagues was John Northrop, who, after training in Thomas Hunt Morgan's group, had turned from genetics to physical and chemical biology because he felt that these disciplines were scientifically safer. A year before Stanley arrived at Princeton, Northrop succeeded in crystallizing an enzymatic protein, pepsin.

Above all, Stanley found that the atmosphere at the Rockefeller Institute and at Princeton was philosophically supportive. As we have seen in the case of Oswald Avery, the dominant biological approach at the Rockefeller Institute was reductionist—researchers sought to explain biological phenomena in physicochemical terms. John B. Sumner's success in crystallizing the first enzyme, urease in 1926, together with Northrop's more recent results had strengthened confidence in this reductionist approach, first championed in the 1920s by the German-born American biologist Jacques Loeb, who was also Northrop's mentor.[6]

Stanley was not the first person to try to purify and crystallize TMV, but he brought to the project his substantial skills as a chemist and, most importantly, his conviction that the virus was a protein and that methods for purifying proteins would enable him to isolate the virus in a pure form. Having shown that the infectivity of TMV was destroyed by pepsin, by extreme pH levels, and by a number of chemicals known to inactivate proteins, he required only a few weeks to complete the purification and crystallization of the virus. In his article published in *Science* in 1935, Stanley stated that the crystalline material he obtained was a protein, and that its infectious properties remained unchanged after more than ten successive crystallizations.[7]

In the years that followed, Stanley discovered more about the properties of the virus. He gave a sample to the Swedish chemist Theodor Svedberg while Svedberg was on a visit to New York. On his return to Uppsala, Svedberg used ultracentrifugation to show that the molecular mass of the virus was 17 million daltons and that it was rod-shaped. This structure was confirmed in 1940 in one of the first studies performed with the electron microscope built in the RCA laboratories at Princeton, close to Stanley's laboratory. Stanley's laboratory was also one of the first in the United States to be equipped with a Tiselius electrophoresis machine (see Chapter 9). Stanley used the new machine to characterize different strains of the TMV as well as other plant viruses.

Stanley's discovery had a huge impact. It was seen as the most important discovery to have been made at the already prestigious Rockefeller Institute. The press gave Stanley's work a great deal of coverage, to which he made no small contribution. It was the crystallization of the virus that most excited the popular imagination. A crystal was a symbol of the material, mineral, mechanical world, but this particular crystal was part of life. As the *New York Times* put it, "in the light of Dr. Stanley's discovery, the old distinction between life and death loses some of its validity." In a way, the crystallization of TMV represented life's Twilight Zone.[8]

Despite this enthusiastic reception, Stanley's work was soon criticized. Some of these criticisms dealt with matters of principle—crystallization does not mean purity—others with the experiments themselves. In England, Frederick Bawden and Norman Pirie tried to reproduce Stanley's results; they managed to crystallize the virus but found that 6 percent of it was composed of a nucleic acid (ribonucleic acid, RNA).[9] Furthermore, they showed that any treatment that inactivated this RNA also altered the infectivity of the virus. Their conclusion was that the virus was not a pure protein but a nucleoprotein. This result fitted in with those models that considered nucleoproteins to be the essential constituent of genes and chromosomes. Although Stanley never replied directly to his critics, he later admitted that he had known that RNA was a constituent of TMV, but he did not give this observation any great weight.[10] It is true that, in 1940, the structure of RNA was as little known as that of DNA, and this did not encourage researchers to think that it could play a role in hereditary transmission.

For all these reasons, the purification and crystallization of TMV did not shed any real light on the chemical nature of the genetic material. In 1956, when there were still a few lingering doubts that nucleic acids might carry hereditary information, Gerhard Schramm and Heinz Fraenkel-Conrat finally provided decisive proof that the RNA in TMV was indeed the infectious material. Avery's experiment thus remained the first to show that the genetic material was formed of a nucleic acid, despite the fact that the link between transformation and gene expression was a lot more tenuous than the link that was thought to exist between viruses and genes.

To leave the matter there, however, would be to underestimate the importance of Stanley's experiment. Despite its fundamental error, on a philosophical level it helped to validate the reductionist, physicochemical approach to biological phenomena. Scientifically speaking, it had an enormous impact, in particular on geneticists. Muller welcomed the "discovery" enthusiastically, while Northrop reoriented part of his work to isolating the bacteriophage. Stanley himself devoted more and more time to genetics and encouraged some of the researchers in his laboratory to enter this new field.

Bit by bit, the gap separating geneticists and biochemists began to close. The techniques Stanley used to purify and above all to characterize "his" virus became the experimental reference for virologists.[11] The Rockefeller Institute was equipped with a new ultracentrifuge, which not only permitted the preparation and purification of viruses on a large scale but, more importantly, opened the way to the fractionation and isolation of cellular constituents. These studies rapidly led to the discovery of microsomes, intracellular particles that were somehow involved with protein synthesis. But it would be another twenty years before the relation between genes and microsomes was finally clarified (see Chapter 13).

7

Enter the Physicists

AS WE HAVE SEEN, a number of physicists played an important role in the birth of molecular biology, despite having no biological training. Strangely enough, some of these physicists did not work very closely with biologists; instead, they chose their own research projects and independently developed the biological "spin-offs" of their results.

The term "physicists" here refers to an ensemble of researchers belonging to different disciplines and subdisciplines—mathematicians, theoretical physicists, specialists in quantum mechanics, and so on—who contributed in different ways to the rise of molecular biology. Despite this diversity, they can be grouped under a single heading because they were all external to biology. They also believed that it was necessary to accelerate the development of biology and to rid it of its lingering metaphysical interpretations through the use of concepts and techniques that were derived from physics.

The movement of physicists into biology involved far more than a handful of specialists—many scientists with long and distinguished careers in physics decided to retrain completely and begin studying biology. One such example was the Hungarian physicist Leo Szilard, who had provided the first physical explanation of Maxwell's demon (a well-known thought experiment) and thus helped found what was to become information theory.[1] As the inventor of the atomic chain reaction, he played a key role in setting up the Manhattan Project to develop the atomic bomb: with two other Hungarian physicists, he alerted Albert Einstein to the risk that Nazi Germany might develop an atom bomb, which led Einstein to write to President Franklin D. Roosevelt. But at the beginning of 1945, once it became apparent that Nazi Germany was on the verge of defeat and had no nuclear bomb, Szilard became one of the first scientists to oppose the military use of atomic weapons. At the end of the war he abandoned physics and turned to the study of bacterial metabolism, playing an important role in the analysis of gene regulation in bacteria (see Chapter 14).

George Gamow, one of the fathers of the "big bang" theory of the origin of the universe, was another physicist who made important contributions to molecular biology. Gamow reacted quickly to Jim Watson and Francis Crick's second 1953 article, on the genetical implications of the double helix structure of DNA. In the days that followed its publication, Gamow wrote to the pair proposing a code that would explain the relation between the sequence of bases in DNA and amino acid chains (see Chapter 12).[2] Gamow's model and its errors will be discussed later; for the moment it is sufficient to note that he was the first to try to decipher the genetic code, and without his prompting Crick might not have turned his attention to the question. Gamow's letter came as a shock to Watson and Crick—despite the fact that they had used the term "code" in their article, they had not taken their model to its logical conclusion and imagined that it would be possible to decipher the code. Gamow later made a number of other contributions to the theoretical study of the structure of the genetic code.[3]

Many other physicists were fascinated by biology. Nicholas Mullins has attempted to quantify their role through a detailed study of the phage group, using data on the scientific origin of the various researchers involved in the group between 1945 and 1966, as measured by their doctoral discipline.[4] Before 1945 three of the six members of the group had obtained a PhD in physics or chemistry. Between 1946 and 1953, ten of the nineteen new members had a PhD in physics, biophysics, or chemistry. The weight of physicists (and chemists) in the group began to decrease only in 1954 (four out of thirteen new members between 1954 and 1962). This shift was no doubt due to the impact of the discovery of the DNA double helix (see Chapter 11), which helped to attract young biologists to molecular biology at a time when it was becoming a separate science with a large number of potential research subjects for graduate students. This attempt at quantification is based only on the phage group, which was only one of the many branches of molecular biology. Furthermore, the structure of the phage group was sufficiently vague for its exact composition to be difficult to determine; data relating to it should therefore be treated with caution.

Whatever its precise value, the numerical importance of the physicists cannot be denied. François Jacob provided a particularly penetrating analysis of the reasons for this intellectual migration and the changes that took place as a result of World War II.[5] Before the war, some young physicists were disappointed by the discipline in which they were training. It was not so much that physics, basking in the glow of the double revolution of quantum mechanics and relativity, was not a brilliant subject—each day these new theories led to new discoveries. Rather, physics had entered a phase of what Thomas Kuhn called "normal science," in which most activity consists of "puzzle-solving" rather than questioning the very bases of the discipline.[6] For the most ambitious young physicists, the prospect of checking or at best slightly improving their elders' models was somewhat uninspiring. Furthermore, the very structure of research in physics was changing. Isolated researchers were being replaced by teams focusing on big projects or massive pieces of apparatus such as particle accelerators. In this kind of work, which is based on the collaboration of a number of specialists, the role and contribution of each individual are reduced, or at least more difficult to determine. Some scientists are too individualistic to accept being near-anonymous members of a large multidisciplinary team.

With the outbreak of World War II, other factors tended to push some scientists away from physics. Many American and British physicists were closely involved in the war effort. For the first time in history—with the exception, perhaps, of the role of scientists in the organization of the armies of the First Republic in France—physicists and mathematicians were called upon to help the military. They developed radar and sonar, improved communications, decoded messages, and developed powerful calculators (of which computers were later a natural development).[7] Physics made major contributions to key technological advances during World War II, which was a war of intelligence, organization, and science. The British were the first to realize the new form that the war had taken and to organize and direct their scientific work accordingly—their geographical and military situation left them little alternative.[8] Their only hope of victory was to shift the terrain of battle.

The teaming up of different scientists and the importance given to information and its exchange led to a new vision of the world. But a negative consequence of scientists' involvement in the war effort was the desire felt by many of them to escape from what they saw as a form of conscription. They wanted to put an end to the guilt they felt at being associated with various military operations at the end of the war, which fundamentally altered what, for a long time, had seemed to be a struggle between good and evil. Unlike the Manhattan Project itself, the atomic bombs dropped on Hiroshima and Nagasaki could not be justified by a potential German nuclear threat—the terrible destruction reminded physicists that the military and the politicians had motives and objectives that were different from theirs and which they could not influence. Research on the atom and on fundamental physics emerged from the war sullied and, for many years, became suspect. Biology appeared as a fresh new field, far from political concerns and sheltered from potential military uses.

Physicists were also attracted to biology because it seemed to harbor a large number of unsolved fundamental problems and to be the new frontier of scientific knowledge. Quantum physics, and the new chemistry it had produced, appeared to be able to provide the tools and concepts necessary for understanding the mysteries of biology. In addition, physicists saw the opportunity to naturalize biological phenomena, ridding biology of all remnants of metaphysics and teleology, as had been done in physics three centuries before. This kind of philosophical motivation was very important for Crick, who had trained as a physicist, and also for biologists like Jacques Monod. In fact, it was the fathers of quantum mechanics, Niels Bohr and Erwin Schrödinger, who were the heralds of this new approach, as can be seen from their writings, which reveal the motivations and hopes they shared with many other physicists.

Of all the leaders of the quantum revolution, Niels Bohr is probably the best known to the general public. All high school students learn about Bohr's conception of the atom, with its electrons orbiting on precise trajectories around a positively charged nucleus. His description of the position of electrons around the nucleus—which went against the dogma of classical physics—was able to explain experimental data on the absorption and emission of light by atoms and laid the basis for quantum mechanics.

Bohr did not, however, play a direct role in the birth of the new physics. He was not involved in the wave approach developed by Louis de Broglie and Erwin Schrödinger, nor in the more mathematical studies of Werner Heisenberg, Paul Dirac, and Wolfgang Pauli, nor did he contribute to

Schrödinger's synthesis. Nevertheless, their results justified his atomic model a posteriori. Furthermore, Bohr was the first to see the radical novelty of the new theories, which put into question the absolute determinism of classical physics and justified new, complementary approaches to reality, abolishing the division between the observer and the observed subatomic world. A small group was formed around Bohr at Copenhagen, pushing the consequences of the new theory to their limits. Many physicists rejected what became known as the Copenhagen interpretation—neither Einstein nor Schrödinger would follow it in abandoning determinism. For the rest of his life, Einstein searched for the hidden variables that would restore determinism to the new physics, while Schrödinger developed his famous thought experiment involving a cat in a box in order to demonstrate the unacceptable implications of the new view.

As well as playing this decisive role in physics, Bohr was also interested in biology. He was the son of Christian Bohr, a famous Danish physiologist who had studied the fixation of oxygen by blood; and as we have seen, Bohr encouraged many physicists, such as Max Delbrück, to turn to biology. Bohr invited George Hevesy to the Institute of Theoretical Physics, which he had founded in Copenhagen, and encouraged Hevesy's research into the use of radioactive isotopes as biological markers, including asking for support and funding from the Rockefeller Foundation.[9] After the war, this new technology was to be of major importance for the development of biochemistry and molecular biology.

In August 1932, Bohr was invited to give the inaugural lecture at the International Congress on Light Therapy at Copenhagen. The following year, this lecture was published in *Nature* under the title "Light and Life."[10] After a long presentation of quantum mechanics, Bohr turned to his main theme: how these recent results on the nature of matter (and of light) change our vision of life. Bohr's aim was not to carry out some kind of takeover of biology by physics through the mere transposition of results from one science to another, but rather to explore the epistemological implications of the new vision of the physical world for perceptions of the biological world.

One of the key results of the new physics was the principle of complementarity. A given object, such as a photon, could and indeed should be studied by approaches that were different but complementary—it had to be studied both as a wave and as a particle. In the same way, Bohr argued, life should be studied by complementary approaches. No one could deny the importance of chemical and atomic phenomena in the functioning of all organisms. Advances made in the study of biological molecules had quashed the idea that there was a vital force and suggested that living matter was no different from inorganic matter. Nevertheless, studies of organic molecules generally required that the organism that contained them be dead—the reductionist study of life meant that life had to be destroyed. To resolve this paradox, Bohr proposed that life should be accepted as an elementary fact that could not be explained, the biological equivalent of the quantum in physics. Bohr asked if it was possible to imagine and even develop another biology alongside the reductionist approach, one that would accept life's teleological, intentional aspects.

In his conclusion, Bohr tried to correct the potentially negative impression given by the body of his lecture. The limits of knowledge of the inanimate world had not prevented quantum mechanics from leading to a far superior understanding of nature and to a massive increase in humanity's power over it. Similarly, renouncing any attempt at explaining life would not be an obstacle to making enormous progress in the future understanding of biology.

This lecture has often been cited but rarely read, and it seems that those who have read it have misunderstood it. Bohr's ideas, and even more so those of Delbrück, who developed Bohr's positions and used them as the basis of his biological research program, were dismissed by their critics. Bohr suggested that physics would never be able to explain the functioning of organisms, and that other principles would have to be discovered before the biological world would become comprehensible. As we have seen, this was one of Delbrück's motivations, which explains why he began to turn away from molecular biology shortly after Watson and Crick reduced the self-replicative abilities of the gene not to new laws but simply to the complementary chemical structure of the double helix. Although it may appear that Bohr was also motivated by the desire to discover new principles, he was more than somewhat reserved when Delbrück later tried to present him as an inspiration.[11]

This apparent misunderstanding between Delbrück and Bohr suggests a different reading of Bohr's Copenhagen lecture. Bohr was quite at home with the idea that life could not be reduced to the material world. But the suggestion that a new approach to biology might be productive implied merely that Bohr was being prudent rather than that he was promoting some kind of antireductionist manifesto. Paradoxically, because this approach rejected all philosophical reductionism, the declaration that the irreducibility of life is the quantum of the new biology opened the road to experimental reductionism. By limiting the philosophical implications of chemical, physical, or biological approaches, scientists could proceed as though life were nothing more than the result of interactions between the molecules of the organism. A similar strategy had been adopted in the nineteenth century by Claude Bernard, who rejected both materialism and vitalism, considering them both to be metaphysical questions that had no place in science and preferring instead to focus on the physicochemical interpretation of biological phenomena.

In the final analysis, Bohr's motives do not matter. What does matter is that his lecture helped to turn the attention of physicists toward biological objects.

In a series of lectures given in Dublin in 1943 and in his book *What Is Life?*, which was based on these lectures and was published in 1944, Schrödinger drew the attention of young physicists to the latest results from genetics and suggested that quantum mechanics would be able to explain them.[12] No sooner was Schrödinger's book published than it was widely seen—unlike Bohr's lecture—as a transfer of the new concepts of physics into biology, or, more polemically, as a takeover bid by physicists on the mechanisms of heredity.

The plan of the book reveals Schrödinger's intentions. He began with a description of the principles upon which order is based in classical physics. For the physicist, thermodynamics—order at the macroscopic level—is merely the statistical result of microscopic disorder. Schrödinger then described the main concepts and findings of genetics, underlining the stability of genes and the fact that this stability is perturbed only by sudden and rare mutations, which then lead to new stable states. Schrödinger also described Hermann Muller's success in inducing mutations by radiation. Schrödinger argued that these two properties of genes—stability and mutability—could not be explained by classical physics: genes are far too small for their stability to be the result of a statistical behavior of their constituent molecules. Instead, claimed Schrödinger, their properties evoked the stable energy levels that quantum mechanisms showed existed in molecules. According to Schrödinger, the molecules that form genes would have to be sufficiently complex for the energy levels that they could attain to have the observed degree of stability. Schrödinger then showed how Delbrück, using this molecular model of the gene, had been able to explain some of the characteristics of mutations and had tried to measure the size of genes. The book continued with two chapters on order in organisms and closed with an epilogue on determinism and the problem of free will.

Schrödinger's book was remarkably influential. Many of the founders of molecular biology claimed that it played an important role in their decision to turn to biology. Gunther Stent, a geneticist (and a historian of genetics), argued that for the new biologists it played a role like that of *Uncle Tom's Cabin* in the run-up to the American Civil War.[13] Schrödinger presented the new results of genetics in a lively, compelling way—much better than the biologists had been able to do. More than seventy years later, the book has lost none of its seductiveness: its clarity and simplicity make it a pleasure to read.

Modern molecular biologists feel quite at home studying the pages of Schrödinger's book. They share Schrödinger's determinist vision of the gene—"In calling the structure of the chromosome fibers a code-script we mean that the all-penetrating mind, once conceived by Laplace, to which every causal connection lay immediately open, could tell from their structure whether the egg would develop, under suitable conditions, into a black cock or into a speckled hen, into a fly or a maize plant, a rhododendron, a beetle, a mouse or a woman."[14] This view can be traced down the centuries, as far back as René Descartes in the 1600s.

Today's molecular biologist would wholeheartedly agree with this research program. Furthermore, Schrödinger was the first to use the word "code" to describe the role of genes. In another part of the book, where he attempted to describe the structure of genes, Schrödinger put forward the hypothesis that they are formed of some kind of aperiodic crystal—a remarkable anticipation of the nonmonotonic polymer structure of chains of nucleic acids.

The idea that Schrödinger's views were original or even coherent has been vigorously contested, and most contemporary historians and philosophers of science are ill at ease with the idea of "precursors" or "founding fathers."[15] Philosophers of science—often rightly—consider this kind of position a consequence of scientists' habit of developing their own historical mythology. It can be argued that Schrödinger's book was fully appreciated only in the 1960s, once the key findings of molecular biology had been discovered. Perhaps this is a case of the new science constructing its own historiography, furnishing itself with prestigious founders such as Niels Bohr and Erwin Schrödinger.[16]

Schrödinger did indeed borrow a number of ideas from other scientists—not only from Delbrück, but also from geneticists such as Muller—and it is easy to point out Schrödinger's blind spots: despite his perceptive guess about the nature of the gene, he was not in fact interested in its chemical nature. He did not mention the early, unpublished results of Oswald Avery on the transforming factor, even though they had already been highlighted by Dobzhansky in his 1941 book *Genetics and the Origin of Species*.[17] Furthermore, although he cited Delbrück, it was for his work on the gene and not as the leader of the embryonic phage group, which at the time was beginning to make its influence felt. Finally, Schrödinger did not discuss the exact role of genes, nor did he mention George Beadle and Ed Tatum's work on genetic physiology.

None of these criticisms, however, detracts from Schrödinger's originality, which flowed from the physicist's farsighted vision, allowing him to see genes merely as a code that determines the formation of the individual.[18] Schrödinger dared to say what no geneticist would have said, that "these chromosomes . . . contain in some kind of code-script the entire pattern of the individual's future development and of its functioning in the mature state."[19] For Schrödinger, genes were no longer merely guarantors of order within the organism, mysterious conductors that ensured its harmonious functioning. Instead, they were musical scores that, down to the smallest detail, determined the functioning and the future of every organism. The implication was that to decipher the molecular code-script contained in chromosomes is to know the organism. Schrödinger transformed the geneticists' conviction that genes were the heart and soul of the organism into a new molecular vision.[20] He thus anticipated the results of molecular biology that showed how genes determine the position and nature of all amino acids, and thus of all proteins within the cell.

Schrödinger's vision was undoubtedly original, but this originality was not necessarily perceived by readers at the time; a survey of the various reviews of Schrödinger's book in journals of the time is particularly revealing in this respect.[21] Most of the reviewers concentrated on the final chapters of the book, devoting most of their attention to his thoughts on order, on what Schrödinger called negentropy—the property of living beings that enables them to increase order at the cost of their surrounding environment—and on the problem of free will. None of the reviews even mentioned the idea of a code, of genes being constituted by an aperiodic crystal, or the new conception of the role of genes. Furthermore, when Schrödinger's readers were interviewed twenty years after the book appeared, none of them remembered having been struck by the novelty of the ideas it contained, but instead claimed to have been seduced by its clarity, by the fact that it made biology attractive to young physics students. One exception was Crick, who in August 1953 sent Schrödinger reprints of the two *Nature* articles on the structure of DNA that he had published with Jim Watson, pointing out the link between Schrödinger's ideas and their findings. There is no record of any reply from Schrödinger.

The failure of early reviewers to recognize Schrödinger's originality is not, however, the end of the story. It simply shows that Schrödinger's book was not, strictly speaking, a scientific work. It was neither a treatise on biology nor a collection of articles. Its aim was not to encourage new experiments nor to present new models of how organisms function, but rather to offer a new vision of biology. But a new vision does not impose itself in the same way that a new theory can. At first it goes unnoticed, and it is never the work of one person. In this respect, Schrödinger was as much the representative of this vision as its author.

To say that this new vision had no direct influence on scientists or on the experiments they carried out does not mean that it had no influence at all. Schrödinger and the new concepts of code and of program were not the source of the new science of molecular biology. But together with the newly emerging idea of information—not mentioned by Schrödinger—they provided a conceptual framework within which experiments were interpreted, and which allowed new experiments and new research projects to be developed.

Although many other physicists were also interested in biology, few of them went so far as Schrödinger—giving popular lectures and publishing them as a book. Although Schrödinger gave the lectures in fulfillment of an obligation that came with his new post in Dublin, he clearly enjoyed giving them, extending the series from one to three lectures, and even taking them on tour to provincial Irish cities. The reasons for this dedication can be found in Schrödinger's personality, his academic training, and his philosophical ideas.[22] Schrödinger was one of the founding fathers of quantum mechanics, but he refused to accept Bohr's nondeterminist explanation. His final years were devoted to an attempt to unify the separate branches of physics. Restoring determinism and reuniting the different domains of physics seem to have been his two key aims throughout his career. In this respect, he was faithful to Ludwig Boltzmann, who, although he was not Schrödinger's teacher, had been his model at the University of Vienna between 1906 and 1910. Boltzmann had unified two branches of physics—thermodynamics and mechanics—that had been separated throughout the nineteenth century, and had shown that there is physical order behind the apparent disorder of atoms.

Although Boltzmann's influence can help us understand Schrödinger's desire to find a principle of order at the center of life, Schrödinger's personality and the environment in which he grew up also explain his highly original approach to biological problems. In 1925, before he made his key discoveries in physics, Schrödinger wrote a philosophical book in which he exposed his "view of the world."[23] This book reveals the philosophical framework within which his scientific thinking was situated.[24] An adept of Hindu philosophy, Schrödinger refused to accept the dualism of life and the world. Each living being, he believed, is merely a facet of the same totality: this explains both how we can all know the world and that this knowledge is the same for each of us. He claimed that organisms differ from inanimate matter only by the existence of memory—the memory of past events, and also the memory of previous generations. Instincts are the mnemonic trace of the behavior of the organisms that preceded us, he argued, while embryonic development is the memory of the evolution of life. Although he did not express it this way, for Schrödinger the existence of a genetic memory constituted the specificity of all life.

This early work thus contains the hazy outline of the research program Schrödinger was to propose to his physicist colleagues some twenty years later in *What Is Life?*—find a principle of order in organisms and link it to the memory of things past. These two themes—memory and the search for a deep, hidden order—were at the heart of early twentieth-century Viennese ideas.[25] Thanks to Schrödinger, molecular biology can be seen as a late flowering of the intellectual effervescence that characterized Vienna in the early years of the century.

Today, with the partial exception of bioinformatics, very few physicists or mathematicians dare to propose new views of organisms, in contrast with the situation in the first half of the twentieth century. The growing specialization of the sciences is not the only factor responsible for this change. Early in the twentieth century it was commonplace to think that biological knowledge lagged behind that of the inanimate world. Bridging that gap by using the recently acquired knowledge in physics and chemistry was a "natural" project, at least for the physicists. Biology appeared to be the new frontier of knowledge.

But the history of science is not only the history of ideas and desires; it is also the history of techniques and financial means. And in that respect one institution played a key role though its desire to regenerate biology by encouraging the use of the tools of physics and chemistry—the Rockefeller Foundation.

8

The Influence of the Rockefeller Foundation

FUNDING PROVIDED by the Rockefeller Foundation played an essential role in the birth of molecular biology.[1] There have been so many studies about this period in the history of science that the underlying driver is clearly not simply a desire to understand the development of molecular biology. The stakes are much higher: the question has very important implications for science policy. If the Rockefeller Foundation did indeed play an essential role, this shows that scientific research can be oriented in a given direction from the outside and implies that a similar policy could be applied today.

In contrast with other countries, in particular those in Europe, in the United States foundations have played an important role in the development of science.[2] The Rockefeller Foundation was created in 1913 by John Rockefeller, the famous petroleum magnate, to contribute to the well-being of humanity. This aim, it was held, could and should be attained by a systematic application of scientific knowledge. Within this rather general framework, the aims of the Foundation were nevertheless relatively diverse. From 1913 to 1923, the Foundation was devoted to general education and public health. From 1923 to 1930, it mainly supported medical and scientific teaching. After World War II an important part of the Rockefeller Foundation's work was devoted to the development of high-yield strains of wheat and their distribution in Mexico, India, and other developing countries. The period that concerns us here stretched from the 1930s to the years that followed World War II, during which the Foundation underwent a major change of orientation.

The new policy was the result of intense discussions that took place after the onset of the Great Depression of 1929. This economic crisis, one of the most serious that America had ever known, came after a long, uninterrupted period of growth. The scale of the slump implied a profound, noncontingent cause. For the Rockefeller Foundation administrators, the origin of the crisis lay in the gulf between humanity's understanding of the productive forces, which had grown continuously in the previous period, and our understanding of ourselves, which had not progressed. A quote from Warren Weaver, a mathematician and physicist by training, who was the head of the natural science division of the Rockefeller Foundation from 1931, illustrates the Foundation's analysis of the crisis and the objectives of their new policy:

> Our understanding and control of inanimate forces has outrun our understanding and control of animate forces. This, in turn, points to the desirability of an increased emphasis, within science, on biology and psychology, and on the special developments in mathematics, physics, and chemistry which are . . . fundamental to biology and psychology.[3]

In consultation with Weaver, the administrators gradually drew up a program for the Foundation that was initially extremely ambitious. Until 1934 an important aspect of Weaver's program was the understanding of humanity through the study of psychology, hormones, and nutrition. Weaver wrote:

> Can man gain an intelligent control of his own power? Can we develop so sound and extensive a genetics that we can hope to breed, in the future, superior men? Can we obtain enough knowledge of the physiology and psychobiology of sex so that man can bring this pervasive, highly important, and dangerous aspect of life under rational control? Can we unravel the tangled problem of the endocrine glands, and develop, before it is too late, a therapy for the whole hideous range of mental and physical disorders which result from glandular disturbances? Can we solve the mysteries of the various vitamins . . . ? Can we release psychology from its present confusion and ineffectiveness and shape it into a tool which every man can use every day? Can man acquire enough knowledge of his own vital

processes so that we can hope to rationalize human behavior? Can we, in short, create a new science of Man?[4]

In 1934, after the program had been subject to internal discussion that focused particularly on its psychobiological aspect, Weaver refocused his attention on fundamental biology and the application of the new techniques of physics to biochemistry, cellular physiology, and genetics. The importance given by Weaver to endocrinology and to the study of vitamins first decreased and then eventually disappeared completely. In 1938, Weaver was the first to use the term "molecular biology" to describe the new approach of applying the techniques of physics and chemistry to biological problems.[5] In this sense, Weaver can be considered the father of molecular biology, even if this expression poorly describes what the new field actually involved at this time. Macromolecular biology would have been a more precise term, although undoubtedly less catchy.

These decisions were not simply the product of internal discussions by Weaver and his fellow administrators. The Foundation's program officers regularly consulted leading scientists, seeking guidance about how best to support promising young researchers, and above all about which new subjects and approaches should be the focus of its largesse. Between 1932 and 1959, the Rockefeller Foundation contributed an estimated $25 million to molecular biology—a small fortune in today's money. In the 1930s, this funding was particularly significant, as federal financing of research was virtually nonexistent. Some of these sums were given as grants to young or experienced researchers to pay, for example, for visits by European scientists to U.S. laboratories; some were used to fund specific research projects. This controlled allocation of grants broke with the prior practice of the Foundation, where important sums had been given to prestigious bodies that would themselves distribute the money and oversee its use. Awarding limited sums for specific projects tended to encourage the acquisition of new pieces of equipment rather than the development of new research groups.

As a result, the money provided by the Foundation played an important role in equipping many laboratories with spectrophotometers, ultracentrifuges, material required for X-ray diffraction, Geiger counters to detect radioactive isotopes, and so on. The Rockefeller Foundation also contributed to the development of these new technologies (see Chapter 9). Finally, the Foundation continued to give substantial sums of money to long-term projects proposed by a handful of high-ranking laboratories, such as Linus Pauling's laboratory at Caltech and William Astbury's laboratory in Leeds, England.

The way in which these grants were awarded was also new: the funding decision was made by Weaver and a handful of his close colleagues. The quality of the project and of the principal investigator was evaluated by a network of correspondents, chosen from the most eminent scientists in the country concerned. This informal network provided Weaver with a massive amount of information and was particularly helpful when it came to reporting how the money had been used and the scientific quality of the research that had been carried out thanks to the Foundation's funding.

European laboratories in particular benefited from this new grant-awarding policy as the Foundation sought to rebalance its previous American-centered approach. Furthermore, the diversity—or disorganization—of European science policies gave the Foundation a field of action that was both larger and freer than in the United States. Some of the laboratories that received Rockefeller grants played an important role in the rise of molecular biology. In the 1930s and 1940s, Thomas Morgan's laboratory, like Pauling's and George Beadle's, was heavily funded by the Foundation. On the advice of the Canadian-born French scientist Louis Rapkine, the administrators of the Rockefeller Foundation made important grants to the Centre National de la Recherche Scientifique (CNRS) in France after World War II. The funds were to be used both for equipping laboratories and for organizing international conferences. Among the key beneficiaries of this money were Boris Ephrussi's group and the laboratories of André Lwoff and Jacques Monod at the Pasteur Institute.[6] Chapter 14 describes the role of these French research groups in the development of a specific school of molecular biology.

By contrast, some groups that were to play a major role in the development of molecular biology in the 1950s and 1960s did not receive any money at all from the Foundation. For example, the various members of the phage group did not receive any Rockefeller money. This is partly explained by the fact that, from the middle of the 1940s on, public funding for molecular biology increased rapidly, in particular through the National Institutes of Health (NIH) in the United States, the Medical Research Council (MRC) in Britain, and, a bit later, the CNRS in France.[7] This initial overview suggests that grants from the Rockefeller Foundation were more important for the prehistory of molecular biology, for the initial steps in its development, than for its subsequent progress.

Most historians of science have argued that the Rockefeller Foundation had an essential role in the paradigm shift that took place in biology in the 1940s. One dissenting voice has been Pnina Abir-Am, who has denied that the Foundation had any positive role, arguing that it awarded grants only to laboratories that were already relatively rich, to well-known research groups, and not to young groups or to truly innovative projects.[8] Grants awarded to physicists tended to further isolate their recipients, preventing them from adopting a genuinely biological approach—or at least not helping them to do so. The real molecular biology, according to Abir-Am, was born out of the close cooperation of biologists, biochemists, geneticists, and physicists, without any help from the Rockefeller Foundation. The valid points in this critique have led historians to clarify the precise role of the Foundation, and the consensus view can be summarized as follows.

The grant-awarding policy followed by Warren Weaver and the Rockefeller Foundation, which prioritized funding for physics and chemistry laboratories that wanted to start studying biology, or biology laboratories that needed buy equipment to make physical studies of organisms, was quite clearly based on a materialist, reductionist philosophy that saw future progress in biology as coming from an understanding of the molecules that constitute all life. Providing laboratories with the ability to use physicochemical techniques undoubtedly played a key role in changing scientists' mentality. A priori, biologists might be expected to think that studies of biological molecules would not be sufficient to fully understand organismal function. If, however, techniques and machines were available that could enable their research to study such molecules, there can be little doubt that this outlook would be shaken. According to the thinking of the Foundation, they would slowly but surely realize that the physicochemical study of biological molecules would be the foundation of the new biology.

Too often, a conceptual presentation of the history of science makes it impossible to see just how much the models and the concepts used by scientists are determined by the technologies employed. These techniques are frequently poorly described in terms of the psychological constraints they induce. The largely experimental origin of the reductionism employed by the early molecular biologists also explains why it was relatively limited.

They only rarely descended to the most fundamental physicochemical level, and often they considered macromolecules to be mere black boxes whose molecular complexity was only moderately interesting. This can be seen in Max Delbrück's striking recollection of how he viewed the significance of identification of DNA as the hereditary material: "It just meant that genetic specificity was carried by some goddamn other macromolecule, instead of proteins."[9]

On a more concrete level, the laboratories supported by the Rockefeller Foundation played an essential role in the development of molecular biology, even if many of them, taken in isolation, were not at the forefront of the new science. Many of these laboratories pursued lines of research that turned out to be mistaken, but to highlight these errors, or the dead-ends they led to, would be to ignore the contingent aspect of scientific work. For example, Pauling failed to discover the double helix structure of DNA, but this is of little importance compared to the essential role he played in the reduction of biological problems to the physicochemical level. Surely the most important point is that in trying to discover the structure of DNA, Pauling, who already had a substantial reputation, showed how important this structure was for the whole of biology (see Chapter 11).

Pauling also played an important role in later developments in molecular biology, in the study of the role of genes in controlling protein structure as well as in the demonstration of the importance of molecular data for evolutionary studies (Chapter 23). His reorientation toward biology from chemistry was partially encouraged by a substantial grant from the Rockefeller Foundation—in the space of approximately twenty years, Pauling's laboratory received nearly 1 million dollars from the Foundation.

One decision by the Rockefeller Foundation has been claimed to have had a negative effect on the development of molecular biology—in 1935 it refused to finance the creation of an Institute of Mathematical and Physico-Chemical Morphology at Cambridge University.[10] The project was intended to use a molecular approach to study embryonic development and morphogenesis. It was supported by the crystallographer J. D. Bernal, by the theoretical biologist Joseph Woodger, by the embryologists Joseph Needham and Conrad Waddington, and by the mathematician Dorothy Wrinch.[11] Without Rockefeller Foundation support, the institute was not set up, and the opportunity to encourage the molecular study of embryogenesis was apparently allowed to slip away. This is not quite the whole story, however—Waddington was awarded a Rockefeller grant to visit U.S. genetics laboratories in 1937 and 1938, and the work of this group of scientists was supported by the Foundation, even if it did not finance the creation of the institute.

Weaver had several reasons for rejecting the proposal. Cambridge University itself was not particularly enthusiastic about the project, and the British biologists consulted by Weaver were also very reserved about both the nature of the project and the personalities involved, three of whom were well-known radicals. The theoretical and antireductionist outlook of two of its members, Woodger and Wrinch, was opposed to that of the Foundation—they sought to use mathematical formulas to understand embryonic development, in particular the biological problem of organization.

Waddington and Needham were well known for their work on what was called "the organizer," which they saw as the central strength of the proposed institute. The nature and function of the organizer was one of the most exciting and fundamental problems of embryonic development. Embryonic induction had been discovered through studies of early development in frogs and toads—in 1924, Hans Spemann and Hilde Mangold had used ablation and transplant experiments to show that a particular part of the amphibian embryo (the mesoderm of the blastoporal lip) could organize embryonic development—for example, by inducing the development of neural cells in non-neural tissues.[12] This suggested that an inducing substance of unknown physicochemical nature was present in the mesoderm. Studies by Spemann's colleagues, and by Waddington and Needham, showed that induction was not altered when the mesoderm was treated with heat or alcohol and that cellular extracts were also active. Research on characterizing the inducer—later rebaptized the evocator—led to a dead end, as a range of components found in tissue extracts (glycogen, nucleic acids) were found to have an inductive power. Even completely synthetic compounds such as methylene blue were active, and, most bewilderingly, inductive activity could be extracted from tissues that had shown no such effect.

Finally, the central idea of a mathematical approach to the structure and functioning of organisms turned out to be a bad choice at this point in the development of the field. Dorothy Wrinch's theory was based on a model of protein structure (the cyclol theory) that was chemically incorrect and was brusquely rejected by Pauling (see Chapter 9).[13]

With the wisdom of hindsight, the proposed project looks decidedly shaky, and the problem of embryonic induction remained unresolved until the 1990s. As Jonathan Slack has put it, the search for the compounds involved in induction was a bit like "looking for a contact lens in a swimming pool, with the added possibility that the contact lens might be soluble in water."[14] There are a number of problems in science that are of considerable theoretical importance but that experience suggests are not ripe for resolution and should therefore be avoided. Few scientists would say so out loud, but avoiding this kind of project is part and parcel of the unwritten training of all researchers.

It would, no doubt, have been highly commendable had Weaver funded researchers with an outlook so far removed from his own, but given the project and some of the people involved, it seems unlikely that such an institute would have contributed significantly to the development of molecular biology.

In sum, the Rockefeller Foundation undoubtedly played an important role in the development of a molecular approach to biology that was based on the use of the techniques of chemistry and physics. This led to the more or less unconscious validation of a reductionist conception of biology. However, there is a flaw in this argument. The Foundation should not be seen as the driving force behind this change. Instead, by its grant-awarding policy, it was both the reflection and the servant of a more general movement within science (described in Chapter 7).[15] This movement, which was not restricted to a single institution or a single country, eventually led to the birth and growth of biotechnology in the 1970s.[16]

The Institut de Biologie Physico-Chimique (IBPC) of Paris, in which Ephrussi and Beadle worked, is one example that reveals how widespread were the ideas supported by the Rockefeller Foundation. The institute was created in 1927—before the Rockefeller Foundation program was launched—by Baron Edmond de Rothschild (the head of an eponymous foundation) and the physicist Jean Perrin. Its objectives were very similar to those that were later adopted by the Rockefeller Foundation. By gathering physicists, chemists, and biologists under one roof, the goal was to develop the least advanced of these three sciences, biology, by the use of concepts and development of technologies that came from physics and chemistry. The ambition of its founders was to return biology to the study of fundamental questions, such as the nature of living phenomena, which

had been partly abandoned after Pasteur's discoveries had pushed biological research toward medical applications. At its creation, the institute
possessed the best equipped laboratories in France. It offered opportunities for foreign scientists, including Beadle and the biochemist Otto
Fritz Meyerhof, which was unusual at the time in France. However, it
was hit hard by the anti-Jewish laws adopted by the Vichy regime during
World War II.[17] Partly as a result of these tragic events, it was at the Institut
Pasteur, not at the IBPC, that research in molecular biology developed
in France.

9

Physical Techniques in
Molecular Biology

THE BEST WAY to understand the nature of molecular biology is not to read popular accounts but, quite simply, to visit laboratories where molecular biological experiments are done. The look and feel—and smell!—of these places is remarkably similar the world over, and there is a striking consistency in the techniques and apparatuses that are used. Some of these have been altered first by the genetic engineering revolution, then by the appearance of laboratories devoted to a single technology such as gene sequencing, and by the increasing role of computers, but the fundamental principles mostly remain the same.

Many molecular biology laboratories employ bacteriological techniques, some of which go back many decades, sometimes for more than a century. The culture media, the form of the instruments, and even the way in which the equipment is handled were all established at the end of the nineteenth century by the French and German schools of bacteriology (Louis Pasteur and Robert Koch, respectively).[1] The phage group—Max Delbrück, Salvador Luria, and their colleagues—adopted and adapted these techniques.

There are also techniques for separating biological macromolecules and their constituents. These methods have been continually improved, and even though these changes have not always been spectacular, they have been part of important advances in our understanding. For the first half of the twentieth century, the most widely used method for fractionating proteins was salt precipitation. It was slowly replaced by chromatography.

In chromatography (long used in organic chemistry and then in biochemistry) a liquid phase is passed through an absorbent, porous, and rigid material. The development of partition chromatography and of new detection methods by the British scientists A. J. P. Martin and R. L. M. Synge increased the sensitivity of this technique and made it possible to analyze the constituents of biological macromolecules, and in particular to separate amino acids.[2] Instead of using a normal solid adsorbent material, Martin and Synge used an inert powder as the material support for the liquid phase. The column was washed with a second liquid phase that was immiscible with the first. In this kind of chromatogram, the partition between the two liquid phases replaces the adsorption on a solid phase. The material support used for the first phase was initially silica, then paper.[3] Another British scientist, Fred Sanger, immediately used this new technique to determine the amino acid sequence of the insulin molecule (see Chapter 12). Intrigued by Oswald Avery's results, Erwin Chargaff used the same method when he tried to test Phoebus A. Levene's tetranucleotide theory by measuring the base composition of different nucleic acids (Chapter 3).[4]

Various other chromatographic techniques were developed or adapted for the separation of biological macromolecules, including chromatography on ion-exchanging resins or dextran beads that separated molecules according to their size.[5] In affinity chromatography a molecule that has a high affinity for the protein to be purified is fixed to an inert material support; the subsequent application of high pressure makes it possible both to speed up the purification steps and to increase their effectiveness by restricting diffusion.[6]

The two techniques that best characterized the rise of molecular biology were ultracentrifugation and electrophoresis. These techniques can be used with two rather different aims: an analytical objective—the characterization of the properties of biological macromolecules (mass, form, electric charge, and so on), or a preparative objective—the purification of these macromolecules by separating them from contaminating molecules. The development and spread of these techniques in biological laboratories were both greatly assisted by the Rockefeller Foundation. These techniques changed the work and objectives of biologists and encouraged the adoption of a molecular vision of biology. They are sufficiently important for their history to be dealt with in some detail.

Ultracentrifugation is probably the best example of a technique that is tightly linked to a place (the University of Uppsala) and to a researcher (Theodor "The" Svedberg).[7] The first centrifuge used to determine the molecular mass and the properties of proteins was developed by Svedberg in 1924, and the first measurements were made in 1926. As noted in Chapter 1, it took some time for the concept of the macromolecule to overcome the influence of colloid theory. According to colloid theory, biological substances were composites formed by the aggregation of small molecules; their molecular mass would thus be the mean of the mass of the different aggregates. According to the macromolecular approach, however, molecular mass had a precise value. Measuring this mass was therefore crucial. Existing techniques, based on the diffusion of light, or variations in osmotic pressure or viscosity, gave only imprecise results. The first experiments aimed at deducing molecular mass from the speed of sedimentation during rapid centrifugation had been performed at the beginning of the century, but the relatively slow speeds obtained had permitted only the mass of microbeads to be measured.

Svedberg had trained in the physicochemical analysis of colloids and had initially studied Brownian motion, becoming interested in ultracentrifugation in the 1920s.[8] During a visit to the University of Wisconsin in 1923, he developed optical methods for observing the sedimentation of proteins during ultracentrifugation. On his return to Sweden, Svedberg built a machine that could produce a centrifugal force 7,000 times that of gravity (g). Two years later, he finished a new machine that could produce a centrifugal force of 100,000 g. With this apparatus, he was able to determine the molecular mass of hemoglobin (the protein that transports oxygen in the blood) and to show that it behaved like a macromolecule with a precise molecular mass. In the years that followed, Svedberg measured the molecular mass of a series of purified proteins, hormones, and serum components. Each time, exact values were obtained. Wendell Stanley sent Svedberg recently crystallized samples of the tobacco mosaic virus so that he could measure their molecular mass (see Chapter 6). Svedberg also selected the proteins he studied as a function of the work being performed at the Rockefeller Institute.

The history of science is never simple. These initial results of ultracentrifugation led to the decline of colloid theory and the triumph of Hermann Staudinger's macromolecular theory. But they also seemed to indicate that protein molecular masses were relatively consistent. Several of the proteins studied by Svedberg turned out to be made up of subunits with a constant molecular mass of around 35,000 daltons (a value that was subsequently reduced to 16,500 daltons).[9] This value took on great significance and led to the development of a number of models of protein structure. On the basis of these data, Dorothy Wrinch, a member of the Cambridge Theoretical Biology Club who worked with Conrad H. Waddington and Joseph Needham on the Institute of Mathematical and Physico-Chemical Morphology project, proposed the cyclol theory.[10] According to this theory, protein structure was stabilized by covalent bonds other than the peptide bond. In a 1939 paper published in *Science,* Linus Pauling and Carl Niemann rejected this model and argued that if proteins have preferential molecular weights, this could not be explained by chemistry but instead had a biological origin that was linked to the evolution of life.[11] The supposed existence of common molecular masses turned out to be an artifact produced by the simple fact that very few proteins had been studied. Nevertheless, the interest stimulated by these preliminary results showed that many biochemists hoped to discover simple rules that could explain the structure of proteins, and above all their formation (see Chapter 12).

Svedberg thus played an essential role in the development of analytical ultracentrifugation. In 1926 he received the Nobel Prize—not for this work but for his studies of Brownian motion, carried out twenty years earlier. The theoretical value of these studies had been strongly criticized by Albert Einstein, and Jean Perrin would later undermine their scientific value. Svedberg found the circumstances surrounding his Nobel Prize embarrassing—during initial discussions in the special commission of the Swedish Academy of Sciences, he refused to allow his name to go forward for the prize. In a subsequent plenary meeting, however, he kept silent, to the great surprise of the other members of the commission. As he wrote in an autobiographical note discovered in the 1980s, "I promised myself to use the following ten years of my life to make myself worthy of the prize."[12] Although the Nobel Prize is usually awarded for a major scientific achievement, in Svedberg's case the award precipitated the achievement. It made his work easier and enabled him to acquire the equipment necessary for his subsequent experiments. The Swedish parliament gave more than 1 million crowns toward the construction of a Department of Physical Chemistry, to which, in 1926, the Rockefeller Foundation added $50,000 to buy equipment.

The links between the Rockefeller Foundation and Svedberg were established early on, at a time when support for groups studying biological problems with physical tools was not yet the Foundation's main priority. These links were maintained over the years that followed. In 1936, the University of Uppsala received more than $250,000 from the Rockefeller Foundation. Warren Weaver was a personal friend of Svedberg's (they met in 1923 at the University of Wisconsin); these links also explain how the Rockefeller Institute at Princeton University, where Wendell Stanley and John Northrop worked, soon acquired an ultracentrifuge as powerful as that developed in Sweden. This centrifuge, obtained with Svedberg's assistance, helped the institute take a lead in virus research, which they maintained in the years that followed (see Chapter 6).[13]

Ultracentrifuges are important not only for measuring molecular mass but also for the preparation of samples—for separating different macromolecules. Beginning in the late 1920s, relatively slow preparative centrifuges were widely used in biochemistry laboratories. High-speed preparative centrifuges—ultracentrifuges—became available only later. Despite being simpler than analytical ultracentrifuges (they did not need the same optical equipment for taking measurements during centrifugation), they were nevertheless subject to the same mechanical constraints.

Protein biochemists and molecular biologists still use preparative centrifuges in various ways, depending on the speeds the machines can attain and thus the centrifugal forces they can produce. With low-speed machines, you can collect raw extracts after the rupture of cells in the sample, and you can isolate proteins or nucleic acids after precipitation by the addition of salts or alcohol. High-speed machines enable researchers to isolate viruses and separate the various cellular structures or organelles, each of which plays a particular role in the life of the cell-energy production, protein synthesis, and so on.[14] With a few experimental tricks, such as the creation of a sucrose gradient in the centrifuge tubes, ultracentrifuges can even be used to separate proteins. Fractionation by centrifugation and ultracentrifugation of intracellular contents led to the isolation of cellular extracts with which the first in vitro protein synthesis was carried out, and which was also the starting point for deciphering the genetic code (see Chapter 12).

For the ultracentrifuge to become reliable and easy to use, some key improvements were necessary, such as putting the centrifugation chamber in a vacuum to avoid overheating and using an electric motor to drive the centrifuge.[15] These modifications were the work of Jesse W. Beams and Edward G. Pickels of the University of Virginia. In the middle of the 1940s, Pickels set up a company called Spinco (SPecialized INstruments COrporation), which, in less than a decade, sold more than 300 examples of its new ultracentrifuge. The history of the installation of these machines in biochemistry and molecular biology laboratories has still not been written, but it would no doubt teach us a great deal about the spread of theoretical models and concepts in both disciplines.

Like ultracentrifugation, electrophoresis had a wide range of uses.[16] During the early years, it was employed as an analytical method for characterizing macromolecules—in particular proteins, but also nucleic acids—to show that these compounds corresponded to unique substances with well-defined properties. It was also used as a preparative technique in the separation of both proteins and their fragments. With the development of genetic engineering, electrophoresis took on a new importance, becoming *the* technique that was used to separate DNA fragments and to determine the sequence of bases in DNA molecules (see Chapter 16).

The idea of using an electric field to separate different biological components is not new. Once again, the first practical application was at least partly the brainchild of The Svedberg, who, after a number of failures, handed the project to his student Ame Tiselius in the early 1930s. In 1948 Tiselius received the Nobel Prize for his development of the apparatus.

The first results were disappointing. Like the other researchers who had tackled the problem, Tiselius encountered difficulties with heating and diffusion, which were so great that he more or less abandoned the project until he visited Princeton University on a Rockefeller Foundation grant in 1934. During his stay he met many U.S. protein researchers as well as the Rockefeller Institute immunochemists—Wendell Stanley, Karl Landsteiner, Leonore Michaelis, and Michael Heidelberger—all of whom emphasized the importance of developing a new method for separating and characterizing proteins. Tiselius therefore returned to the project, concentrating on eliminating those sources of variation that prevented the technique from being a precision tool. A new, extremely large apparatus that resolved the problems of heating and diffusion was completed in 1936 with the aid of a

Rockefeller Foundation grant. The optical system employed in the ultracentrifuges was used, so direct observations could be made of the different components in the sample being analyzed. It was even possible to remove parts of the sample during electrophoresis, thus making the apparatus both an analytical and a preparative tool.

Tiselius's first results with the new apparatus were spectacular: he was able to separate the different proteins present in serum. Together with Elvin Kabat and Michael Heidelberger, he showed that antibodies were found in the γ-globulin fraction. Tiselius's apparatus was subsequently copied and improved, but even in its initial version it enabled L. G. Longworth at the Rockefeller Institute to show that there were differences in the serum protein composition of healthy and sick subjects, thus opening the road to medical applications of the new technology.

However, the apparatus was cumbersome, expensive, and difficult to use, requiring great skill. By 1939, only fourteen such machines were in use in the United States, most of them bought with Rockefeller Foundation money, five of them in Rockefeller Institute laboratories. In 1945, the American company Klett made the first commercial electrophoresis machine. By 1950 four different companies were making them, selling them at a considerably reduced cost. But it was only in the early 1950s, when optical observation was replaced by easily detectable labeled molecules, that the use of electrophoresis became truly widespread.

In addition to liquid phase electrophoresis, the technique could also employ silica gel (or paper).[17] From the 1940s this method was widely used to separate amino acids and nucleotides. It was eventually replaced by electrophoresis on agar (a solid sugar polymer made from seaweed), then starch, which in turn was replaced by polyacrylamide gel. From the 1960s, polyacrylamide gel electrophoresis became the main technique used in molecular biology for separating macromolecules.[18] In 1970, Ulrich K. Laemmli used a reducing agent to disrupt the disulfide bonds present in proteins, and a denaturing agent (sodium dodecyl sulfate, SDS) that unfolded proteins by breaking the weak bonds that stabilize their three-dimensional structure. SDS binds to the denatured polypeptides, giving them an electrical charge proportional to their mass; as a result, migration depends only on the volume occupied by the denatured polypeptides, with the small polypeptides migrating faster. These innovations made it possible, in a single electrophoresis experiment on polyacrylamide gels, to separate polypeptidic chains according to their molecular mass and to measure this mass.[19]

It took many years for electrophoresis to have the same impact on the development of molecular biology as ultracentrifugation had. Nevertheless, according to the U.S. historian Lily E. Kay, electrophoresis led to a situation where "vital processes . . . were increasingly probed through systematic applications of tools from the physical sciences. This trend altered the nature of biological knowledge, the organization of research," in particular by involving more and more physicists in biological research.[20]

PHYSICS ALSO MADE a decisive contribution to the new biology through the creation of labeled molecules—molecules that behaved chemically like normal molecules but could be distinguished by physical criteria (mass or radioactivity). Detecting these differences made it relatively easy to follow the fate of these molecules, both in the test tube and in vivo. One of the differences between Avery's experiment and the study by Alfred Hershey and Martha Chase was that Hershey and Chase used radiolabeling to distinguish DNA from proteins. This new technology increased the impact of their experiment.

George Hevesy was one of the first scientists to use isotopes in biology when in 1923 he measured the uptake of a lead isotope by plant roots. As we have already seen, Hevesy joined Niels Bohr's group in Copenhagen to continue his work on the use of isotopes in biology and medicine.

Most of the molecules that were initially labeled were not radioactive but contained heavy isotopes of certain elements.[21] Greater sensitivity of detection techniques meant that radioactive isotopes later almost completely replaced heavy isotopes, to the extent that it has been completely forgotten that for ten years or so heavy isotopes enabled biology to make considerable advances. This forgotten chapter in the history of biology is worth studying in detail.

In 1931, in the Chemistry Department of Columbia University, Harold Urey discovered deuterium, a heavy isotope of hydrogen. This discovery quickly led to the use of isotopes in biological research. The first description of the use of deuterated compounds for the study of intermediate metabolism was published in 1935 by Rudolph Schoenheimer and David Rittenberg of the Department of Biochemistry at Columbia University. In the same year, the Rockefeller Foundation set up a special fund intended for

chemists who knew how to handle deuterium and who wanted to incorporate it into biologically interesting molecules.

These studies developed on an even larger scale when Urey obtained
compounds enriched by the heavy isotope of nitrogen, ^{15}N, which were immediately used by Schoenheimer and his coworkers. Deuterated compounds were well-adapted to the study of lipid metabolism, whereas ^{15}N was
particularly well-suited to the study of amino acid and protein metabolism.
But though deuterium could be relatively easily detected (it simply involved
a few measures of density or refraction), the study of molecules containing
^{15}N required far more complex equipment—a kind of miniature particle ac-
celerator called a mass spectrometer. This restricted the use of this isotope
to a handful of well-equipped laboratories.

Schoenheimer extended his studies first to proteins and then to all biochemical substances. After working with Hevesy, in 1933 he was forced to
flee Germany for the United States, where he obtained a position in the Department of Biochemistry at Columbia University. Schoenheimer's research convinced him that all the constituents of the organism were in a
state of dynamic instability. His conclusions were described in an important book published posthumously in 1942, *The Dynamic State of Body
Constituents:*

> The large and complex molecules and their component units, fatty acids,
> amino acids, and nucleotides, are constantly involved in rapid chemical
> reactions. . . . The free amino acids are deaminated, and the nitrogen lib-
> erated is transferred to other previously deaminated molecules to form new
> amino acids. Part of the pool of newly formed small molecules constantly
> re-enters vacant places in the large molecules to restore the fats, the pro-
> teins and the nucleoproteins.[22]

This view had an important impact on biochemistry, in particular on the
analysis of metabolism. Nevertheless, the influence of Schoenheimer's book
was ambiguous: by arguing that proteins had the same metabolic instability
as amino acids, and by suggesting that the amino acids that form proteins
could be continually added and subtracted, Schoenheimer conflated the
biosynthetic and degradation pathways of proteins with those of amino
acids. He thus placed the problem of protein synthesis in the context of
metabolism in general. His solution was the opposite of that which was

subsequently proposed by molecular biology (see Chapter 12). In the 1960s, the French molecular biologist Jacques Monod acknowledged that Schoenheimer's book had an important influence on him. Armed with Schoenheimer's ideas, Monod and his colleagues doubted that proteins were synthesized in a single step and once synthesized were metabolically stable—an issue that was significant when they had to interpret their results on β-galactosidase (see Chapter 14).[23]

Once again, the Rockefeller Foundation played a key role. The work of Schoenheimer, Urey, and others was largely financed by the Rockefeller Foundation, which, among other things, helped them buy mass spectrometers. Indeed, Warren Weaver directly oversaw the development of their research.

From the beginning of the 1940s, radioactive isotopes gradually replaced heavy isotopes for labeling molecules. The first radioisotopes produced by particle accelerators had a very short half-life and thus had little effect on the domination of the heavy isotopes. Only radioactive phosphorus, as used by George Hevesy, became an important tool for biochemistry. In February 1940, in the Radiation Laboratory at Berkeley, Martin Kamen and Samuel Ruben discovered the radioisotope ^{14}C, which had a very long half-life. This heralded the end of the golden age of heavy isotopes.[24] The use of heavy isotopes involved a combination of skills in physics, physical chemistry, and chemistry. The use of radioisotopes, on the contrary, merely required the acquisition of a Geiger counter: their use therefore spread rapidly between 1940 and 1950.

The techniques of ultracentrifugation, electrophoresis, and isotopes all show the same process of transformation: initially extremely cumbersome and requiring the close collaboration of physicists, chemists, and biologists, these methods were gradually simplified enough to be employed in all biology laboratories.

The same process can be seen in the use of other physical techniques such as spectroscopy, the study of the absorption of light—visible or ultraviolet— by molecules.[25] In this case, too, the Rockefeller Foundation played a very important role in developing the first apparatuses and in funding experiments that showed the importance of this new technique in the physicochemical study of biological phenomena, and in quantitative measures of biological components.[26] It is interesting to note that, more than half a century later, the rapid entrance of computers into biological laboratories was characterized by the same process of speedy simplification / despecial-ization (Chapters 27 and 28).

One final technique from physics that contributed to the growth of mo-lecular biology was the electron microscope, which made a direct contri-bution to two major discoveries: the structure of the bacteriophage (Chapter 4) and the existence of ribosomes (Chapters 12 and 13).[27] Apart from these examples, however, the electron microscope played a secondary role, even in the identification of intracellular structures (Chapters 12 and 13), despite what a naively realist conception of science might suggest. The images produced by the electron microscope were in fact too complex; to correctly interpret the images, scientists needed some kind of preexisting model of what they were looking at. Furthermore, the energy from the elec-tron beam was so great and the chemical treatments involved in stabilizing the sample and increasing the contrast were so extreme that the technique produced a large number of artifacts and made pseudo structures appear where none were present.[28] Images produced by electron microscopy no doubt had a certain pedagogic value and were useful as documentary proof, but they led to few if any major discoveries. Instead, electron microscopy simply confirmed the reality of results that had generally been obtained using other techniques. More recently, cryoelectron microscopy extended the power of electron microscopy and, in combination with X-ray diffrac-tion studies, made it possible to determine the structure of high-molecular-weight macromolecules (Chapter 21).

10

The Role of Physics

SCIENTISTS AND TECHNICIANS from the world of physics played a fundamental role in the birth of molecular biology, but historians disagree as to the nature and extent of this influence. For example, Horace Judson has argued that the concept of information, thought by many to be inherited from physics, is virtually indispensable for explaining the main discoveries of molecular biology. But although this concept is of fundamental importance today, it was not so at the time: it played no role in the early development of molecular biology.[1] For Arthur Kornberg, physicists who turned to biology introduced many of the strategies of physics into their new field, but few of its tactics.[2] In other words, they changed the form of biological research without altering its fundamentals. For Alain Prochiantz, a neurobiologist, molecular biology is simply an avatar of biology. According to Prochiantz, Schrödinger in no way favored the growth of molecular biology—the subject existed long before the publication of *What Is Life?,* and the book's interest is rather to be found in the analogies it suggested.[3]

Although I cannot provide a definitive answer to this debate, a number of points can be highlighted. First, although physicists did influence the development of biology, biology was nevertheless not reduced to either chemistry or physics. There is no trace of any physical formalism in contemporary biology (except very recently in systems and synthetic biology; see Chapter 28). To use the words of Nils Roll-Hansen, biological phenomena were not reduced to physics and chemistry—they were explained by physics and chemistry.[4] One of the arguments put forward to refute the idea that physicists had any real influence in the birth of molecular biology is that during this period biological research showed an apparent continuity and there was no sudden change in orientation. But this continuity, which was very real, does not imply that there was not a molecular revolution—in fact, such continuity is required for old and new scientific visions to be confronted, which is one of the hallmarks of a revolution in science.[5]

Ironically, both those who too easily accept and those who too easily reject the influence of physicists on the development of molecular biology generally share the same mistaken vision of scientific progress. To find the key influences on a science, they look to its origins, to its infancy—this is a kind of preformationist view of the history of science. Horace Judson's comments are particularly striking in this respect, because they inadvertently reveal that it is impossible to detect any influence of the concepts of physics on the early experiments of molecular biology. And yet these concepts—in particular those based on information—have become indispensable for describing the discipline. This paradox can be explained simply by the fact that today's molecular biology was not present in the initial experiments that founded the science—it has only gradually taken the form we know today. Those concepts that are so useful for understanding and explaining the results of modern biological research (information, feedback, programs, and so on) were not borrowed from physics (or from computing). Instead, they were common both to these disciplines and to biology and were at the heart of the new worldview that developed during and after World War II. They constituted the framework within which the experiments, theories, and models of molecular biology took shape.

The migration of scientists from Europe to Britain and above all to the United States as a consequence of the rise of fascist and Nazi regimes also played an important role in the birth of the new worldview. The forced confrontation of different scientific traditions and the forging of links between scientists working in very different domains but sharing the same difficult exile status were important aspects of a rich cultural mix that helped to shape the new science.[6]

The development of the techniques used in genetic engineering shows that the molecular understanding of biology acquired between 1940 and 1965 was an *operational* understanding.[7] Today both molecular biologists and physicists share a scientific worldview in which knowledge and action are intimately linked. Physicists played an important role in this change in the form of biological knowledge, by the way they conceived of and carried out their experiments. By following Delbrück and asking simple questions of biological objects, they obliged these objects to reply in the same simple language.

However, the most important contribution of the physicists was perhaps simply to have been convinced (with a certain dose of naiveté) that the secret of life was not an eternal mystery but was within reach, and above all to have convinced the biologists that this was the case.

Historians of science have a great deal of work to do before we can hope to understand what happened in the period from the 1940s to the 1960s, when two fundamental disciplines—computer science and molecular biology[8]—were born and developed, when a new worldview appeared that in many respects saw information and logic as being more fundamental than energy and matter.[9]

PART TWO

THE DEVELOPMENT OF MOLECULAR BIOLOGY

11

The Discovery of the Double Helix

FEW SCIENTIFIC FINDINGS are as well known as that of Jim Watson and Francis Crick, who in 1953 discovered the double helix structure of DNA.[1] The discovery was not the work of a coordinated group or a laboratory but a result of the meeting of two extraordinary personalities who had been trained in different scientific disciplines but had complementary abilities. Crick, a physicist by training, had begun his research at University College, London, where he studied the viscosity of water heated at high pressure.[2] Having worked for the Admiralty on the development of magnetic mines during the war, after 1945 he decided to turn to biology.

Recruited by the Medical Research Council (MRC) to study the movement of small magnetic particles within cells, he heard about a new research group being set up at the Cavendish Laboratory at the University of Cambridge. This small team, led by Max Perutz and supervised by Sir Lawrence Bragg, was working on the description of protein structures by X-ray diffraction.[3] Crick asked to be transferred to this group and got his wish.

The protein Perutz decided to study was hemoglobin. The work involved purifying the substance, obtaining crystals, then directing a beam of X-rays at the crystals. The diffraction of the beam and its decomposition into a number of different beams left a trace on a photographic plate placed behind the crystals, forming a diffraction pattern. The theory of diffraction, developed by Lawrence Bragg and his father William, described the distribution and intensity of the diffraction pattern as a consequence of the structure of the molecules in the crystal.

Research carried out before the war by Bragg's group in England and by Linus Pauling in California had shown that, by using crystals of small molecules, it was possible to deduce the three-dimensional structure of the molecules from the diffraction pattern. Bragg's aim was to study more complex molecules and to show that the diffraction technique could help determine the structure of molecules as complex as proteins.[4] This was not an easy task. In the first place, the photographic diffraction images were not very good. There were also a number of major theoretical problems. The intensity of the diffraction spots was the result of the diffraction of a wave with a given intensity and a given phase, but the observer had access to only one value when it came to determining the two factors. Added to these theoretical problems was the fact that the calculations required to interpret the images were extremely complex and had to be carried out with the aid of slide rules and primitive calculators (computers were not yet available for this kind of work).

Protein crystallographers were thus unable to interpret the diffraction images directly. They could propose only vague models of the form of the proteins they were studying, theoretically determine the diffraction spectrum of these models, and see whether the theoretical results were compatible with the observed images. One of the characteristics of a diffraction image, or of some of the mathematical functions that can be derived from an image such as a Patterson curve, is that they tend to be symmetrical and regular. The symmetry is a reflection of repetitions and regularities in the crystals and in the molecules that form them.

During his doctoral research, Crick became familiar with the complexity of analyzing diffraction images and quickly realized the limitations of the tools available. His generally negative attitude was not appreciated by his laboratory colleagues.[5] A thirty-something who still had not finished his thesis, he was widely seen as competent but somewhat left field.

Jim Watson had been a brilliant student. Having completed his studies in biology at an early age, he was Salvador Luria's first graduate student. For his doctoral research Watson had studied the effect of X-rays on the development of bacteriophage, but the data were difficult to interpret. Impressed by the results obtained by Seymour Cohen on the biochemical characterization of the bacteriophage, Luria and Watson were convinced that to understand their results and, more generally, to understand experiments on the bacteriophage, it was necessary to have a better understanding of the chemical constituents of the phage, and in particular of DNA. The phage group increasingly suspected that DNA played an important role in phage replication: the end of Watson's doctoral studies preceded by a mere two years the publication of the experiments by Alfred Hershey and Martha Chase (see Chapter 4).

Luria and Max Delbrück thought that European science was more imaginative than its American counterpart and that it would be ideal for a brilliant young researcher like Watson to do his postdoctoral research in Europe. Luria decided to send Watson to Copenhagen to work with Herman Kalckar, a specialist in nucleic acid metabolism whom he had met at the Cold Spring Harbor phage course.

Watson's visit to Kalckar's laboratory turned out to be relatively unproductive, apart from a study on phage that he carried out with Ole Maaløe in another laboratory in Copenhagen. While on a visit with Kalckar to the Naples Zoological Station, Watson attended a conference on macromolecules, almost by accident. He was struck by Maurice Wilkins's talk on the diffraction of X-rays by DNA fibers and decided to prolong his stay in Europe in order to work in a crystallography laboratory. Luria contacted John Kendrew, who worked with Max Perutz; with Perutz's agreement, Watson joined the team at the Cavendish Laboratory. Watson's official project was to purify and crystallize a protein that was analogous to hemoglobin but simpler (myoglobin), but as soon as he arrived at Cambridge, Watson began a close collaboration with Crick aimed at determining the structure of DNA. For Watson and Crick, understanding this structure became fundamental for understanding genes, and above all their ability to replicate.

At first glance, Crick and Watson did not have much going for them. Although their training was complementary—Crick could provide Watson with the information necessary for understanding the principles and interpreting the results of crystallography, whereas Watson could inform Crick about developments in bacterial genetics and the latest results from the phage group—their handicaps were far greater. Crick had to focus on his doctoral thesis, while Watson had to devote time to crystallizing myoglobin, the project for which he had been recruited. But the biggest obstacle was that they did not have direct access to diffraction spectra of DNA fibers— they had to rely on results previously published by William Astbury in 1938 and especially on data from the group at King's College, London, which was provided by Maurice Wilkins.

During the war Maurice Wilkins—who, like Crick, had been trained as a physicist—had worked at Berkeley on the separation of uranium isotopes as part of the Manhattan Project. Like Crick, he had turned to biology after the war. After working on ultraviolet microscopy, he began studying DNA fibers by X-ray diffraction. Using high-quality DNA samples obtained from the Swiss scientist Rudolf Signer, Wilkins was able to obtain very precise diffraction images and to show that the degree of hydration played an important role in the structure of DNA molecules.

Wilkins had recently been joined by Rosalind Franklin, a physical chemist who had an excellent knowledge of X-ray diffraction, which she had previously used in a study of the structure of carbons that she did in Paris.[6] Franklin was an exceptional experimentalist: she rapidly produced images that were far better than any previously obtained.

Wilkins's group apparently had all that was required to make rapid progress in the study of DNA structure. But a fundamental problem between Franklin and Wilkins made it impossible for them to work together productively. When Franklin arrived at King's she assumed, on the basis of a letter from the director of the laboratory, John Randall, that she would have her own research project on DNA. Wilkins, however, thought she had been recruited simply to provide the technical expertise necessary for him to carry out his research successfully. This major misunderstanding limited the exchanges between members of the group and prevented the elaboration of a clear research strategy.

On the basis of data presented by Wilkins in informal discussions and at scientific conferences, Watson and Crick began to build models of the DNA molecule. They had very little to work on: the proportion of water present in DNA was not known, and the structure of the bases was still a matter of debate. The only thing everyone agreed on was the structure of the polynucleotide chains. Their model, consisting of three chains of polynucleotides, was instantly dismissed by Franklin and Wilkins. This incident led to a conflict between the groups in London and Cambridge, and between Randall and Bragg. The upshot was that Watson and Crick were not allowed to continue working on the structure of DNA because all the experimental data were produced by King's College in London.[7] DNA structure was King's baby.

Although he did not know it, Linus Pauling played a decisive role—both directly and indirectly—in the development of Crick and Watson's work. In a series of articles published in 1950 and 1951, Pauling proposed the structure of some elementary motifs that were present in proteins—the first X-ray diffraction images of proteins had revealed that these molecules contained repeated regular structures.[8] As early as 1933 the British crystallographer Astbury, working at the University of Leeds, had proposed models for these structures on the basis of his study of the proteins that form wool and silk—α- and β-keratin.[9] More recently, models of polypeptide chains with a regular helical conformation had been presented by Lawrence Bragg, Max Perutz, and John Kendrew.[10] Pauling completed and corrected these initial attempts. The superiority of Pauling's work was a result of his excellent knowledge of the stereochemistry of the peptide bond and of the hydrogen bonds that could play a role in stabilizing protein conformation.[11] This knowledge came both from a simple use of quantum mechanics and from the direct measurement of angles and bonds by X-ray diffraction on crystals of small, "model" molecules. If structural constraints were respected, there was only a limited range of possible conformations for polypeptide chains. By using simple but precise molecular models, Pauling proposed that there were regular structures in proteins—these were soon found to exist, in the shape of α-helices and β-sheets.

Pauling's work showed that the helix was an important structure in macromolecules. This idea was already widely accepted by crystallographers and by specialists in the structure of macromolecules; Pauling's results thus merely confirmed something they were already convinced of. More significant was the lesson of Pauling's approach: valid predictions could be made only if the stereochemical constraints of the molecules were rigorously respected.

But Pauling's most important contributions to solving the structure of DNA were probably simply his decision to study the question and his publication of an initial model that turned out to be completely wrong.[12] For Watson and Crick, Pauling's interest confirmed the importance of their previous work and strengthened their desire to decipher the structure of DNA. Bragg eventually agreed, fearing that Pauling might repeat his success in the α-helix story and discover the structure of DNA before the British groups. Watson and Crick had first tried different models of double and triple helixes in which the bases were situated on the outside of the molecule.

Then, partly in desperation but perhaps also influenced by the work of John Masson Gulland, who in 1947 had suggested that DNA bases interacted through hydrogen bonds, they opted for the opposite model, with the bases on the inside of the molecule.[13] Thanks to the work of the chemist June Broomhead, Watson had a thorough understanding of the hydrogen bonds that could be formed between the bases. Watson and Crick also had access to the recent data of Rosalind Franklin (including the famous photo 51 of the B structure of DNA, although this was far less significant than Watson subsequently claimed). If correctly interpreted, these data showed that the structure was helical, and that the two chains of polynucleotides ran in opposite directions, with ten residues per turn. Above all, the pair could use Crick's complex method for interpreting X-ray diffraction data from helical molecules, which he had developed as part of his doctoral work and had published in 1952.

Watson first made a double helix model of DNA in which the bases paired with themselves—adenine with adenine, thymine with thymine, and so on. This model immediately suggested how the strands of the DNA molecule could replicate, but it was soon abandoned when Jeremy Donohue, who had worked with Pauling for several years, pointed out to Watson that he had chosen the wrong tautomeric forms for the bases—that is, he had put the hydrogen atoms in the wrong position on the bases because the cardboard templates he was working with were literally upside down. Having jettisoned both the external model and the pairing of identical bases, Watson had the idea of pairing different bases. He discovered that the base pairs A-T and G-C had the same spatial structure, thus enabling the construction of a perfectly regular double helix. He realized that the pairing of A with T and G with C explained the existence of roughly equal concentrations of A and T and of G and C, which had been discovered by Erwin Chargaff and of which they had been completely unaware until Chargaff's recent visit to Cambridge. The significance of this 1:1 ratio—and even its reality—had until that point been completely obscure (see Chapter 3).

Watson and Crick quickly wrote an article describing their model, which they initially intended to send as a standalone paper to *Nature*. After protests by the King's group, it was published together with an article by Maurice Wilkins, Alexander Stokes, and H. R. Wilson, and another by Rosalind Franklin and Raymond Gosling, on April 25, 1953.[14] These last two articles contained X-ray diffraction data that supported the model, most of which were already known to Watson and Crick, as they indicated in the acknowledgments to their paper. They had only been able to make the double helix model because they had access to an MRC report from the King's laboratory that contained key data from Rosalind Franklin.

Taken together, the three articles showed how, on the basis of the double helix structure, it was possible to propose a very simple model for the replication of DNA, and thus explain the autoreplicative power of genes. The two strands of DNA could separate, and the bases on each strand would associate with the complementary bases present in the cellular medium in the form of free nucleotides. All that would be required would be for the nucleotides to bond together to produce two daughter DNA molecules, each of which would be identical to the parent molecule. Watson and Crick also suggested that mutations resulted from rare transient forms of bases that were responsible for incorrect pairing during replication.

A few weeks later, Watson and Crick sent a second article to *Nature,* this time dealing with the genetic implications of the double helix structure of DNA. This article, which was even more speculative than the first, was published on May 30, 1953, and contained the decisive phrase "the precise sequence of the bases is the code which carries the genetical information."[15]

Watson presented the structure, and its genetic implications, at the Cold Spring Harbor Symposium in the summer of 1953—Max Delbrück gave photocopies of the *Nature* articles to all the participants. Despite some criticisms and a lack of absolute proof (the model was still just a model, even with the accompanying data from Franklin and Wilkins), the double helix structure of DNA was rapidly accepted by the scientific community. At one point, historians claimed that the double helix had little immediate impact, but a precise bibliometric analysis has since demonstrated that this was not the case.[16] A long series of visitors came to admire the model at the Cavendish Laboratory. A page had been turned in the history of the young science of molecular biology. The structure of DNA, of the gene, had been discovered, and this structure explained the autoreplicative properties of life that had so fascinated biologists since the beginning of the twentieth century and even before.

IT IS DIFFICULT for the historian to know how to deal with Watson and Crick's discovery. It appears that everything that could be said about it has been said, that there is no unknown element that could falsify this or that aspect of the story, or at the very least provide a new and original point of view. Paradoxically, this information overload weakens our understanding. The apparent transparency of the discovery of the double helix erases the very object of historical research: nothing remains to be explained because everything is already known.

In reality, the fact that the discovery of the double helix is so well-known is a golden opportunity for the historian of molecular biology: using details and anecdotes, it is possible to re-create the scientific context and the social forces that were at work in the development of the new discipline.

The scientific careers and personalities of the codiscoverers of the double helix are particularly revealing. Francis Crick and Maurice Wilkins were members of the group of physicists who turned toward biology in the anticipation of major new discoveries. They discovered biology through genetics, or rather through the distorted image of genetics presented by Erwin Schrödinger in *What Is Life?* Their change of direction was encouraged by research administration policy decisions—the MRC had set aside a number of positions for physicists.

In Britain, an obvious area of research for such scientists was the study of the structure of macromolecules by X-ray diffraction. This method had been developed by William Bragg, Lawrence Bragg's father, and had grown as a result of the work of his son and other researchers, such as J. D. Bernal (first at Cambridge, then at London) and by William Astbury (at Leeds) whose contributions have recently been intensively studied.[17] With Dorothy Crowfoot (later Hodgkin), Bernal was the first crystallographer to obtain clear X-ray diffraction images of proteins. To do this, he had placed the protein crystals in a capillary tube that was closed at both ends, thus avoiding dehydration during the experiment. Although the first images, obtained in 1934, did not lead to a three-dimensional description of protein structure, they were nevertheless perfectly clear, confirming the macromolecular nature of proteins.[18] Resolving the three-dimensional structure thus became a "doable" project.

The British school of crystallography was without doubt the best in the world. But this strength was also its weakness. The crystallographers were convinced of their ultimate success: continued progress in X-ray diffraction showed that the determination of the structure of biological macromolecules—DNA and proteins—was quite possible. But armed with this certainty and unaware of the importance of these structures for molecular biologists, for whom the chemical nature and structure of genes had become a key problem, the crystallographers were in no particular hurry. Crick and Watson thought that Wilkins and Franklin worked very carefully but far too slowly because the structure of DNA was of only minor importance for them—a small step in the long march of scientific knowledge.

Watson's impatience was representative of a change in the attitude of the phage group. In the initial phases of his research, Delbrück had hoped to explain the self-replication of genes and bacteriophages without having to understand their chemical composition. All the research carried out in this initial period had shown that the phage was extremely complex and that it was hopeless to expect to understand how it replicated without opening the black box—without exploring the nature, structure, and interaction of its component parts. Irradiating bacteria during the replication of the phage, as Watson had done during his doctoral studies, was not enough to reveal the mechanisms of replication.

Luria was well aware of this, so he encouraged Watson to work on the biochemistry of nucleic acids. This work, however, contained a trap that Watson only narrowly escaped thanks to a series of lucky chances—he ran the risk of becoming embroiled in the complexity of the metabolism of nucleotides and bases, losing his way in the cellular pathways of biosynthesis and degradation.

Watson and Crick's discovery presents us with an analytical problem similar to that encountered earlier with Osward Avery's work. Both Avery and his readers were astonished by the result he obtained because they were convinced that DNA was not an important, specific molecule. On reading Watson and Crick's articles and studying the reactions to them, the surprise is exactly the opposite. According to the textbooks, by this stage everyone was convinced of the role played by DNA in heredity, and no one felt it necessary to justify this conviction. That is not quite accurate—as late as 1960 *Nature* published an article speculating that proteins might be the hereditary material. But it is clear that in the period around the discovery of the double helix there was a sea-change in the views of many biologists, with the widespread adoption of what was frequently called the working hypothesis that DNA was the genetic material (remember that there was still no proof whatsoever of the role of DNA in anything more complex than a bacterium).

Several theories have been suggested to account for the rapid adoption of this new view that DNA was fundamental to heredity, but none of them fully explain a shift in opinion that for some resembled a conversion. We should reject one idea that is widely held by both the admirers and the detractors of molecular biology, according to which Watson and Crick inaugurated a new style of research in which discussions and theories became more important than experiments and observations. It is true that Watson and Crick determined the double helix structure of DNA without having carried out a single experiment on the molecule. It is also true that Delbrück had introduced a new style of work into the phage group in which discussion of experiments and articles was brought to the fore.[19] Bacterial genetics involved experiments that were difficult to think up but relatively simple to carry out.

But Watson and Crick were exceptions. Experiments in both crystallography (determining protein structure) and biochemistry (studying protein synthesis), which were essential components of molecular biology, required both time and care. In the 1960s, some specialists in bacterial genetics would abandon the subject in order to perform molecular studies of multicellular organisms. They had to leave their offices, where they had spent most of their time working out subtle genetic crosses, to spend many hours in cold rooms trying to isolate minute quantities of proteins or RNA.

A close study of Watson and Crick's approach also reveals that their research, like that of their colleagues, was more akin to tinkering than to the work of an engineer or an architect.[20] A builder chooses elements in which they have confidence and, on that basis, constructs a stable building. Researchers, by contrast, have only rotten planks available, with quite a high probability that they might give way. From these rotten planks they choose a few that they will use to build a new edifice. In most cases, the building collapses, but occasionally it holds. The rotten planks then become more and more solid as building progresses.

This is what makes historical analysis of science so difficult. There is a clear tendency, encouraged by scientists themselves, to justify the choice of the planks used to build the "edifice" (the right theory), to suggest that these planks were clearly the best from the outset.[21] This is rarely the case. The reason certain planks are chosen and subsequently become more solid is the product of a combination of chance and scientific flair. Watson and Crick were not certain of the structure of the bases, the number of polynucleotide chains, the interpretation of the diffraction patterns, or even of the role of DNA in heredity. And yet, despite all that, they proposed a structure that was so elegant that it convinced everyone who studied it.

The geneticist and historian Gunther Stent has suggested that Watson and Crick's discovery was akin to a work of art, and he has used the example to point out similarities between scientific and artistic creation.[22] These two activities are often presented as opposites—in a work of art, the artist is supposedly completely free; they invent something that has never existed. The scientist, however, is not free—he or she merely reveals concealed truths that, up until now, have been hidden from human eyes. The work of art is unique, inextricably attached to the identity of its creator. Scientific discovery is the product of a community, and scientific results are often anonymous because the name of their author is of no importance.

Stent is particularly convincing when he argues that in reality the artist is no freer than the scientist in their artistic choices: though it is true that all artistic creation is unique, nevertheless artistic schools do exist, within which all works bear a certain resemblance. By contrast, taking the example of the double helix, Stent claimed that this was a unique discovery that belonged to Watson and Crick. If they had not existed, the double helix structure would certainly have been discovered, but probably less elegantly and more gradually. Yet Crick argued that the importance of the discovery came above all from the object discovered—the double helix—and its remarkably simple structure.[23]

Some models of the history of science have equated research with other human activities. They argue that scientific theories are a reflection not of reality but simply human constructions, having the same status as literary or artistic works. This constructivist vision of science may be seductive when applied to the highly abstract models of particle physics, but it is difficult to see how the double helix could be one of many possible human constructions rather than a reflection of reality.

The story of the discovery of the double helix structure of DNA would not be complete without acknowledging the tragic fate of Rosalind Franklin. Franklin, who died from cancer a few years after the discovery of the double helix, got short shrift from Watson in the main body of his 1968 novelized account of the discovery, *The Double Helix*, although Watson's epilogue balances the picture quite substantially. More recently, she has been presented as a female researcher whose problems were typical of those encountered by women in the male-dominated research milieu. Franklin, according to this analysis, was deprived of *her* discovery of the double helix.[24]

This is wrong. The life of a woman scientist was certainly difficult— Franklin harked back longingly to her stay in Paris, where she considered relations between the sexes to be simpler. But, as outlined by her biographer Brenda Maddox, her difficulties in Britain stemmed mainly from her upper-class intellectual origins, which were very different from those of her collaborators.[25] The problems that existed between Franklin and Wilkins, which caused her such unhappiness and conceivably prevented the pair from discovering the double helix, had an even more trivial explanation— confusion over their respective roles, and a profound clash of personalities. In the years that followed the discovery she was happy at Birkbeck College, and she completed a very important crystallographic study of the structure of the tobacco mosaic virus that could conceivably have won her a Nobel Prize.[26] She was close friends with Crick and his wife Odile, going on holiday with the couple, and also maintained cordial relations with Watson.

Franklin's crystallographic data were very important to the discovery of the double helix structure of DNA, but it is also clear that Franklin did not initially share Watson and Crick's keen awareness of the biological importance of the structure of DNA, nor of the need to explain the autoreplicative ability of genes on the basis of this structure.[27] Furthermore, her hostility to speculative model-building and her insistence that the data should "speak for themselves" meant that she did not attempt to interpret the key findings that were literally under her nose.

Nevertheless, in the mythology that has developed around the discovery of the double helix, Franklin has taken on the role of a martyr. Watson and Crick are claimed by some to have stolen her data; in less serious versions of history, she is even said to have discovered the double helix herself. The enthusiasm of sectors of the public for a female protagonist is understandable, but it is less clear why some scientists might adopt such a position. One explanation is the well-studied need felt by scientists—not just molecular biologists—to ascribe a mythical origin to new disciplines.[28]

In the case of molecular biology, Pnina Abir-Am has shown how such a mythical vision can distort the historical account.[29] Such a distortion is not particularly problematic when it rectifies scientists' personal failings in order to turn them into heroes. It is, however, much more unhelpful when it twists historical reality and renders the history of a discipline incomprehensible. In the case of molecular biology, this mythical vision has a tendency to concentrate the entire history of the discipline into a single discovery. Previous results were merely steps on "the path to the double helix" (this is title of a major history of the period), from which flowed all subsequent discoveries, including messenger RNA and the genetic code.[30] This is mistaken: in reality, the discovery of the double helix played an extremely ambiguous role in the subsequent development of molecular biology.

Although the discovery of the double helix showed that the sequence of bases was fundamental for the genetic function of DNA and its replication, it also suggested that knowledge of the structure of molecules alone could lead to an understanding of their function. As will be seen (Chapters 12 and 13), ribonucleic acid (RNA) also plays an essential role in protein synthesis. Many researchers, including Watson, therefore turned to the structural study of this molecule.[31] Despite the "genetical implications" of the double helix structure of DNA outlined by Watson and Crick in their second *Nature* article of 1953, the true meaning of the existence of a genetic code, which was that only the information carried by nucleic acid molecules was important and not their structure, was not properly understood by most scientists.

Over the subsequent decade, the discovery of the double helix had an enormous impact on the general public, including architects and painters such as Salvador Dali. Paul Doty, a chemist specializing in DNA, recalled that, at the end of the 1950s, he saw a button with the letters "DNA" on it. When he asked the vendor what it meant, he was told that it referred, of course, to the gene and that he had better get wise![32] In many respects it was the discovery of the double helix, not either Avery's experiment or Hershey and Chase's, that convinced the biological community that genes were almost certainly composed of DNA, and that it was thus the basis of heredity. Similarly, it was through the double helix that the genetic role of DNA entered the textbooks.[33]

This weight given to the double helix is highly paradoxical—despite the widespread enthusiasm for Watson and Crick's discovery, in reality it was extremely fragile. The double helix was merely a hypothesis, a model. Despite the empirical data in the two accompanying reports from Franklin and Wilkins, there was no proof. Although the notion of base complementarity was widely accepted, the idea that DNA was a helix was quite difficult to believe. The helix would have to unwind during replication, and this posed a major biochemical (and topological) problem that Watson and Crick had barely touched on.

Other models of DNA were proposed, in which the two polynucleotide chains were in the same plane or turned differently than in the Watson-Crick model (such as in the Z-DNA structure described in 1979).[34] Indeed some crystallographic results appeared to support these heterodox models. It was only in the 1980s, twenty-five years after Watson and Crick's discovery, that the double helix structure was unambiguously demonstrated, and the role of DNA in multicellular organisms was not finally proved until the late 1970s.

Watson and Crick thought that during DNA replication the two strands separated and that each would gather bases from its nucleotide-rich surroundings to synthesize a complementary strand. Such a model of replication is called semiconservative because half the DNA is totally conserved, whereas the other half is synthesized de novo.

To clarify matters, but also because they thought that Watson and Crick's model was too simple, Delbrück and Stent outlined the other possible models of replication: dispersive replication (in which nothing was conserved, the parent DNA molecules were degraded, and the new DNA molecules were synthesized) and conservative replication, in which the two original strands of DNA remained intact and the two strands of the daughter molecule were synthesized *ex nihilo* using available nucleotides.[35]

In 1941—before the exact chemical nature of the gene was known—J. B. S. Haldane, a renowned enzymologist and one of the founders of population genetics, had suggested that gene replication could be studied through the use of heavy isotopes such as nitrogen 15 (^{15}N). The idea was that by suddenly changing the composition of the milieu, scientists could distinguish old and new copies of genes.[36] As seen in Chapter 9, radioactive isotopes had already replaced heavy isotopes in the arsenal of biochemists and molecular biologists. The initial experiments to test the semiconservative model of DNA replication were thus carried out with radioactive isotopes without much success.

In 1954 Matthew Meselson (one of Pauling's ex-students) and Franklin Stahl decided to address the idea of separating newly synthesized molecules of DNA according to density. After a number of trials, they decided to label DNA with nitrogen 15 (^{15}N), but separating DNA molecules of slightly different densities posed a problem. The medium needed to have a density very close to that of DNA (this could be obtained by dissolving a heavy salt in water; Meselson and Stahl used cesium chloride) and a physical method—ultracentrifugation—would have to be used to create a density gradient.

In 1957, Meselson and Stahl finally carried out their experiment, before which they allowed bacteria to reproduce for several generations in a medium containing ammonium chloride labeled with ^{15}N.[37] At the beginning of the experiment, the bacteria were diluted in a normal medium containing nitrogen 14 (^{14}N). Meselson and Stahl took samples of the bacteria during the course of the experiment, extracted the DNA, then mixed it with cesium chloride. After twenty hours of centrifugation at 45,000 rpm (revolutions per minute), the position of the bacterial DNA in the centrifuge cell was noted.

At the beginning of the experiment, the very dense DNA formed a band at the bottom of the cell. After a while another, less dense, band appeared. After one cycle of bacterial reproduction, this latter band was the only one that was visible. It corresponded to the density of a molecule of DNA formed by a heavy chain labeled with ^{15}N and a light chain. This result was exactly that predicted by the semiconservative replication model. Meselson and Stahl confirmed that the medium-density DNA band detected after a single generation was in fact composed of one light and one heavy strand. They heat-denatured the DNA molecule and obtained two bands of slightly different molecular weights, one with a density corresponding to that of a heavy DNA, the other to the density of a light DNA.

The results were clear, clean, and in perfect agreement with the predictions of the semiconservative model.[38] Furthermore, the value of the result was reinforced by the elegance of the method—the density gradient—which was subsequently widely used in molecular biology for separating DNA molecules.[39] In the words of the historian Frederic L. Holmes, it was "the most beautiful experiment in biology."

Meselson and Stahl's finding was particularly striking because it had been obtained with *Escherichia coli* and not bacteriophage, thus showing that the semiconservative model of DNA replication was also valid for whole chromosomes.

12

Deciphering the Genetic Code

THE PATH THAT LED from the discovery of the double helix structure to the deciphering of the genetic code was extremely convoluted. To resolve the problem of the role of genes in protein synthesis, two experimental approaches were possible. Researchers inspired by genetics could try to deduce the role of DNA in protein synthesis on the basis of the structure of the DNA molecule and the action of genes in the cell. A more biochemical line of attack was to try to create in vitro systems that could synthesize proteins—it was this that ultimately led to the deciphering of the genetic code. A third approach, which preoccupied a large number of very clever people in the 1950s, was to attempt to crack the code by using theoretical, mathematical methods. This ended in complete failure.

Many accounts, including Jim Watson's and Francis Crick's autobiographical versions, support the view that the revelation of the structure of DNA almost naturally gave rise to the idea of a genetic code. According to this hypothesis, the succession of nucleotides (bases) in DNA codes for different amino acids, which in turn are chained together to form proteins. In their second *Nature* article, dated May 30, 1953, Watson and Crick did indeed write that "the precise sequence of the bases is the code that carries the genetical information."[1] But this striking and novel formulation reflected the fact that the ideas of code and information had recently entered the scientific mainstream—it did not imply that Watson and Crick had a clear idea of the existence of a genetic code, nor that they imagined that trying to understand that code would be a viable research program. They were therefore surprised to receive, after the publication of their articles in *Nature,* a letter from George Gamow, a Russian-born physicist working in the United States, that contained a concrete proposal for a genetic code: a direct correspondence between bases and amino acids.[2] Gamow's idea was that the DNA double helix contained twenty spaces, the form of which depended on the nature of the surrounding nucleotides. These spaces could be "containers" for amino acids. For a sequence of bases to correspond to a protein, covalent bonds would simply have to link the different amino acids.[3]

Gamow's letter was a surprise to Watson and Crick. At the time, knowledge of protein structure was far too vague for biologists to be confident that the sequence of amino acids was genetically determined, and although the detail of protein synthesis remained mysterious it was known that RNA was somehow involved. Gamow's view that DNA was the direct physical template for protein assembly was evidently incorrect. Gamow had been able to accept such a simple hypothesis only because of his ignorance of the field.

Crick's first reaction to Gamow's letter was to ignore it—written in childish handwriting, it seemed to be the work of a crank. But four months later a second, more detailed letter arrived, prompting Crick (who was now in New York) to show that Gamow was wrong. One weak point of Gamow's model was his choice of twenty amino acids—more than one hundred different amino acids are found in organisms. Some of these are present in all proteins; others are relatively rare and are produced by the modification of other amino acids after protein synthesis. Gamow's division of amino acids into fundamental and derived compounds was thus nevertheless perceptive. Crick, using his long experience in the study of proteins, proposed another list of twenty fundamental amino acids that, remarkably, turned out to be right.

Despite its mistakes, Gamow's proposal was extremely fruitful. Crick adopted the idea of a literal genetic code—of a correspondence between nucleotides and amino acids, so technically a cipher—whereas previously he had used the term very loosely. Gamow's suggestion implied that the code might be deciphered directly, without any experiments, without having to study all the intermediate biochemical steps leading from DNA to proteins.

In a later version of Gamow's code, each group of three bases (what we now call a triplet or a codon) coded for a different amino acid, but triplets coding for successive amino acids overlapped. This type of code limited the number of possible amino acid sequences. The rare protein sequences that were known at the time soon enabled Crick and Sydney Brenner to dismiss this idea. Gamow and Martynas Yčas then devised a new version, the combination code, in which the order of bases was unimportant; what counted was their combination. This code easily allowed for twenty ways of combining four objects—the four bases taken in groups of three—but it was difficult to imagine how the cellular machinery could recognize a combination of bases.

Although the idea of an overlapping code was rejected, the problem with a simple code was that two letters provided only sixteen possibilities (fewer than the likely number of amino acids), whereas a code with three letters gave sixty-four possibilities—much greater than the number of amino acids used in protein synthesis. Furthermore, a nonoverlapping code raised the problem of what was later called the reading frame. How could the cell know where one triplet ended and the next one began?

In 1957 Crick and Leslie Orgel proposed a solution to these problems.[4] If each nucleotide triplet was read in the right reading frame it would code for an amino acid, but if the reading frame were shifted one base, all the triplets would lose their meaning. This meant that there were only twenty meaningful triplets. This perfect relationship between the number of codons and the number of amino acids appeared too good not to be true.

Many other codes were proposed, all of which sought to reduce the number of significant codons to twenty or to find a simple method of determining the reading frame. When the genetic code was finally cracked, beginning in 1961, all these attempts were shown to have been wrong and ultimately pointless.[5]

Most of the proposed codes, and in particular that of Crick and Orgel, were based on the assumption that if proteins had a given amino acid composition, nucleic acids had to have a given nucleotide composition. The more that was known of the DNA composition of different organisms, the more it appeared that the amino acid composition of the same protein in different organisms was more or less identical, whereas the nucleotide composition of these different organisms showed substantial variations. In a despairing lecture given in 1959 at the Brookhaven National Laboratory near New York, Crick listed all the difficulties encountered by solutions to the coding problem, which at the time he was ready to abandon.[6]

ALTHOUGH THE IDEA of a genetic code—and especially the theoretical approaches that were adopted to characterize that code—led to a dead end, the years following the discovery of the double helix reinforced the idea that genes closely controlled the nature of the amino acids that form proteins.

The decisive steps in the characterization of proteins that had taken place in the first half of the century have already been described. The concept of the macromolecule decisively vanquished colloid theory, and, following Linus Pauling's work, researchers generally accepted that proteins were formed by making linear chains of amino acids linked by peptide bonds. The protein chain would then fold into a precise conformation that was stabilized by the formation of weak bonds (hydrogen bonds).

This quick overview is not meant to suggest that in 1950 scientists held today's view that, whatever its amino acid sequence, a polypeptide chain will spontaneously fold into a stable conformation. All biochemists at the time felt that there must be rules that would simplify the problem of protein synthesis. Some biochemists felt that these rules must exist at the level of the chaining of amino acids. In 1937, Max Bergmann and Carl Niemann of the Rockefeller Institute proposed that each protein was formed by $2n \times 3m$ amino acids, each being present at precise positions in $2n' \times 3m'$ copies, where n, m, n', and m' were whole numbers.[7] This mysterious arithmetic was the result of the action of proteinases (what we would call proteases), which were thought to be responsible for protein biosynthesis (see the description of this model later in this chapter). Not all biochemists had such a precise view of protein structure. Many of them, however, agreed that there must be rules for assembling amino acids.

Other scientists hoped to find general principles that would govern the three-dimensional structure or folding of proteins. In this context, Pauling's 1950 discovery of the secondary structures of polypeptides, and in particular of the α-helix (see Chapter 11), was especially important. For Pauling, the significance of his finding was not so much that he had described several possible models of how proteins folded, but rather that he had revealed the fundamental structures upon which folding took place.

This conviction that there was a simple explanation of how proteins formed had its roots in the biochemists' view of the role of genes in protein synthesis. Contemporary theories of protein synthesis were relatively vague: several models existed, all of which were at least partly mutually incompatible. Nonetheless, they all gave an essential but limited role to genes, implying that protein synthesis should be a relatively simple biochemical process.

The hopes of finding any regularity in the chaining of amino acids evaporated, however, when the first protein sequences were established. Only a few years earlier, Richard Synge, Archer Martin, and their colleagues in the Leeds Wool Industries Research Association had developed chromatographic techniques for separating amino acids and peptides, first on silica, then on paper (see Chapter 9).[8] The British biochemist Frederick Sanger had subsequently developed methods for fragmenting polypeptide chains under the action first of acids, then of proteases (and in particular trypsin), followed by their separation and sequencing.[9] Insulin was the first protein sequence to be completed.[10] This sequence showed that amino acids had precise positions in the polypeptide chain, but did not reveal any regularity to their order.

The virtually simultaneous demonstration that genes control both the nature and the position of the amino acids in polypeptide chains was made in two steps, separated by more than seven years, both involving the same human disease. The first was made by Linus Pauling. In a 1949 article published in *Science,* Pauling had shown that sickle cell anemia—so called because patients suffering from the disease have sickle-shaped red blood cells—was linked to an abnormal structure of hemoglobin, the protein that transports oxygen in the blood.[11] Pauling reported that the hemoglobin of sickle cell anemia patients had an electric charge that was different from that of normal hemoglobin. Furthermore, at low oxygen pressures it was less soluble, precipitating in the form of long needles, deforming the red blood cells and giving them their characteristic sickle shape. These red blood cells could not fit properly in the patient's capillaries and thus tended to block the oxygenation of the tissues. More fragile than normal red blood cells, they produced anemia in the patient. Shortly before Pauling's discovery, James Neel, a specialist in human genetics at the University of Michigan, had reported that sickle cell anemia was a genetic disorder displaying Mendelian inheritance.[12]

Pauling's article was extremely important: by showing how a molecular disorder can explain the symptoms of an illness, it founded molecular medicine. Pauling and his colleagues wanted to go on to characterize the modification(s) of hemoglobin that resulted in the observed new properties, but preliminary studies wrongly suggested that normal and mutant hemoglobin had the same amino acid composition, so they abandoned their attempts.

The second step involved in demonstrating that genes control both the nature and the position of the amino acids in polypeptide chains was made by the British scientist Vernon Ingram, at Crick's suggestion. After a visit to the Rockefeller Institute, Ingram began work with Crick at the Cavendish Laboratory in Cambridge, using Frederick Sanger's primitive protein sequencing method. In 1956 he showed that the mutation responsible for sickle cell anemia was linked to a change in a single peptide and to the replacement of a single amino acid in this peptide.[13] Although Crick was heavily involved in the work, he did not sign Ingram's article.

Ingram's discovery had an important impact on biochemists: it showed that genes intervened directly in protein structure and could influence such apparently unimportant details as the nature and position of a single amino acid.

Up until this point the biochemical problems associated with understanding protein synthesis had only been hinted at. The question of which mechanisms are involved in protein synthesis was initially considered to be the opposite problem to degradation. Because enzymes—proteases—were able to cleave proteins at specific sites, identical or analogous enzymes were thought to carry out the opposite operation. Furthermore, in addition to direct synthesis reactions involving polypeptide chains, there might be reactions involving the exchange of amino acids between peptides, which were energetically more favorable. Rudolph Schoenheimer's experiments using nitrogen 15 had shown that proteins were in an unstable metabolic state, and that amino acids were continually added to or subtracted from them (see Chapter 9). These results agreed with the proposed role of proteases in protein synthesis.

Between 1940 and 1955 the multi-enzyme model of protein synthesis was widely accepted.[14] It was, however, gradually undermined by a series of difficulties. First, it raised a major theoretical problem: if each protein, each enzyme, was synthesized by a multi-enzyme complex, how were the enzymes in the multi-enzyme complex synthesized? If they were themselves synthesized by multi-enzyme complexes, it was difficult to see how the whole process could ever begin.[15] Furthermore, the one protein–one multi-enzyme complex model did not fit with George Beadle and Ed Tatum's results and the one gene–one enzyme hypothesis (Chapter 2).

In reality, the multi-enzyme model of protein synthesis left no place for nucleic acids. Although the precise role of nucleic acids remained unclear, they nevertheless appeared to intervene at two levels. From the late 1940s molecular biologists knew that, in one way or another, genes controlled protein synthesis and that they were formed, at least in part, of nucleic acids. By contrast, researchers on protein synthesis in eukaryotic organisms (rather than in bacteria), in which chromosomes are isolated from the rest of the cell by the nuclear membrane, had shown that protein synthesis takes place in the cytoplasm and not in the nucleus, where the genes are.[16] At the beginning of the 1940s, research by the Swede Torbjörn Caspersson and by the Belgian Jean Brachet had shown a correlation between the level of protein synthesis and the quantity of RNA in the cytoplasm (see Chapter 13).[17] Once again, nucleic acids were implicated in protein synthesis.

In 1952 Alexander Dounce of the University of Rochester, well aware of these problems of the multi-enzyme approach, proposed the radically different template model.[18] According to this idea, nucleic acids were the scaffolding on which the amino acids were assembled in order to form proteins. P1 enzymes made the link between nucleotides and amino acids, whereas nonspecific P2 enzymes formed the peptide bonds between the different amino acids that had been brought together. Despite its appeal, this model left a fundamental question unanswered: what was the template? If it was DNA, why were proteins synthesized in the cytoplasm, whereas genes—DNA—were found in the nucleus? And if it was RNA, how could the role of genes in protein synthesis be explained?

To answer these questions, in 1953 Dounce suggested that DNA might itself serve as the template for the synthesis of RNA, and that RNA was in turn the template for protein synthesis.[19] This was a remarkable anticipation of what was to become the central dogma of molecular biology (see Chapter 13). But the theory was not new—it had first been proposed in 1947 by André Boivin and Roger Vendrely (this work was not cited by Dounce).[20] Furthermore, the nature of the RNAs involved in protein synthesis remained unclear, and the significance of Dounce's model was not widely recognized. For Dounce, Pl enzymes alone were responsible for making the link between an amino acid and a given nucleotide in a polynucleotide sequence. He was the first person to consider the relation between proteins and nucleic acids to be indirect and thus nonstereospecific (see Chapter 1) and structurally arbitrary. But this was not how the template model was understood. Scientists interpreted it in the light of biochemical tradition, according to which the formation of biological macromolecules was the result of their stereospecific interaction with other molecules. The archetype of such models was Pauling's 1940 explanation of antibody formation.[21] A description of this model reveals how molecular biology differed from previous approaches.

When a foreign molecule—an antigen—enters a vertebrate, the organism reacts by synthesizing proteins called antibodies, in virtually all cases and whatever the precise chemical nature of the antigen.[22] The antibodies fix on the antigen, thus forming a complex that can be eliminated. Chapter 9 showed how this understanding of antibodies had developed rapidly through the use of the new physical techniques of electrophoresis and ultracentrifugation. These methods had shown that antibody molecules were γ-globulins and all had more or less the same amino acid composition.

Pauling, using detailed studies on the effects of the chemical modifications of antigens carried out by the Austrian-born immunologist Karl Landsteiner at the Rockefeller Institute, was able to precisely describe the chemical nature of the bonds that formed between the antibody and the antigen (see Chapter 1).[23] These studies explained how the interaction between the antibody and the antigen took place but said nothing about the origin of antibodies. How could the organism produce antibodies with molecular structures that complemented literally any foreign molecule?

The simplest solution had been suggested as early as 1930 by Stuart Mudd, Jerome Alexander, Friedrich Breinl, and Felix Haurowitz: antibodies were able to interact with antigens because the antigen guided the formation of the antibody.[24] The antigen had what was termed an instructive role. In 1940, Pauling adopted this model and gave it the chemical precision it lacked. For Pauling, all antibody molecules had identical amino acid chains. They were, however, very unusual proteins: the central part was highly structured, but the two ends appeared not to have a definite form and "hesitated" between various conformations. This effect might be due to the presence of a particular amino acid, proline, in these regions. Newly synthesized antibody molecules do not immediately take on their final conformation—when in the presence of an antigen each extremity folds around the foreign molecule and adopts the definitive conformation, which is complementary to that of the antigen. The antigen then escapes from the antibody molecule, thus resulting in the production of antibodies ready to interact with any new molecule of the antigen that they might encounter.

This theory was thoroughly scientific in that it led to a certain number of predictions that could be quite simply experimentally tested.[25] For example, because an antibody adopts its final conformation only in the presence of an antigen, the synthesis of specific antibodies should cease if the antigen is eliminated from the organism. But it was already known that immunity could persist for many years, while, as Schoenheimer had shown, proteins—which included globulins—had a rapid turnover. This suggested that the antigen was stored somewhere in the organism. By using a radioactively labeled antigen, scientists could track its fate following injection. Experiments revealed that though most of the antigen rapidly disappeared, a small quantity remained.[26]

Furthermore, antibodies were bivalent, with two antigen recognition sites; if an animal was simultaneously immunized with two different antigens, *a* and *b*, one would expect to obtain antibodies with a dual specificity, directed against both *a* and *b*. Pauling cited a number of results that partly agreed with this prediction. Above all, Pauling's model suggested a way of making antibodies in vitro. These experiments, which Pauling announced in 1940, were carried out with his colleagues and published in 1942.[27] They succeeded beyond his wildest dreams.

The procedure Pauling employed was as follows. First, he used a dye as an antigen. Various immunoglobulins from a nonimmunized animal were then added to the dye (he denatured the immunoglobulins by passing them through an alkaline medium). The globulin-dye mixture was then slowly brought to a neutral pH, forming a precipitate of the antigen and immunoglobulin. Pauling then used a milder procedure for denaturing the immunoglobulins—heating to 10°C below the antibody denaturation temperature. This should have speeded up the conformational changes in the immunoglobulin molecule. More than 40 percent of the globulins developed an antibody function directed against the antigen. Finally, still using this less vigorous denaturing procedure, but this time with a polysaccharide extracted from a pneumococcus as an antigen, Pauling obtained virtually identical results, although there were problems with the experiment owing to difficulties in eliminating the antigen from the precipitate without denaturing the antibodies.

The antibodies obtained in these three experiments precipitated the antigens in conditions that were slightly different from those required by natural antibodies extracted from an animal, but this small difference appeared relatively minor compared with the overall outcome, which completely agreed with Pauling's model. Pauling was sufficiently encouraged by these results to consider producing specific antibodies on an industrial scale. The Rockefeller Foundation clearly understood the importance of these results and provided Pauling with substantial financial support.

Felix Haurowitz, who worked in Istanbul, criticized one of Pauling's experiments by showing that precipitation of globulin molecules and dyes coupled with ovalbumin could occur spontaneously, as a result of the attraction of opposite electrical charges.[28] However, Haurowitz's critique was valid for only one of Pauling's many experiments. The rest of his data, which showed the in vitro formation of antibodies against a single dye molecule or against pneumococcal polysaccharides, were not in question. Despite his criticisms, Haurowitz in fact agreed with Pauling's theory of antibody formation.

Pauling's initial results could not be replicated; as is often the case in science, these negative data were not published.[29] More curiously, Pauling's model is cited in the scientific literature as if it remained purely theoretical.[30] The experiments, which nevertheless provided a striking confirmation of the model, are never mentioned, even by their authors.[31] Pauling's 1940 model of antibody synthesis was later replaced by the model of clonal selection developed by Frank MacFarlane Burnet in 1957 (see Chapter 17). Nevertheless, the 1940 model was of fundamental importance for biochemistry.

Pauling believed that his model was valid only for antibodies, which he thought were the only proteins to show conformational flexibility, the form of other proteins being strictly determined by their amino acid sequence. But his model was implicitly or explicitly extended to all proteins and enzymes. For many biochemists, the substrate of an enzyme—the chemical compound that the enzyme transforms—intervened in its synthesis: enzymes took on their definitive conformation by folding around a substrate molecule.[32]

This theory was particularly well suited to the adaptive enzymes studied by Jacques Monod and Sol Spiegelman (see Chapter 14). These enzymes were produced only when organisms were in contact with the relevant substrates—it was thus logical to imagine that the substrate played a role in synthesis. In 1947 Monod proposed a model that showed how the substrate converted an inactive protein into an active enzyme, almost certainly by conformational change.[33] For Monod, this induction was a general phenomenon, closely linked to protein synthesis.[34] All enzymes could be induced; those enzymes that appeared to have a constitutional activity were, in fact, synthesized under the action of endogenous inducers that were always present in the cell. The model proposed for inducible enzymes was thus generally valid.

The template model and Pauling's model might appear to be different, if not incompatible: the first suggested that protein structure is the product of a kind of copy, the second that this structure is obtained when the antibody molds around another, "negative" molecule. But these models were never formally confronted. The term "template" was sufficiently vague to mean both "pattern," suggesting a process in which an identical copy was produced, and "mold," implying a process in which a complementary copy was made. For most biochemists and geneticists, the two models were part of the same biological vision, according to which form played a fundamental role. As seen earlier, this idea of form, of specificity, taken from enzymology and immunochemistry, played a major and complex role in the birth and development of molecular biology.[35]

The study of these two models is particularly difficult because, although they were omnipresent in the thinking of biochemists from the 1940s to the 1960s, they were almost never explicitly formulated. It is as though scientists thought that if their models were set down on paper, their fragility would be exposed. Sometimes, however, in interviews or popular articles the scientists expressed themselves more freely. For example, in a 1949 interview with *Scientific American,* George Beadle described the development of molecular biology at Caltech in the following terms:

> We are seeking to uncover the principles that govern fundamental processes of life . . . the investigations tend to show that the molecular form

known as protein is the key structure. . . . The genes, we believe, exercise an overruling control on all these activities. They do this, we think, by serving as the master patterns for the many proteins which function in the processes of life. Thus, there is probably a gene which serves as the template for the body's manufacture of insulin, another which provides the mold for pepsin, and so for albumin, fibrinogen, the polypeptide chain that forms antibodies, and all the rest.[36]

A final example will give an idea of the close links that existed between the concepts and tools of immunology and those of other biological disciplines—biochemistry and even genetics. In 1944 the geneticist Alfred Sturtevant, who had worked for many years with Thomas Morgan, linked two sets of data.[37] One was the work of immunologists, who had shown that some antigens, such as blood groups, were the direct product of gene action.[38] The other was the work of Landsteiner (followed by Pauling), who had shown that antibodies and antigens had complementary structures. Sturtevant deduced that antibodies directed against an antigen would probably interact with the gene responsible for the synthesis of the antigen and that it would thus be possible to induce specific mutations. This is what Sterling Emerson had just succeeded in doing in *Neurospora*.[39] The data were extremely important, but when they were not confirmed they were forgotten, together with the models that they supported.

FORTUNATELY, biology is a science, and all models or hypotheses fade away in the glare of contradictory experimental results. In the case of protein synthesis, such findings were the product of the patient work of biochemists who tried to reproduce protein synthesis in vitro.

The active contribution of biochemists to what was to be one of the most remarkable discoveries of molecular biology—the deciphering of the genetic code—is not widely known.[40] This fact, and the consequent resentment felt by biochemists, can be traced to the first descriptions of the birth of molecular biology, particularly those written by the contributors to a 1966 Festschrift for Max Delbrück, in which molecular biology was presented as the fruit of the collaboration of the phage group geneticists and the Cambridge crystallographers and structural chemists.[41] The role played by Watson and Crick in the discovery of the double helix structure of DNA was taken as the symbol of this collaboration.

After the war, using the new experimental possibilities provided by the production of radioactive compounds, several groups directly studied the mechanisms of protein synthesis. One of the most active laboratories was led by Paul Zamecnik at Harvard.[42] The initial aim of Zamecnik's group was to compare protein synthesis in normal and cancerous cells. The early experiments were carried out on slices of rat liver, some of which had been treated with a cancer-inducing agent that provoked the formation of hepatomas. This required the development of some very precise techniques for obtaining radioactive amino acids and measuring the radioactivity incorporated into proteins but led to some very disappointing results. Although there were differences between normal and cancerous tissues, they were apparently merely quantitative and thus did not alter the nature of the proteins that were synthesized.

These results led to a reorientation of the research carried out in the Zamecnik laboratory: the group now sought to open the black box of protein synthesis. This change was stimulated by a novel observation made by Zamecnik's group: protein synthesis appeared to depend on the energetic state of the cell—something that did not agree with the multi-enzyme model of protein synthesis. Furthermore, in 1950, Henry Borsook showed that protein synthesis took place in a particular cell structure, the microsome. Borsook's observation allowed the comparison of in vitro and in vivo data, thus providing a criterion for testing the validity of the former.

As early as 1951, Zamecnik succeeded in obtaining a coarse fractionation of a homogenate of liver cells in which the incorporation of radioactive amino acids into proteins could be observed in vitro. In the years that followed, Zamecnik's group developed protocols that made it possible to distinguish between radioactive amino acids that were actually involved in protein synthesis and those that might be incorporated due to some artifact. After a series of increasingly powerful fractionations, the system became sensitive enough to be useful in studying the key stages of protein synthesis.

In 1954 Zamecnik developed an active in vitro system that contained only amino acids, a donor molecule to provide energy (adenosine triphosphate: ATP), microsomes, and a supernatant from a high-speed ultracentrifugation of a raw cellular extract. This system showed that ATP was necessary for protein synthesis. For biochemists trained in the study of metabolism, the importance of ATP implied that there must be an amino acid form that was activated through a reaction with ATP. The Belgian bio-chemist Hubert Chantrenne proposed a structure for this activated form, which was rapidly confirmed by Mahlon Hoagland in Zamecnik's labora-tory.[43] But this was merely the first step. Hoagland showed that the amino acid was subsequently loaded onto a small RNA, which he called soluble RNA. The same enzymes were responsible first for the reaction of the amino acid with ATP, then for its fixation to soluble RNA. It seemed that there were as many enzymes as amino acids.

These biochemical discoveries validated the hypothesis that Crick had put forward in 1955, according to which there must be small RNA mole-cules, called adaptor molecules, able to fix amino acids and to interact with the nucleic acid matrix.[44] This hypothesis, developed with Brenner, had been inspired by Gamow's model of the genetic code. One of the weak-nesses of this model was that it could not really describe in chemical terms how a nucleic acid could form a container for the lateral chains of amino acids. On the other hand, the discovery of the DNA double helix had shown that one nucleic acid could combine with another nucleic acid by forming a group of hydrogen bonds. Crick hypothesized that there were twenty adaptor RNAs, one per amino acid, and twenty enzymes able to bind spe-cifically one type of amino acid to an adaptor RNA. The adaptor RNA would then bind to the matrix of nucleic acids. Although the soluble RNA mole-cule that was observed was much bigger (several dozen nucleotides) than Crick had imagined, this difference did not raise any serious problems.

The convergence between the most theoretical approach to deciphering the genetic code and the biochemists' ground-level methods confirmed the experimental value of in vitro models of protein synthesis. Shortly after-ward, Zamecnik developed a system of in vitro protein synthesis using bacterial extracts that was subsequently improved by Alfred Tissières.[45] Everything was in place for the deciphering of the genetic code.

On Monday, May 22, 1961, at 3:30 P.M., J. Heinrich Matthaei, a German biologist working in the United States, took a test tube and mixed a ground-up extract of bacteria that had been centrifuged, a fraction containing small molecules of soluble RNA, the twenty amino acids that form pro-teins (sixteen of which were radioactively labeled), ATP as an energy source, salts, and a buffer that kept the pH of the mixture constant.[46] He then added a few micrograms of an artificial molecule of ribonucleic acid that had been synthesized in vitro, consisting of the simple repetition of one type of base, uracil. After incubating the mixture for one hour at 35°C, Matthaei precipitated the proteins from the mixture using trichloroacetic acid, washed the precipitate, and placed it in a radioactivity counter. The result was clear: in the presence of the poly-U nucleic acid, but not in its absence, amino acids were incorporated into a material that could be precipitated in an acid medium—that is, into proteins. The poly-U had led to the synthesis of a protein.

For the rest of the week, Matthaei worked day and night to determine which amino acid(s) had been incorporated into proteins in the presence of poly-U. On Saturday, May 27, at 6:00 in the morning, he finally had the answer: the poly-U coded for a monotonous protein consisting of a chain of a single amino acid—phenylalanine. In less than a week, Matthaei had identified the first word in the genetic code.

Marshall Nirenberg, who directed the small group at the National Institute of Arthritis and Metabolic Diseases in Washington, DC, in which Matthaei worked and who had conceived of the experiment, came back from a four-week visit to Berkeley. Because Matthaei had to stop work for two weeks to attend the course on bacterial genetics at Cold Spring Harbor, Nirenberg finished characterizing the product of the experiment himself. The two researchers then quickly wrote two articles: one describing the modifications they had made to the system of in vitro protein synthesis, the other presenting the results obtained with different RNAs, including the poly-U molecule.[47] These two articles, sent on August 3, 1961, to the *Proceedings of the National Academy of Sciences of the United States of America,* were published in November of the same year.

Even prior to publication, Nirenberg and Matthaei's results had made a splash. They became the main event of the Fifth International Congress of Biochemistry, held in Moscow in August 1961. Neither Matthaei nor Nirenberg had been well known, and the institute where they worked was not particularly prestigious; for both these reasons Nirenberg had not been accepted as a participant at the annual Cold Spring Harbor symposium held in June 1961. But international congresses such as the Moscow meeting were more open to scientists when compared with the relatively select meetings such as those at Cold Spring Harbor, and young researchers could present their results. Nirenberg had fifteen minutes to present his data. His talk was scheduled for a small room, and virtually nobody was there to hear it. One of the few people in the room—Matthew Meselson, who saw the talk quite by chance—told Crick of Nirenberg and Matthaei's discovery. After discussion with Nirenberg, Crick invited him to present his data again the next day, but this time in the main conference hall. The whole conference was electrified by their discovery. In the weeks that followed, other groups reproduced Nirenberg and Matthaei's results and extended them to other synthetic RNAs.

The results showed the superiority of the experimental method over theoretical approaches: one of the strongest predictions of Crick and Orgel's 1957 model was that units of the code formed by a single letter (only one kind of base, for example, U) would not code for anything—they would be nonsense.

Matthaei and Nirenberg's experiment was extremely important. Their study was part of a biochemical tradition aimed at showing that biological phenomena are merely extraordinarily complex physicochemical events that can be understood when they are reproduced in vitro. Zamecnik had been the first to develop cell-free systems of protein synthesis, whereas most of the changes introduced by Nirenberg and Matthaei had already been developed by Tissières, although Nirenberg and Matthaei were able to reduce the noise in the system.

Matthaei and Nirenberg had dared to take the idea of a genetic code to its logical conclusion and to try to determine this code experimentally, without worrying about the precise nature of the RNA involved in protein synthesis (see Chapter 13). They also rejected the idea, deeply rooted in the biochemists' view of the world but rarely expressed openly, that the shape of RNA molecules played an essential role in protein synthesis.

There has been a persistent rumor among molecular biologists, encouraged by a version of history told by Gunter Stent of the phage group, that Matthaei and Nirenberg's success was in fact pure luck. The poly-U RNA was supposedly used in the in vitro synthesis medium as a negative control: artificial, formless RNA was assumed to be noncoding. However, there is no evidence to support this claim, and Matthaei and Nirenberg protested against this view of their work. Historians of science have supported their account, citing Nirenberg's detailed laboratory diary, which shows that the experiment was carried out quite deliberately.[48] Whatever the carping by those who had failed to come up with the experiment themselves,

Nirenberg and Matthaei immediately and correctly interpreted the data they obtained with poly-U.

The rest of the code was deciphered within five years. By 1966 all the codons (the word was coined in 1962)—the base triplets that correspond to the different amino acids, as well as the punctuation signs—had been characterized. The deciphering of the genetic code elicited great enthusiasm and was followed by the major U.S. newspapers virtually day by day.

This rapid progress involved a great deal of ingenious hard work by a number of groups. Following Nirenberg and Matthaei's breakthrough, Severo Ochoa's laboratory, already a leader in the study of polynucleotides, threw itself into the race. Other laboratories soon joined in, and the data accumulated, although the quality of the results did not match the speed with which they were obtained.[49] The first polynucleotides used were composed of either a single nucleotide or several nucleotides that, under the action of an enzyme (polynucleotide phosphorylase), were randomly distributed in the polynucleotide chain. On the basis of such polynucleotides, however, a given codon could not be paired with a given amino acid. This problem was overcome by Har Gobind Khorana, who developed an effective chemical method for synthesizing polynucleotides with a given sequence.[50] Nirenberg also showed that very short polynucleotides (containing only three nucleotides) were sufficient to fix onto the microsomes (now called ribosomes) and to bind the soluble RNA (now called transfer RNA) to them, together with the attached amino acid, thus providing a very simple way of deciphering the code.[51]

Crick and Brenner's elegant experimental approach to the problem of the genetic code, published in late 1961, came too late.[52] The principle of their experiments was to use drugs that led to the insertion of a single nucleotide. It was expected that one or two insertions would have a dramatic effect on the protein that was synthesized by changing the reading frame (and therefore all the amino acids downstream of the insertion). But as soon as the number of insertions was equal to the number of nucleotides forming a codon, the effect on protein synthesis would be far less dramatic because the reading frame was rapidly restored. Their experiment, completed after Nirenberg and Matthaei's announcement, showed that the code was almost certainly composed of three-letter codons (there remained the theoretical possibility that they were composed of some multiple of three) and that the genetic message was linear.[53] These genetic techniques were useful in confirming that certain codons were indeed nonsense codons, their only effect being to interrupt protein synthesis.

When researchers began deciphering the genetic code, the links between DNA, RNA, and proteins, and the role of the different RNAs in protein synthesis, were all unknown. Scientists had to discover messenger RNA before they could begin to probe these mysteries.

13

The Discovery of Messenger RNA

IN 1953, George Gamow suggested that each amino acid fitted into a corresponding "hole" in the DNA molecule. He began his exploration of the issue by looking for a structural relationship between genes (formed of DNA) and proteins. In contrast, biochemical studies of protein synthesis tended to emphasize the role of ribonucleic acid (RNA). The coexistence of these two lines of research shows that the roles of DNA and RNA were not clearly defined. Francis Crick subsequently claimed that in 1953 when he and Watson discovered the DNA double helix they already had a relatively clear idea of the sequence DNA → RNA→ protein.[1] Articles and documents from the time, however, show that this was not the case and that right up until 1960 the relations between the three kinds of macromolecule remained extremely vague.

In 1957 Crick gave a lecture to the Society of Experimental Biology in London entitled "On Protein Synthesis," which was subsequently considered the key exposition of what became known as the dogmas of molecular biology (see Chapter 15). In this talk, Crick argued for the first time that protein folding was a spontaneous process and that the final conformation was simply a function of the amino acid sequence.[2] He also proposed the sequence hypothesis, according to which the specificity of a nucleic acid resided only in its sequence of bases—the code that determines the amino acid sequence in the protein. Most famously, he used the lecture to state what he called the central dogma of molecular biology: sequence information can go from a nucleic acid to a nucleic acid, and from a nucleic acid to a protein, but not from a protein to a protein, or from a protein to a nucleic acid. Finally, he presented the adaptor hypothesis in public for the first time (see Chapter 12) and set out the coding problem and the various theoretical attempts that were being made to resolve it.

From a biochemical point of view, Crick emphasized the importance of RNA-rich microsomal structures (now called ribosomes) in protein synthesis. But though he devoted several pages to proving that genes—DNA—control the amino acid sequence, and spent some time explaining the role of microsomes in protein synthesis, he devoted only one line to the relation between DNA and RNA: "the synthesis of at least some of the microsomal RNA must be under the control of the DNA of the nucleus."[3]

Before adding natural or synthetic polynucleotides to their cell-free system, Heinrich Matthaei and Marshall Nirenberg had tried to see if the spontaneous protein synthesis that occurred in these systems required the presence of DNA or of RNA. To each system they had thus added either DNase, an enzyme that could degrade DNA, or RNase, an enzyme that could degrade RNA. Both enzymes inhibited protein synthesis. The action of RNase was quicker, perhaps suggesting a passage from DNA to RNA.[4] Matthaei and Nirenberg were very cautious in their conclusions as to the roles of the two macromolecules.

The proof that RNA was involved in protein synthesis went back many years and was the result of a set of observations made on very different systems. The first experiments were carried out in the 1940s by Torbjörn Caspersson and Jean Brachet. Caspersson had developed some extremely sensitive spectroscopic methods that enabled him to measure the absorption of ultraviolet light by different components of the cell.[5] Linked to the action of different degrading enzymes and dyes, these studies had shown that chromosomes were composed of nucleoproteins and that the cytoplasm was rich in ribonucleic acids. The level of RNA was proportional to the metabolic activity of the cell—to protein synthesis.

Brachet's approach was different in terms of both the problem he sought to unravel and the methods he employed.[6] Brachet wanted to understand the role of nucleic acids in embryonic development by use of biochemical and histochemical methods. His experiments gave the same result as Caspersson's: the amount of DNA increased in proportion to cell division, whereas the amount of RNA was related to the protein synthesis activity in the cells.

The experiments that most clearly showed that RNA was sufficient for protein synthesis were Brachet's 1955 studies involving the removal of the cell nucleus. Protein synthesis continued for several days in anucleate cells, in the absence of DNA.[7] Parallel to these studies, researchers at the Rockefeller Institute had carried out experiments involving cell fractionation by ultracentrifugation, focusing on problems far removed from protein synthesis.[8] In 1911, Peyton Rous of the Rockefeller Institute had transferred a cancer from one chicken to another via a cell-free extract he called a virus. The American James Murphy and the Belgian Albert Claude tried to purify Rous's tumoral agent; although they succeeded, in a control experiment they also isolated small particles—apparently identical to the virus—from normal, noninfected cells. They hypothesized that Rous's tumoral agent was the result of the autocatalytic transformation of an endogenous component of the cell. They thus adopted John Northrop's theory of bacteriophage replication (see Chapter 4). These endogenous particles were distinct from the mitochondria and were called microsomes.

In 1943 Claude abandoned the idea that microsomes were related to tumoral agents, arguing that they were instead self-replicative particles that were present in the cytoplasm. He dropped all reference to his previous views and subsequently presented his work as being aimed at understanding the complex inner workings of the cell.

The scientist responsible for the characterization of microsomes was the Romanian biologist George Palade, who used the electron microscope to study cell cultures and thin sections of tissues, together with biochemical analyses and ultracentrifugation.[9] After extensive comparison of the results from these different techniques, which were often contradictory, Palade was able to distinguish particles—which in 1957 were named ribosomes—from the membranes with which they were associated. Despite this finding, the two terms ribosome and microsome continued to be used indiscriminately for several years; the RNA that makes up ribosomes was often called microsomic RNA. Each ribosome was shown to be around 250 angstroms in diameter, and to contain equal amounts of RNA and proteins. Ribosomes were found in all tissues, in quantities that were proportional to the amount of protein synthesis. In 1950 Henry Borsook and his colleagues discovered that protein synthesis takes place in microsomes, and in 1955, using a cell-free system, Paul Zamecnik showed that the incorporation of amino acids into proteins takes place on ribosomes.

In addition to these experiments, studies on the tobacco mosaic virus confirmed the role of RNA in protein synthesis. In 1956 Heinz Fraenkel-Conrat and Gerhard Schramm (finally) showed that it was the viral RNA that was infectious and not the protein, as claimed by Wendell Stanley (see Chapter 6). Fraenkel-Conrat separated RNA from viral proteins; by combining them again, he was able to obtain an infectious virus. Using the same strategy, but combining RNAs and proteins from different viral strains, he showed that the virus produced by an infected plant was of the same strain as the RNA, not the proteins. RNA clearly controlled the synthesis and nature of viral proteins.[10]

Messenger RNA (mRNA), at the time called X (pronounced in the French manner "eex" by all researchers) was noticed by François Jacob and Jacques Monod during a study of an inducible enzyme, β-galactosidase (see Chapter 14). Monod had characterized two classes of mutations that affected the production of this enzyme—one prevented the synthesis of an active enzyme, while the other led to permanent, constitutional synthesis of the enzyme, irrespective of whether the inducer was present.

To understand how the second class of mutations produced their effects, Jacob and Monod used bacterial sexual reproduction, discovered ten years earlier by Joshua Lederberg. In crosses between bacteria, the "male" transmits a chromosomal fragment but no other cellular components (see Chapter 5).

An observation made during these experiments was to prove fundamental for the birth of the concept of mRNA. These experiments clearly showed that as soon as the β-galactosidase gene entered a bacterium, enzyme synthesis began at a maximum rate.[11] But the male bacterium was thought to transmit no microsomal particles to the female. The result of the experiment was a surprise, contradicting everything that was known about protein synthesis and the role of microsomal particles.

Arthur Pardee, François Jacob, and Jacques Monod carried out a large number of control experiments in order to eliminate any other explanation. Pardee, who had spent a sabbatical year in France, returned to the United States, where, together with Monica Riley, he conducted an experiment that gave a complementary result. Pardee and Riley introduced radioactive phosphorus into the bacteria; as the phosphorus decayed, it inactivated the bacterial genes. They showed that as soon as the gene was destroyed, β-galactosidase synthesis stopped. The gene thus closely controlled protein synthesis.[12] Despite this result, Jacob and Monod did not think that the protein was directly produced on the gene. Although this was theoretically possible in bacteria, it was known that in eukaryotes genes were confined to the nucleus, whereas proteins were synthesized in the cytoplasm. Jacob and Monod therefore hypothesized that there existed a short-lived intermediary between genes and microsomal particles. They called this intermediary X (the name "messenger RNA" was coined only in the fall of 1960).[13] The data from the PaJaMo experiments (named after their authors—Arthur Pardee, François Jacob, and Jacques Monod) agreed with results obtained at the same time by François Gros, who was also working in Monod's laboratory but using a completely different experimental approach. Gros had shown that the addition of a base analog, 5′-fluorouracil, virtually instantaneously blocked β-galactosidase synthesis.[14]

The eureka episode in the discovery of mRNA has been amply described by the actors, and by historians.[15] During a visit to Cambridge on Good Friday, 1960, Jacob described his data to Crick and Sydney Brenner. As the discussion went on, Crick and Brenner suddenly realized that Jacob and Monod's results were identical to those previously obtained by Elliot Volkin and Lazarus Astrachan, which up until then had remained unexplained.[16] Volkin and Astrachan had studied replication of bacteriophage T2 in *Escherichia coli* and had shown that an RNA, comparable to the phage DNA in its base composition, was synthesized shortly after infection. The results of the Pasteur Institute group shed new light on these data and provided a precise interpretation: the RNA that Volkin and Astrachan had observed was a form of RNA that acted as a "messenger" that controlled the synthesis of proteins required for the phage to replicate.

Everything appeared straightforward: the RNA contained in the microsomal particles was not the RNA that controlled protein synthesis. Ribosomes were merely "reading heads" to which mRNAs would bind in order to be translated into proteins. This kind of RNA had hitherto gone unnoticed; it was short-lived and of varying sizes, and represented only a small fraction of cellular RNAs—it was far less abundant than ribosomal RNA.

The discussions continued into the evening and even through a party held by Crick and his wife, Odile. While bright young things danced in the adjoining room, Jacob and Brenner developed an experimental strategy for proving the existence of this new type of RNA and for distinguishing it from ribosomal RNA. These experiments were performed in the summer of 1960, in Meselson's laboratory in California.[17] Complementary experiments using a different approach were carried out by François Gros and collaborators in Watson's laboratory in Harvard.[18]

Jacob and Brenner showed that RNA synthesized by bacteriophages that infected a bacterium would bind to the ribosomes that were present in the cell prior to infection. To carry out this experiment, they used the density-gradient technique developed by Meselson a few years earlier (see Chapter 11). The bacteria were grown for several generations in a medium rich in nitrogen 15 and carbon 13. The bacteria incorporated these two heavy isotopes in particular into their ribosomes, making the particles denser. At the beginning of the experiment, the bacteria were placed in a medium of normal density that contained phosphorus 32 and would label the nucleic acids, and they were then infected with phage. The experiment showed that the newly synthesized labeled RNA molecules bound to the heavy ribosomes. The ribosomes thus merely played a passive role in the synthesis of phage proteins: they were the material support upon which the short-lived RNA molecules were bound in order to be translated into proteins.

François Gros's experiment was technologically less sophisticated, but it had the advantage of being carried out on normal, noninfected bacteria. After adding radioactive elements that could be incorporated into the nucleic acids, Gros fractionated the RNA molecules by sucrose gradient centrifugation, which separated molecules according to their size. When the radioactive components were added for only a short time, a new family of RNA molecules was found, different from ribosomal RNA and of varying sizes. But these RNA molecules were still attached to the ribosomes. These rapid-turnover RNA molecules—they were strongly labeled despite the short duration of the experiment—had all the expected properties of mRNA.

According to Francis Crick, the discovery of mRNA was "postmature."[19] But for an unfortunate series of circumstances it would have happened earlier: the fact that the most abundant RNA in cells is ribosomal RNA and that this RNA has a structural (and catalytic) function was "bad luck." If ribosomes were made of proteins—this was an unlikely possibility (see Chapter 25)—and if the only cellular RNA, apart from the small molecules of soluble RNA, had been mRNA, it would have been a lot easier to discover!

To understand why it took so long to discover mRNA, we need to appreciate the problems faced by molecular biologists at the end of the 1950s. The confusing difference between the expected properties of the RNA involved in protein synthesis and those of microsomal RNA had begun to shake the fragile edifice of molecular biology and to cause even its most ardent supporters to question the idea of a genetic code. In a letter published in *Nature* on July 12, 1958, the Russians Andrei Belozersky and Alexander Spirin reported a chromatographic analysis of the base composition of DNA and RNA from nineteen different species of bacteria. The composition of DNA showed strong interspecific variation, but the RNA tended to be remarkably constant from species to species. According to the letter, "The greater part of the ribonucleic acid of the cell appeared to be independent of the deoxyribonucleic acid."[20]

UNDERLYING THIS UNCERTAINTY were important changes that had taken place during the early development of molecular biology. The study of metabolic pathways had revealed many possible ways in which macromolecules and their components might be related. One temporary solution to this complexity was to simplify this problem to the extreme—this was the approach taken by Crick in his 1957 lecture, and by Jacob and Monod in their experiments. A counterexample of what happened in the absence of such a simplification can be seen in the case of Brachet and the members of the Rouge-Cloître laboratory in Brussels. Despite their important contributions, they did not fully succeed in extricating themselves from the overwhelming biochemical complexity of the system they were studying and were unable to make the decisive breakthrough.[21] And yet, from a chemical point of view, the direct conversion of DNA into RNA and of RNA into DNA are both relatively simple: it was easy to imagine that cells were able to perform these reactions.

Another major conceptual problem was the role attributed to genes. Classical genetics saw genes as having a distant control function in the cell, whereas molecular biologists believed that the gene determined protein structure down to the finest detail. At the end of the 1950s, the classic conception of the role of genes was still dominant and agreed with what little biochemical data existed at the time. For example, Jean Brachet showed that protein synthesis continued for hours, if not days, in anucleate cells. The fact that in eukaryotes the genes are physically separated from the cytoplasm also tended to reinforce this idea that genes intervened in a decisive but parsimonious manner, operating only transiently. Many scientists thought that the cytoplasm also contained self-replicative particles that were involved in protein synthesis.

This hypothesis—that some cytoplasmic components had genetic continuity and were distinct from nuclear genes although related to them—had haunted biologists (especially in Germany) from the beginning of the twentieth century and continued to do so in the 1960s. Indeed, in some respects history has proved it right: organelles such as mitochondria and chloroplasts do indeed contain their own DNA molecules.

At the end of the 1950s, paragenetic cytoplasmic components, known as plasmagenes, were believed to play a fundamental role in most of the events of cell differentiation. A whole series of experimental results was more or less strictly interpreted in the light of this plasmagene hypothesis: André Lwoff's studies of the morphogenesis of ciliates, Tracy Sonneborn's work on paramecia, Sol Spiegelman's investigation of adaptive enzymes, Jean Brachet's data on cytoplasmic RNA, and even the first experiments on microsomes by Albert Claude and George Palade.[22] The plasmagene hypothesis filled a gap in genetic theory—the gap created by the inability of genetics to explain morphogenesis and embryonic development. This gap would begin to be bridged only with Jacob and Monod's work and their models of gene regulation (see Chapter 14). Plasmagene theory was also favored by the opponents of genetics—those scientists who thought that the cytoplasm played an active role in morphogenesis while the external medium played an important role in evolution.[23]

One of the key accomplishments of the group at the Pasteur Institute was to bring genes and proteins closer together, to show that genes intervened permanently and directly in protein synthesis. Genes were no longer seen as isolated particles; despite their metabolic stability, they actively participated in the life of the cell.

The distant conception of the gene explains why it took so long for researchers to address the question of the precise structural relation between DNA and RNA. In 1959 Crick was still speaking of the translation of DNA into RNA and seemed to be equating two processes (translation and transcription) that today are thought to be completely different.[24] He also argued that DNA "controls" RNA—a vague term that merely re-flected the vagueness of his idea. Even more curiously, a number of bio-chemical studies had begun to show that RNA was synthesized only in the presence of DNA.[25] Today it seems quite obvious that these studies showed that information is transferred from DNA to RNA. But the articles barely mention any potential biological significance of these phenomena. They do not even discuss the problem of protein synthesis. For many scientists, DNA seems to have merely had the function of stimulating the produc-tion of RNA. None of these studies explored the base composition of the newly synthesized RNA to see if there was a link with that of DNA.

Despite the fact that researchers had known since 1952 that RNA could coil into a helix,[26] prior to 1959 there was no mention of the idea that DNA and RNA could form a double heterohelix.[27] The fact that this hypothesis was never suggested shows that the synthesis of RNA from DNA was not considered analogous to the replication of DNA. Even in the first articles that showed the existence of short-lived mRNA, the only data that suggested that these RNA molecules were derived from DNA were measures of base composition.[28] There was no direct proof that RNA was an exact copy of DNA.

The first clear description of the idea that DNA was copied into RNA can be found in a 1959 article published by Mahlon Hoagland in *Scientific Amer-ican*.[29] Even though Hoagland understandably confused mRNA and ribo-somal RNA, this article explicitly expressed the idea of complementarity between the base sequence of DNA and the base sequence of RNA. The first experiment proving that RNA is complementary to DNA was carried out by Sol Spiegelman at the end of 1960, when he showed that during infec-tion of *E. coli* by bacteriophage T2, the synthesized molecules were com-plementary to the phage DNA.[30] This experiment closely followed Paul Doty and Julius Marmur's demonstration that the two strands of DNA, after separation by heating, could reanneal if cooled slowly enough.[31]

Why did it take so long to make this link between DNA and RNA? Why can no clear trace of it be found in the scientific literature prior to 1959? The simplest explanation would be that the link was known and that it was ob-vious to all concerned that RNA was an exact copy of DNA. But this was not so: the confusion in the articles of the time reflected the uncertainty of the scientific community. The most curious point, however, is not so much that people did not realize the exact relation between DNA and RNA, but rather that no one appeared to be interested in the question. Researchers seemed satisfied with phrases like "DNA controls the fabrication of RNA."

Before the relations of DNA, RNA, and proteins could be precisely understood, the notion of specificity and the belief in the existence of a "template" in protein synthesis both had to be abandoned. Although it might seem that the death knell of these ideas sounded in 1953, when Watson and Crick first put forward the idea of a genetic code (see Chapter 12), this was not the case. In fact, the first code—proposed by Gamow—implied a precise structural relation between DNA and the amino acids that were coded by the DNA.

Following from Gamow's model, a number of other genetic codes were proposed, none of which had anything to say about the structural relation between nucleic acids and proteins. The adaptor hypothesis, put forward by Crick in 1955, explicitly excluded all direct interaction between bases and amino acids. It would, however, be wrong to conclude that the idea of form had completely disappeared from the minds of biochemists and molecular biologists: the substantial attention devoted between 1950 and 1960 to the three-dimensional structure of RNA in the hope that this structure would reveal the mechanisms of protein synthesis (Chapter 11) shows that this was not the case. Without doubt, the final manifestation of this was the importance given to microsomal particles, which were seen as the "workbench" on which proteins were "formed."

By giving microsomal particles a secondary role in protein synthesis, the work of the French school represented a final break between form and information and made it possible at last for molecular biology to come of age.

Who discovered mRNA? As Matthew Cobb has shown, it's complicated![32] This complexity probably explains why no Nobel Prize was awarded for the discovery of this key element in our understanding of gene function. The PaJaMo experiment was extremely significant because it made it possible to reinterpret an ensemble of earlier observations. But the mRNA of 1961 was not exactly the mRNA that biologists now have in mind. The supposed instability of mRNA that Brenner, Crick, and Jacob focused on was in fact a consequence of the particular experimental systems in which it was discovered—phage replication and the synthesis of inducible enzymes, where rapid changes in protein synthesis occur. Although this transience was essential for the decisive experiments to be able to distinguish mRNA from ribosomal RNA, it is not a characteristic that is now considered as a fundamental feature of this class of molecule.

14

The French School

IN 1965 **FRANÇOIS JACOB**, André Lwoff, and Jacques Monod received the Nobel Prize in Physiology or Medicine for their research on the mechanisms of gene regulation in microorganisms. Other researchers had already determined the chemical nature of the gene, the molecular structure of DNA, and the correspondence between genes and proteins—the genetic code. The work of Jacob, Lwoff, and Monod provided the final step in the circle of information exchange within organisms. They explained how regulatory proteins controlled gene expression—that is, the synthesis of the protein coded for by the gene—by binding to the gene itself.

The work of the French group was universally praised for its elegance and its association of biochemical techniques with the most advanced tools of bacterial genetics.[1] The group's results opened the way to the understanding of embryonic development in animals (Chapter 22). The study of biological deregulation (for example, in cancer) also benefited from their work.

The success of the Pasteur Institute group was the result of an "unexpected" convergence between Monod's biochemical approach and the genetic outlook of Lwoff and Jacob.[2] The origins of their research can be traced to the traditions of the Pasteur Institute.

Jacques Monod was born into an old Protestant family. He began his research career at the Sorbonne, where he worked on the nutritional requirements of microorganisms.[3] In 1936, together with Boris Ephrussi, he visited Thomas Hunt Morgan's laboratory in the United States, where he learned about genetics. During that trip, he also became aware of the relative backwardness of French biology, which was partly due to France's overly rigid university system.

During his research at the Sorbonne, Monod discovered and characterized the phenomenon of diauxie: when two food sources were added to a microbial culture, the bacteria ate first one type of food and then, after a latency period, the other. This effect produced biphasic growth curves. André Lwoff was the first to provide an interpretation of this phenomenon; in so doing he pushed Monod onto the road he was to follow for more than twenty years.[4] Lwoff realized that Monod's observation corresponded to enzymatic adaptation, a phenomenon that had been described at the Pasteur Institute by Louis Pasteur, Frédéric Dienert, and Emile Duclaux, and given its name in 1930 by the Finnish biologist Henning Karström. When microorganisms are put in the presence of a new food source, they are able to synthesize the enzymes they need to use the new food. The system Monod chose was that of adaptation to lactose, which led to the synthesis of the degrading enzyme for this sugar, β-galactosidase.

These adaptive enzymes were ideal for studying protein synthesis.[5] The experimenter could easily induce the synthesis of an enzyme by adding a simple chemical compound to the culture medium. Such systems appeared to make it possible to determine the relative roles of genes and environment in the biosynthesis of proteins and enzymes.

As seen earlier, George Beadle and Ed Tatum's work had confirmed that enzyme synthesis was under genetic control. In the case of adaptive enzymes, the added food source—lactose in Monod's system—also played a key role. The simplest explanation of the experimental data was presented implicitly in the various articles written by Monod after his arrival at the Pasteur Institute (see Chapter 12). According to this view, under the action of genes, adaptive enzymes were synthesized in the form of inactive protein precursors.[6] The inducer—lactose—formed a stereospecific complex with this protein precursor and converted it into a molecule of the active enzyme, β-galactosidase.[7] Developments in the decade that followed led Monod to question the assumptions of this view and to approach the problem of adaptation from a completely new angle.[8]

Monod's initial studies appeared to confirm the model. In 1952 he succeeded in purifying β-galactosidase and obtaining antibodies against it. With these antibodies he was able to show that the precursor—called the

Pz protein—was present before the addition of the inducer and appeared to diminish following induction. The next year, 1953, the problems began. Experiments using a radiolabeled amino acid that was a component of β-galactosidase clearly showed that this enzyme was synthesized *ex nihilo* after addition of the inducer. The supposed Pz protein disappeared into the ether of failed or misinterpreted experiments, and even its name was expunged from subsequent accounts.[9]

At around the same time, Monod and his colleagues separated the inducing function and the substrate function: a chemical analog of lactose could be an excellent inducer without being metabolized by β-galactosidase. The existence of these free inducers raised fundamental questions about the evolutionary significance of the system. In a letter sent to *Nature* in 1953, Monod and the other key researchers who were studying these adaptive enzymes proposed to change their name to "inducible enzymes," a term that described the effect without ascribing any finality to it.[10]

Such brusque changes of vocabulary are extremely rare in science, especially when they involve a very common term. Of Monod's six key articles published between 1944 and 1947, four contained the word "adaptation" in the title. Scientists generally admit that their vocabulary is not entirely precise, and that it may sometimes hide confusion; far rarer are the cases where the dropping of a scientific term is decided *ex cathedra*.

In *A History of Molecular Biology,* I suggested that political and philosophical motivations lay behind this change. The change of nomenclature ensured that Monod's research could not be seen as part of the neo-Lamarckian current that was still thriving in France, and even less so as linked to the contemporary version of this view, as represented by the Soviet agronomist Trofim Lysenko. Monod sharply criticized Lysenko's theory in a long unpublished manuscript, while in an article published on September 15, 1948, in the newspaper *Combat,* he argued that the theories of Lysenko and Ivan Vladimirovich Michurin had no scientific value whatsoever. He saw the support given to these ideas as the most visible sign of the intellectual bankruptcy of the Communist system in the USSR. According to this interpretation, by changing the terminology in his experiments, Monod crowned his break with the French Communist Party, which had called on intellectuals and scientists to back the new Soviet theories.[11] While this explanation fits with Monod's intellectual evolution, there is no direct evidence to support it.

In the years that followed, Monod showed that other proteins besides β-galactosidase were induced by the same inducer (lactose). One of them, lactose permease (studied by Georges Cohen), enabled the lactose sugar to penetrate the bacterium. Monod had obtained mutant bacteria, some of which showed altered functions of β-galactosidase or of lactose permease; others showed changes in induction itself. Some of these mutants exhibited a constitutive expression of β-galactosidase and lactose permease in the absence of an inducer. Monod did not take the characterization and localization of these mutations very far because this required complex genetic techniques that could not be performed in his laboratory. He therefore began a joint project with Jacob, a geneticist. This work quickly became a close intellectual collaboration, with the first joint experiments revealing significant similarities between the systems that the two groups had previously studied separately.

Jacob's research was part of another "Pasteurian" project. Bacteriophages had been discovered in 1915 by Frederick Twort in Great Britain, and independently two years later by Félix d'Hérelle at the Pasteur Institute in Paris. Their discovery led to the hope that it would be possible to fight pathogenic bacteria by using specific bacteriophages. These initial hopes rapidly vanished, and the therapeutic use of bacteriophages remained limited, in the West at least.

In 1925, Oskar Bail discovered the phenomenon of "lysogeny," which had probably been observed by Frederick Twort in 1915: some bacterial strains were resistant to the destructive action of bacteriophages, but within these strains some bacteria would spontaneously lyse and release bacteriophages into the medium.[12] These lysogenic bacteria thus probably contained an inactive form of the bacteriophage.

Lysogeny had been widely studied during the interwar years, in particular at the Pasteur Institute by Eugène and Elisabeth Wollman.[13] However, it had not attracted the attention of the American phage group (Max Delbrück, ever the skeptic, thought that the observations on lysogeny were worthless, perhaps the result of contamination). Whatever the case, the phenomenon was uncontrollable because the induction of lysogenic bacteria took place spontaneously, thereby limiting its experimental utility.

Partly in response to Delbrück's criticisms, Lwoff began work on lysogeny after the end of World War II. Using a micromanipulator developed by Pierre de Fonbrune at the Pasteur Institute and a culture of lysogenic bacteria, Lwoff isolated individual bacteria that he observed undergoing successive divisions. He was thus able to demonstrate that "lysogenic power"—the ability to release bacteriophages—was transmitted, implying that the bacteriophage was present in the bacteria in a cryptic form, which Lwoff called the prophage. Lwoff was able to induce this prophage in a reproducible way by irradiating the bacteria with ultraviolet light or through the addition of various chemicals. This discovery was of fundamental importance because it made the study of lysogeny much easier. It also led Lwoff to recruit François Jacob, a young doctor who had recently come out of the army and knew a bit about biological research, having been inspired by reading Schrödinger's *What Is Life?*[14] In the space of a few years, using all the tools of bacterial genetics, François Jacob and Elie Wollman gradually revealed the complex mechanism of lysogeny.[15]

They confirmed a previous observation of Lederberg that, at the prophage stage, the *Escherichia coli* bacteriophage called λ (lambda) was closely associated with the bacterial chromosome. They were able to precisely localize the prophage on the chromosome by using high frequency of recombination (Hfr) strains of bacteria that showed high rates of conjugation and a Waring blender (the same kind of device used in the experiments by Alfred Hershey and Martha Chase) to interrupt conjugation at different times (an idea dreamed up by Elie Wollman). Strangely enough, the bacteriophage was induced during conjugation, as soon as it entered into the bacterium—Jacob and Wollman initially called this phenomenon erotic induction.

It was already known that lysogenic bacteria that carried a prophage were immune to infection by other bacteriophages. This immunity was the result of the action of a single dominant bacteriophage gene, the *C* locus. Induction and lysogeny remained separate phenomena until one of the earliest results of the Jacob and Monod collaboration put them in a new light.

Jacob and Monod did their experiments in a loft in an old Pasteur Institute building (hence the laboratory's nickname *le grenier,* or attic); they soon showed that the same principles of regulation were at work in the two systems. The generality of the phenomenon made Jacob and Monod's discovery particularly important. Remarkably, both groups now used the term "induction" to refer to different phenomena but did not consider this to be significant.[16] In fact, Lwoff formed the living link between the two research projects, both of which dealt with phenomena that appeared inexplicable according to strict genetic determinism. As will be seen, this was a unique characteristic of French biology in the 1930s to 1950s.

The first experiments were simply intended to apply the technique of conjugation to the study of β-galactosidase induction. These PaJaMo (or sometimes "pyjama") experiments produced some surprising results.[17] When a male bacterium that could produce β-galactosidase only in the presence of an inducer (lactose) was crossed to a mutated female bacterium that constitutively synthesized an abnormal form of the enzyme, normal β-galactosidase was rapidly produced in the absence of the inducer. This synthesis was transitory and would cease after a few hours.

These experiments showed that the action of the regulator gene required the formation of a cytoplasmic product that must inhibit β-galactosidase synthesis. The product of this regulator gene was called the repressor. The induction of β-galactosidase during mating was analogous to the erotic induction of the bacteriophage λ already described: the same mechanisms should explain lysogeny and the control of inducible enzymes. Monod and Jacob later showed that the products of regulator genes—repressors—were proteins. Naturally active, these repressors were inhibited by lactose (or, more precisely, by a substance derived from lactose) and, in the case of the bacteriophage, by ultraviolet light. For some time it was unclear whether repressors acted at the level of the gene, by controlling its copy or transcription into ribonucleic acid (RNA), or at the level of protein synthesis. Jacob and Monod opted for the first hypothesis: repressors blocked the transcription of genes into RNA by binding to the DNA sequences, called operators, situated upstream from the regulated genes.[18]

Jacob and Monod's experiments also showed that as soon as the repression of a gene stopped, protein synthesis was quickly induced. Similarly, once a gene was inactivated, synthesis rapidly came to an end. This implied that there must be a short-lived intermediary between genes and proteins that Jacob and Monod later called messenger RNA (see Chapter 13).

Few experiments have been so rich and have led to so many discoveries as the PaJaMo experiment: the discovery of messenger RNA, the proof of the existence of repressors, and the elaboration of a general schema of negative regulation of gene expression. But who should be credited with these discoveries, Jacob or Monod? Following his observation of erotic induction, Jacob was certainly the better placed to interpret the results of the PaJaMo experiment. Furthermore, Monod had not completely abandoned the idea that the inducer played a role in protein folding and in the conformation of inducible enzymes.[19] He was thus not prepared to accept that the inducer acted on a protein—the repressor—that was distinct from the inducible proteins β-galactosidase and lactose permease.

The analogy of the erotic induction of the prophage and the transitory activation of β-galactosidase in the PaJaMo experiment was clearly stated in Jacob's Harvey Lecture, given in New York in the summer of 1958. In his autobiography, Jacob suggested that the similarity between the two systems sprang into his mind when watching a movie in a cinema near Montparnasse. It seems probable that the trigger was his desire to describe the results that he had recently obtained with Monod in a lecture that was nominally focused on lysogeny and conjugation. Whatever the case, the similarities between the observations made in the two setups suddenly appeared in his mind: constraints often stimulate creativity![20] In order to explain how the phage repressor could simultaneously inactivate *all* viral functions, Jacob put forward the hypothesis that the repressors had to act directly on DNA. This was particularly problematic for those such as Monod who thought that genes were distant, isolated, and untouchable structures.

These different views reflected the different systems being studied. Bacterial geneticists like Jacob had come close to the gene and had even begun to fragment it. The classical geneticists and other biologists, such as Monod, still operated with a much less realistic view of the nature and role of genes. If Jacob played an essential role in the development of the regulation model, Monod was key to the detailed elaboration of the model, to the definition of its different components, and to the design of experiments that would prove its validity in an extremely elegant manner.

Jacob and Monod confirmed that several structural genes could be controlled by a single regulator gene. These genes were generally grouped together in a structure called an operon and were transcribed into a single messenger RNA from a special DNA sequence called a promoter. Jacob and Monod's model of regulation was thus known as the operon model.[21]

One characteristic of this model, the idea of a repressor, requires closer historical analysis. It was several months before Monod accepted that induction could be caused by the inhibition of repression, that is, that a *positive* phenomenon could be caused by a *double negative*. For a long time he had considered induction—adaptation—to be a positive phenomenon, the result of a complex interaction between the cell and the perturbation that was imposed upon it. This kept him from accepting the inducer's minimal role in the new model. It was only after a seminar given by Leo Szilard that Monod became convinced that regulator genes coded for repressors.[22]

After the war, Szilard had abandoned his research in physics and began to study bacterial metabolism, helping to discover the inhibition of metabolic pathways by a feedback effect of the end product. The characterization of the mechanisms involved was one of the most effective applications of cybernetics to biology and led to the regulatory vision of molecular biology and to the study of what are called allosteric enzymes.[23] This work put Szilard into contact with Monod and led him to play an active role in developing the conception of induction as a derepression. Szilard had an unusual career, often abandoning a subject as soon as it had been even partly resolved. Like Delbrück, toward the end of his life Szilard became interested in what seemed to be the final frontier—the study of the nervous system—and proposed a number of models to explain memory. An ideas man rather than an experimentalist, Szilard, like many of his physicist colleagues, played an essential role in the clarification of biological problems.[24]

The outcome of this work was the conclusion that, in addition to genes that code for enzymes or structural proteins, there are regulator genes, whose sole function is to control the activity of other genes. The impact of this fundamental idea on the development of molecular biology has not been sufficiently emphasized.[25] It is particularly interesting to note that the rare experimental data available to Jacob and Monod did not justify either this clear distinction between structural genes and regulator genes, or the generalization of the concept of a regulator gene.

According to Jacob and Monod, this distinction, which was clearly set out in an article published in the *Comptes Rendus de l'Académie des Sciences* in October 1959, rapidly became the "first postulate" of their models of gene regulation.[26] It appears that it was Jacob who, inspired by studies of the bacteriophage λ that showed the importance of a single gene in the control of lysogeny, first conceived of the distinction between regulator genes and structural genes. This distinction introduced a hierarchy of genes. It also changed the aim of research in molecular biology.

For the molecular biologists of the 1960s, one of the major objectives became to understand the functioning and development of animals. How could a single cell lead to the appearance of the two hundred or so different types of cells required to form a multicellular organism, such as a human?

What mechanisms were involved in making skin or muscle cells, which possess the same genes but express these genes differentially and thus synthesize different proteins? The discovery of regulator genes implied that the work of molecular biologists should center on the study of these genes (Chapter 22).

The idea of the regulator gene also played an important role in cancer research—a cancerous cell is a cell in which gene expression is deregulated (Chapter 18)—and in studies of molecular evolution (see Chapter 23).[27] The concept of the regulator gene is important not only because it oriented molecular biological research programs for more than thirty years, but also because it closed—virtually definitively—another line of research that had been partly opened by the German-born geneticist Richard Goldschmidt and had been developed in particular by the American geneticist Barbara McClintock's work on maize.[28] According to McClintock, the differential expression of genes resulted from their physical movement in the genome. As they moved, genes were placed under the control of different regulator elements that modulated their expression. This genetic transposition was in turn under the control of another gene. McClintock had in fact anticipated the discovery of regulator genes, but the complexity of the system she studied, difficulties in explaining her results to other scientists, and above all the key importance she gave to the physical movement of genes in regulation all limited the impact of her findings.

The work of François Jacob and Elie Wollman on the bacteriophage and conjugation showed that the movement of genes in bacteria could also alter gene function. Two models of genetic regulation were thus possible: one saw this regulation in terms of a stable genome, under the control of networks of regulator genes; the other linked the regulation of gene expression to a structural modification of the genome that took place during the lifetime of the organism.

In June 1960 Jacob published an article in *Cancer Research* in which he explained to cancer specialists the mechanisms of regulation that existed in microorganisms, placing the two possible forms of regulation on an equal footing.[29] In the book Jacob had published with Elie Wollman a year earlier *(La sexualité des bactéries),* they argued that mobile genetic elements, which they called episomes, could intervene during cell differentiation to modify nuclear potential.[30] Jacob eventually opted for a model of stable regulation with no physical alteration of the genome, first in his lecture to the 1961 Cold Spring Harbor Symposium.[31] In a 1963 article written with Monod, he more clearly wrote, "most of the known facts support the view that the genetic potentialities of differentiated cells have not been fundamentally altered, lost, or distributed."[32] (Jacob's choice might have reflected the influence of Monod, who, considering nature to be straightforward and Cartesian, was always inclined to choose between two competing hypotheses.)

The distinction between regulator and structural genes, indeed the very notion of regulator genes, had a historical and philosophical importance. It brought genetics closer to embryology and implied that it would be merely a matter of time and hard work before they fused definitively. Philosophically speaking, it represented the final step in the development of a new vision of biology that had begun in the 1940s, which made the question of the information contained in genes an ordering principle of all life. Jacob and Monod went further, distinguishing two kinds of information in organisms: structural information that was necessary for the formation of the components of the organism, and regulatory information that was responsible for the gradual spatial organization of these structural components during development. This distinction had been outlined by Schrödinger in *What Is Life?* and found its clearest expression in the work of Jacob and Monod.

The historian Edward Yoxen has shown that Schrödinger's book, which had been influential in the early days of molecular biology but had subsequently been largely forgotten, was rediscovered in the 1960s.[33] Jacob and Monod's articles contain formulations similar to those used by Schrödinger, although it remains unclear whether this was the result of a direct influence or an indication of the continuing relevance of Schrödinger's ideas and of a strange resonance between his speculations and the models developed by the molecular biologists.[34]

After more than five years of joint work, Jacob and Monod went their separate ways. Monod turned to the study of a class of proteins—regulatory proteins—for which the repressor was the model. The repressor protein interacts with DNA, but that interaction may be inhibited by a breakdown product of lactose binding at another site on the repressor molecule. Many other proteins or enzymes, and in particular those subject to negative feedback control, have similar properties: their activity can be modulated by activators or inhibitors binding to sites other than the catalytic site.

Monod, together with Jean-Pierre Changeux and Jeffries Wyman, developed the "allosteric model," which explained the properties of these proteins and enzymes.[35] This model was largely inspired by work on hemoglobin, from Linus Pauling's pioneer studies in the 1930s to the most recent results from crystallography.[36] Other models of regulation were proposed by competing groups—such as the induced-fit model of Daniel Koshland—and for several years there was a lively debate over allosteric theory.[37] History—and the structural data obtained by X-ray diffraction— has generally supported Monod's hypothesis, although it did not turn out to have the universality that he hoped for.[38]

In retrospect, two characteristics of the allosteric model are particularly striking. The first is that its dogmatic presentation, in the form of a series of postulates, was overly rigid and in some cases unnecessary. This kind of dogmatism was probably reinforced by the absence of a strong tradition of structural chemistry in France. These postulates, however, were extremely rich in terms of the information they provided about the enzyme and its structure. The other models, which were much less restrictive, described the kinetic behavior of enzymes without seeking to explain their structural bases.

The second characteristic of the model that distinguished it from its competitors was linked to the first. Whereas the other models were almost always able to describe the behavior of enzymes, the allosteric model made a number of easily falsifiable predictions. These two characteristics of the allosteric model were the result of Monod's persistence in trying to impose some kind of order on the anarchy of life. They were also the product of his adoption of Karl Popper's philosophy of science, according to which science advances through easily falsifiable models and not through theories of everything.[39] In the context of contemporary molecular biology, allosteric theory appeared as an island of rigid Cartesianism in a sea of Anglo-Saxon pragmatism. It is only recently that the scientific potential of the allosteric theory has been fully acknowledged, with the number of articles devoted to allostery increasing each year (Chapter 21).

François Jacob remained faithful to the research program outlined in his 1963 article with Monod; he studied increasingly complex bacterial regulatory circuits before turning to the regulation of gene expression in multicellular organisms. Continuing to study the bacteriophage λ, Jacob and his coworkers described complex regulatory circuits that controlled the

transition between infection and lysogeny. They also studied the regulation of cell division, an essential step in the bacterial life cycle. Division
takes place only when the bacteria have become large enough and when
there is sufficient food available. Jacob, Sydney Brenner, and François
Cuzin developed a model of regulation called the replicon based around a
positive regulatory gene, which still serves as a reference for researchers.[40]
Finally, Jacob bridged the gap that separates bacteria and multicellular organisms, and in 1970 he began to study the embryonic development of
the mouse (Chapter 22).[41]

Jacob and Monod's research put the finishing touches to the basic principles of molecular biology and laid the foundations of future molecular biological research programs. But French researchers had been virtually absent
from the early stages of the development of this new science. Furthermore, the two roots of molecular biology—biochemistry and genetics—
developed much later in France than in other countries.

There is a striking contrast between the brilliant interwar developments
in biochemistry that took place in Germany, then in Britain and America,
and the virtual nonexistence of French biochemistry in the same period.
This delay can mainly be explained by the actions of one man, Gabriel Bertrand, the chair of biological chemistry at the University of Paris and the
director of one of the main biochemistry laboratories at the Pasteur Institute.[42] After a series of striking studies on enzymes, Bertrand catastrophi-
cally decided that proteins played no real part in them and that the key catalytic role was played by the metal ions that are often found associated
with enzymes. The orientation of Bertrand's research and its decisive influence meant that French biochemistry played no part in the subsequent
characterization of the various major metabolic cycles or in the study of proteins, despite the medical and agricultural significance of Bertrand's
research.

The reasons for France's backwardness in genetics are more complicated.
Some historians have argued that the slow rate at which chairs in genetics
were created in French universities was a result of the top-heavy and overly
rigid French university system, a sign of its fossilization. Other reasons
can equally well explain the delay: at the end of the nineteenth century
France was particularly slow in accepting—or even discussing—Darwin's
theory.[43] In the first half of the twentieth century neo-Lamarckism was
very strong among French biologists.[44] They did not directly reject the

results of Mendelian genetics, but they nevertheless tended to minimize the role of genes in the functioning and development of organisms while emphasizing the role of the cytoplasm.[45]

To understand how, despite these apparent handicaps, France eventually played a major role in the development of molecular biology, we must see this weakness of biochemistry and classical genetics as an advantage for the French researchers.[46] As shown earlier, molecular biology often developed in opposition to biochemistry, by showing that proteins and enzymes were not at the heart of self-replicative phenomena, by opposing the idea that biological facts could be explained by thermodynamic rules linked to the totality of chemical reactions that take place in the organism. In a way, the French molecular biologists were freer than their British and American counterparts because they spent less of their time fighting the biochemists.

Relations with French geneticists were both more complex and more productive than those with the biochemists. The genetics laboratories that were set up in France after World War II, such as Boris Ephrussi's laboratory at Gif-sur-Yvette outside Paris, had an unusual approach to the subject.[47] Turning away from the characterization of chromosomal genes, which was dominant in the U.S. school of genetics, the French tended to concentrate on the study of hereditary phenomena that were apparently independent of the nucleus. This was the continuation of a tradition begun with André Lwoff's work on the genetic continuity of intracellular organelles in unicellular organisms called ciliates and with the work of Elisabeth and Eugène Wollman on the fate of phage in lysogenic bacteria.[48]

In addition to this concentration on paragenetic phenomena, there was also the influence of the great French physiologist Claude Bernard. As Jean Gayon, Richard Burian, and Laurent Loison have pointed out, Bernard's approach emphasized physiology rather than morphology.[49] For Bernard, organisms were characterized by their ability to adapt continually to physiological changes rather than by their structure or form. Whereas the geneticists of Morgan's group were interested in the genes controlling the form of organisms (the wing structure of flies, for example), the French groups were interested in the genetic control of organic adaptability.[50] They were able to show that an adaptive process, conditioned by the external medium, is nevertheless subject to strict genetic control.

But more than any of these French particularities, it was the very special framework of the Pasteur Institute that offered Lwoff, Jacob, and Monod a favorable environment for their research. The institute allowed its groups complete scientific autonomy and total independence from the university system. Despite the lethargy that had overtaken the institute in the first half of the century, during the long directorship of Emile Roux, its prestige remained intact, and it continued to attract large numbers of visiting foreign scientists from both Europe and America. This independence and prestige enabled Lwoff's group to be fully integrated into the burgeoning international network of molecular biology.

The 1965 Nobel Prize had a major impact in France, extending far beyond the few French people able to understand the significance of the discoveries that had been made. This impact was the consequence of the personalities of, particularly, Jacob and Monod, and of their heroic actions during World War II as well as their involvement in politics and philosophy.

Monod had been highly active in the Resistance against the Nazi occupation.[51] In the 1950s, he joined the efforts made to reform French universities and research, and he would actively support the student protest movement of May 1968 and fight for the rights to abortion and euthanasia. In 1970, he published *Chance and Necessity,* a book describing the recent developments in biology and exploring their philosophical implications.[52]

Jacob's life was no less adventurous.[53] After facing anti-Semitism in his early years, he left France in June 1940, joined de Gaulle in London, and for four years was a medic with the Free French armed forces in Africa. He was severely wounded in Normandy in August 1944, and he had six difficult years before his recruitment by Lwoff. In 1970 he published *The Logic of Life*.[54] In the following years, Jacob wrote various historical and philosophical books on science, which were influenced by Michel Foucault and Thomas Kuhn, as well as an autobiography, *The Statue Within.* As politically active as Monod although less visibly so, Jacob consistently defended the freedom of research and emphasized the ethical responsibility of scientists.

Jacob, Lwoff, and Monod, the three musketeers of molecular biology, affected French society as much as they did French biological research. Their influence would continue to be felt for many decades.

PART THREE

THE EXPANSION OF MOLECULAR BIOLOGY

15

Normal Science

ONCE RESEARCHERS had deciphered the genetic code and described regulatory mechanisms in microorganisms, molecular biology entered what the historian and philosopher of science Thomas Kuhn called a period of normal science.[1] Research no longer involved testing global models but puzzle solving within the framework of existing theories. Molecular biologists did not think they had solved all the mysteries of biology, but their understanding of fundamental molecular mechanisms appeared sufficient to imagine how the unresolved problems (development, the origin of life) might be approached. This impression of completeness was shared by most molecular biologists (as it had been by physicists in the 1930s), and it led some of them to turn to what appeared to be the final frontier of human knowledge—the brain.[2]

In the 1950s, Max Delbrück, the founder of the phage group, was disappointed that the study of genes had not revealed new physical principles and turned instead to the study of a fungus that is sensitive to light and other stimuli, as a model for studying sensory physiology at the molecular level.[3] Seymour Benzer, a specialist in bacteriophage genetics, turned to *Drosophila* behavior genetics, while his good friend Sydney Brenner sought to understand development by studying the tiny roundworm, *Caenorhabditis elegans*.[4] A decade later, Francis Crick followed them into the nervous system, ambitiously focusing on humans.[5]

Other molecular biologists, such as Gunther Stent and François Jacob, wrote the history of the subject, reinforcing the impression that a key phase in the discipline's history had come to an end.[6] Finally, in *Chance and Necessity*, Jacques Monod set out the principles of the new ethics that seemed to be prompted by molecular biology. The book was the consequence of the new view of the biological world and an eyewitness account of its development:

> The theory of the genetic code constitutes the fundamental basis of biology. This does not mean, of course, that the complex structures and functions of organisms can be deduced from it, nor even that they are always directly analyzable on the molecular level. (Nor can everything in chemistry be predicted or resolved by means of the quantum theory which, without question, underlies all chemistry.) But although the molecular theory of the genetic code cannot now—and will doubtless never be able to—predict and resolve the whole of the biosphere, it does today constitute a general theory of living systems.[7]

Monod's book brought the principles of the new science to the attention of the general public in France and elsewhere and led to one of the last great French intellectual debates of the twentieth century. Dozens of books were written, all replying in slightly different ways to Monod's views. The heart of the debate was the role of chance in evolution, and thus the place left for a religious conception of the universe by the new scientific view.[8]

THIS CONVICTION that the essentials of the new science had been discovered inevitably led to dogmatism and a refusal to accept any results that tended to question—no matter how slightly—what was considered already proven. Two examples will suffice to illustrate how molecular biology was affected by such narrow-minded reactions. The first is taken from the work of Jacques Monod and his colleagues on gene regulation in microorganisms. We saw in Chapter 14 that Monod found it difficult to accept the existence of negative feedback control. Other systems of bacterial gene regulation, such as that studied in Monod's laboratory by Maxime Schwartz and Maurice Hofnung, did not follow the model of negative regulation. Other groups had already proposed models of positive regulation, which Monod rejected. It was only after a great deal of effort that Schwartz and Hofnung were able to convince Monod that positive regulation did indeed exist.[9] In fact, this concept was to be of great significance in bacteria and especially in multicellular organisms (see Chapter 20).

The other oft-cited example of the resistance of molecular biologists to even a partial questioning of their results was the highly significant discovery of an enzyme that could copy ribonucleic acid (RNA) into DNA. The existence of such an enzyme had been proposed by Howard Temin in 1962, but was finally accepted only eight years later.[10] To understand the importance of this enzyme and the questions raised by its discovery, it is necessary to return to Francis Crick's declaration of the central dogma in 1957, in his famous lecture to the Society of Experimental Biology (see Chapter 13).[11] This lecture was important because, for the first time, Crick gave a synthetic presentation of the principles of the new science.

Crick's lecture discussed the nature of the genetic code and proposed the adaptor hypothesis. Most importantly, he set out the philosophy of the new science: all the information required to create organisms can be found in molecules of DNA. This information goes into proteins, where it determines the order of amino acids. No transfer of information in the other direction—from proteins to DNA—is possible. On the one hand, no part of the cellular machinery appeared able to convert the three-dimensional information contained in proteins into the one-dimensional information contained in a DNA molecule. On the other hand, the existence of such a transfer would call into question the fundamental principles of genetics. Although this was not pointed out by Crick, allowing the possibility of a transfer from proteins to DNA would open the door to neo-Lamarckism.

Although the study of regulation in microorganisms had shown that the medium could have an effect on protein synthesis, if proteins could change the message contained in DNA, then the environment could change DNA, and the inheritance of acquired characters would be possible. For many biologists, the rejection of any transfer of information from proteins to DNA was written in stone—more because of the traditions inherited from genetics and neo-Darwinism than because of the biochemical data that were available to Crick and his contemporaries.

As noted earlier, in 1957 Crick was particularly vague about the role of RNA. The hypothesis of messenger RNA and the experimental proof of its existence were three years in the future. Many experiments had suggested that RNA might be the precursor of proteins and be derived from genes (DNA), but the possibility of a transfer from RNA to DNA was by no means

excluded; some models, in particular those from the Rouge-Cloître laboratory in Brussels, even proposed that RNA might be a common precursor
of both DNA and proteins.[12] Strikingly, in a preliminary, unpublished sketch
of the central dogma, drawn in 1956, Crick included a potential transfer of
information from RNA to DNA. He theoretically foresaw this possibility
six years before Howard Martin Temin. However, none of this was used in
the published version of the lecture.

The totality of relations between biological macromolecules formed what
Crick somewhat unfortunately called the central dogma of molecular biology. Defending himself later, Crick explained that he was unaware of the
theological meaning of the word dogma and that he had confused it with
axiom.[13] After the discovery of messenger RNA, the central dogma took on
its canonical form in the book *Molecular Biology of the Gene* published by
Jim Watson: DNA makes RNA makes proteins.[14] The potential passage from
RNA → DNA was removed from the models without any justification beyond the fact that it made things simpler and that there was no positive evidence for such a transfer.

In science in general, and in molecular biology in particular, models have
their own weight. As a result, the passage from RNA to DNA was generally
considered to be quite simply impossible. But Howard Temin's experiments
on retroviruses soon made it necessary to reassess what had truly become
a dogma. Retroviruses like the Rous sarcoma virus are tumor viruses (oncogenic viruses), but they differ from the other tumor viruses in that their
genetic material is RNA. By the end of the 1960s, the study of oncogenic
viruses had become a major research topic in molecular biology. Many scientists thought that oncogenic viruses would provide a royal road to the understanding of multicellular organisms, much as the study of bacteriophages had helped in the understanding of bacteria.

Renato Dulbecco played a very important role in developing quantitative methods for studying animal viruses.[15] He worked first with Salvador
Luria, then with Delbrück, on the rather strange phenomenon of the reac-
tivation by visible light of phages that had been inactivated by ultraviolet
light. In 1950 Max Delbrück asked Dulbecco to interrupt this work and visit
a series of U.S. laboratories that were working on cell culture and the multiplication of viruses in these cultures. Thanks to the work of John Enders,
Thomas Weller, Frederick Robins, and Wilton Earle, cell culture technique
had recently been improved through the use of antibiotics and changes in

medium composition, to the extent that it had become a powerful tool for the study and production of viruses.[16] On his return to the California Institute of Technology (Caltech) in Pasadena, it took Dulbecco only a few months' work with the horse encephalitis virus to develop a lysis plaque technique, similar to that used to detect bacteriophages, which made it possible to quantitatively measure the number of viruses. Dulbecco first applied this new approach to the poliomyelitis virus, then in 1960 he turned to the study of a small oncogenic DNA virus, polyomavirus. Instead of the plaques of lysed cells, plaques of transformed cells appeared on the surface of the dish.

Dulbecco's work shows that Delbrück's influence extended far beyond the world of the phage. It also shows how molecular biologists were able to benefit from new technological tools being developed in other areas of biology. Above all it provides a very concrete example of the influence of research funding on scientific developments. Delbrück's approach can be explained by two very large grants he received, one from the National Foundation for Infantile Paralysis (set up by President Franklin D. Roosevelt to support research into polio, it provided substantial funding to virology laboratories), and the other a sizable donation to Caltech by the American millionaire James G. Boswell, who suffered from shingles.

Studies of small oncogenic viruses such as polyoma and SV40 had shown that their DNA was integrated into the genome of infected cells and that this process was linked to the transformation of the cells—their passage from a normal to a cancerous state. Temin had previously proposed the same hypothesis to explain transformation by retroviruses. In his model, Temin was obliged to imagine the existence of a conversion step from RNA to DNA, preceding the integration of the retroviral genetic information into the genome, which he felt had been proved by the effect of different inhibitors on transformation.

Nevertheless, it was more than eight years before Temin convinced the rest of the scientific world and the central dogma was "reversed," as an editorial in *Nature* proclaimed in 1970.[17] The scientific community was finally convinced when Temin and David Baltimore independently isolated an enzyme, associated with the viral particle, that was able to convert RNA into DNA and was thus called reverse transcriptase. There can be no doubt that this discovery was made more difficult because of the prevailing dogmatism of molecular biology. But the picture is not entirely black and white: the experiments reported by Temin in 1962 were not entirely convincing, as they had a low signal to noise ratio.[18] And opposition to his claims stopped immediately following the discovery of reverse transcriptase.

In hindsight, the whole controversy over the existence of reverse transcriptase seems ridiculous. After all, its discovery did not shake molecular biology to its foundations. As will be seen, it in fact furnished one of the most important tools used in genetic engineering, a development that in turn reinforced the models of molecular biology (Chapter 16).

But if Temin did not destabilize molecular biology, it was not for want of trying. No sooner had reverse transcriptase been discovered in retroviruses than Temin proposed a protovirus model, according to which retroviruses were merely the pathological products of the normal cell machinery, which was able to copy RNA into DNA. Temin argued that the amount of DNA in different cells in the organism varied, depending on the quantity of RNA present in the cells and their functioning. He put forward the hypothesis that this reverse transcription machinery played an essential role in development, in complete opposition to the models of Jacob and Monod (see Chapter 14).[19] However, a reverse transcriptase-type activity in healthy, noninfected cells has still not been found, and eventually Temin was obliged to abandon his protovirus theory (see also Chapter 23).

THE PERIOD FROM 1965 TO 1972 saw an increase in the number of groups and laboratories working in molecular biology, the creation of institutes devoted to the subject, and above all an expansion of the molecular vision beyond its original field, leading to a takeover of other biological disciplines by molecular biologists. In Europe, this expansion led not only to the spread of molecular biology beyond its original centers (Cambridge, Paris, and Geneva), but also to the creation in 1964 of the European Molecular Biology Organization (EMBO) and of the European Molecular Biology Laboratory (EMBL) in Heidelberg in 1978.[20] Although this molecularization was generally gradual, it sometimes took place swiftly: a single article could push a whole discipline into the molecular field.

One of the best examples is that of biological membranes. Both physicochemical and morphological (electron microscopy) studies had suggested that membranes were formed of a lipid bilayer, one side of which was covered with proteins. In 1972, Jonathan Singer and Garth Nicholson proposed a very different model, according to which proteins were inserted fully into the membrane, implying that they played an essential role in the structural and functional properties of membranes.[21] The study of membrane properties thus escaped from the lipid biochemists and became the domain of protein specialists, and in particular of molecular biologists.

These takeovers occurred in a number of ways. Molecular biologists received important funding, often at the expense of other researchers. They took control of scientific journals, either by creating new publications devoted to the new molecularized sciences, or by changing the outlook of existing journals.[22] Molecular biologists also took over in the universities, either by introducing molecular biology into the curricula or, more often, by updating biochemistry or genetics courses. Particularly important in this struggle were job descriptions in both teaching and research, and the choice of candidates. In both respects, the molecular biologists clearly came out on top.

Later chapters describe the results of the introduction of the new science into the main biological disciplines. At this point, however, we need to understand how the molecular biologists were able to take control and to impose their new vision of biology onto the whole of the biological community. This study will be limited to France, where the belated introduction of molecular biology and the isolation of the pioneering groups of André Lwoff, Jacob, and Monod make this seizure of power and the subsequent molecularization of biology all the more spectacular.[23]

For the Whig historian of science, focused on interpreting science as endless progress, none of this would make much sense. They would argue that molecular biology succeeded simply because of its worth, because of the beauty of its results, and because of the new areas of research that it opened up. By contrast, for sociologists of science (or at least for the most dogmatic of them), new science can have no objective superiority. Superiority is merely the outcome of confrontation, not the reason for victory. It is thus necessary to think in terms of the strategies of disciplines, and not of the value of scientific theories. Max Planck argued that a new theory is accepted only when the partisans of the old theory have died. The triumph of molecular biology was far too rapid for Planck's explanation to be true!

No one can deny that molecular biology is a compelling science. The primary reason for this lies in its simple and pedagogic models (Chapter 29). These models are not a way of popularizing the science but form one of the basic tools used by molecular biologists in their daily work. They make molecular biology attractive. Take Jacob and Monod's models of regulation: how many young biologists have been impressed for life by these models, which seem to clarify the complexity of biological phenomena? As early as 1944 the clarity and simplicity of Erwin Schrödinger's book was attracting many young physicists to biology. Twenty years later, the inspirational brilliance of Jacob and Monod's models would lure young biologists away from the traditional disciplines.

These models are persuasive because the logic that inspired them and the image of the biological world that they reveal are in harmony with the picture of the world that is presented both by the media and by other scientific disciplines. Explaining the functioning of organisms in terms of information, memory, code, message, or negative feedback involves using a language and a set of images that everyone now knows.

Despite what you might think, the results produced by molecular biology did not immediately convince everyone that this approach was worthwhile. As we will see, extending the new science to the study of multicellular organisms proved a lot more difficult than expected. Few new results were obtained, and the onward march of knowledge proved to be difficult. The sociologists of science are at least partly right: molecular biology did not win simply on the battlefield of facts.

In France the rise to power of the molecular biologists was assisted by the Nobel Prize that was awarded to Jacob, Monod, and Lwoff in 1965. The Nobel Prize confers enormous prestige and gives the winner influence at all levels of the decision-making process in his or her field—recruitment, financing, and the construction of new laboratories. Monod, who was appointed professor at the prestigious Collège de France, became director of the Pasteur Institute in 1971, and was able to intervene directly in the development of molecular biology in France. But this political power was a consequence of a scientific judgment by the Nobel committee: in the case of Jacob and Monod, the attribution of the Nobel Prize sanctioned a discovery that the scientific community already considered to be fundamental. This is shown by the fact that Jacob and Monod were given pride of place at the 1961 Cold Spring Harbor Symposium, long before they received the Nobel Prize.

Jean-Paul Gaudillière has explained how the intervention of Monod and the Pasteur group in the development of molecular biology in France in fact predated their Nobel Prize and the national and international recognition of their work that this prompted. The main force behind the development of molecular biology in France was not the Centre National de la Recherche Scientifique (CNRS) or the universities but the Délégation Générale à la Recherche Scientifique et Technique (DGRST). The DGRST, which became operational in 1961, was under the direct control of the prime minister, and was given the task of making up for France's scientific backwardness. A creation of the Gaullist regime, it was intended to contribute to the renewal of French science, bypassing the cumbersome administrative machinery of the French research organizations and university system.

One of the main tasks of the DGRST was to promote molecular biology. The priority given to this discipline was the result of the personal links that existed between the molecular biologists and those in General de Gaulle's entourage who were responsible for developing the new science policy. These links had been forged in the Resistance during World War II and had been reinforced between 1955 and 1960 at a series of meetings and conferences, such as the one that took place in Caen (Normandy), all of which were aimed at reorganizing French research and universities. The political contacts that made the action of the DGRST possible had thus been established long before the work of the French school of molecular biology reached its peak.

BETWEEN 1965 AND 1972, some of the dogmas of molecular biology were confirmed. Crick's 1957 sequence hypothesis, according to which the structure of a protein depends only on the order of its amino acids (Chapter 13), was demonstrated by the in vitro synthesis of an active enzyme—ribonuclease A—solely from amino acids.[24] The most striking development of this period was the description of the three-dimensional structures of proteins through the use of X-ray analysis of crystals.[25] The first crystallographic analyses of enzymes led researchers to propose a number of models of enzymatic catalysis (see Chapter 21).[26]

These developments were not necessarily the result of molecular biology. In reality they were additional steps on a long road that began many years earlier with the attempts of William Astbury, Linus Pauling, and J. D. Bernal to decipher the structure of biological molecules. These results did not relate to the heart of the new discipline—the mechanisms of information exchange within cells. No dramatic progress was made in this key respect. Further investigation of the mechanisms of the regulation of gene expression only complicated Jacob and Monod's seductively simple model. In addition to negative control, other mechanisms, such as positive regulation, turned out to play a role. The molecular mechanisms of the major biological processes—replication, transcription, and translation—were all clarified, at least in bacteria. After the discovery of the first enzyme that could replicate DNA in vitro, DNA polymerase I, another enzyme, DNA polymerase III, was found to be responsible in vivo for the replication of DNA in bacterial cells (see Chapter 19).[27] Similarly, RNA polymerase—the enzyme that copies DNA into RNA—was purified, and its action was described: a small protein subunit, the σ factor, was found to be responsible for beginning transcription by precisely positioning the RNA polymerase upstream of the gene.[28]

During this period, researchers characterized many of the factors necessary for protein synthesis and began to understand ribosomal structure more fully; the experiments initiated by Masayasu Nomura to reconstitute ribosomes from their RNA and protein components were particularly striking.[29] Molecular biologists also crystallized transfer RNA molecules and began determining their three-dimensional structure.[30] Meanwhile, the structures of messenger RNA and of viral RNA were studied using various techniques but without any decisive breakthroughs.

Molecular biologists also determined the first nucleic acid sequences, isolated and purified the first genes from bacteria and eventually from animals, and chemically synthesized a gene corresponding to a transfer RNA.[31] Studies involving gene manipulation were extremely complex—as shown by the number of authors involved—and difficult because of the primitive technology employed.

Many molecular biologists were aware that the new frontier of their discipline was the study of multicellular organisms—their functioning, their development, and their pathologies. But molecular studies of multicellular organisms proved extremely difficult. A mouse has around a thousand times more DNA in the nucleus of each of its cells than does a bacterium, with a concomitant increase in the complexity of the genetic information. Furthermore, because of the large number of different cells present in a multicellular organism, it is much more difficult to purify a mouse protein than a bacterial protein. The isolation of the proteins that regulate transcription is difficult enough in *E. coli* (the isolation of the repressors of the lactose operon and of the phage λ took more than five years).[32] It was virtually impossible in mice. Combined with the feeling that molecular biology was now entering a period of normal science, these difficulties led to a kind of crisis among molecular biologists that has been well described by François Gros.[33]

All was not bleak during the years of 1965 to 1972, however. Of course, it took some time for researchers who had never studied anything other than *E. coli* and the bacteriophage to get used to the complexity of worms, flies, and mice. But this period was useful for developing methods for extracting RNA and DNA for the molecular study of multicellular organisms. Links were forged between specialists from different disciplines—embryologists, cell biologists, molecular biologists—informal networks were established, and different laboratories became focused on similar objects. These years were clearly necessary for the molecularization of biology.

During the 1960s, the use of the word "molecular" spread like wildfire. Everything became molecular: pharmacology, neurobiology, endocrinology, and even medicine, if several popularizers of the time are to be believed.[34] A wave of young researchers trained in molecular biology appeared at the same time as new techniques from the discipline were introduced into biological laboratories: the use of labeled molecules to follow RNA or protein synthesis, ultracentrifugation and chromatography to purify RNA or protein molecules, and electrophoresis to characterize them. This introduction of researchers and techniques was a consequence of the invasion of various disciplines by the concepts and models of molecular biology. Whereas a traditional biologist would characterize a function in physiological or cellular terms, the molecular biologist—or the biologist recently retrained in the new discipline—would isolate the molecules (generally proteins) responsible for that function.

In endocrinology (the science of hormones and their mechanism of action) or in neurobiology, the aim was henceforth to isolate the proteins onto which the hormone or the neurotransmitter bound (the receptors) and that were responsible for the observed cellular effect. Similarly, pharmacologists tried to isolate the macromolecules that were the target of drugs (these macromolecules were also called receptors).

The notion of a receptor predated molecular biology, but it fitted perfectly into the models of the new science. Allosteric theory explained how the binding of a small molecule such as a hormone onto a receptor could alter the receptor's conformation and enable it to interact with other molecules and transport the hormonal or pharmacological signal to the inside of the cell. In medicine, the aim was to describe pathologies in molecular terms. As early as 1949 Linus Pauling had set the example, when he described the causal chain that linked the slight changes in the structure of the hemoglobin molecule observed in sickle cell anemia with the macroscopic changes in red blood cells that were responsible for the disease (see Chapter 12).[35]

Nevertheless, the extent of this molecularization of biology should not be overestimated. Many laboratories—now sporting the term "molecular" in their title—carried on as before, with the same research objectives and using the same techniques. Jean-Paul Gaudillière has shown that, in France, the DGRST's "concerted action in favor of molecular biology" often had little effect on the research performed by the laboratories involved.[36]

Furthermore, the molecularization of biology often involved the introduction of the techniques and models of biochemistry rather than those of molecular biology. Only a few groups began studying genes and gene regulation directly. In some cases, this can be explained by the persistence of research traditions and experimental systems that predated molecular biology. In other cases, the absence of appropriate molecular techniques stopped researchers from turning to molecular studies or, when initial attempts were unsuccessful, led them to return to a biochemical approach.[37]

Such prudence was probably preferable to the molecular excesses that characterized some areas of research such as embryology, neurobiology, and immunology. All three disciplines almost simultaneously took up the idea that RNA might be responsible for intercellular information transfer.

In embryology, several groups argued that inductive effects—the possibility that cells of a given type could induce other cells to undergo differentiation—were produced by the transfer of RNA molecules from the inducing cells to the induced cells. Embryonic induction, which in the 1930s had led to a dead end, thus received a new lease on life (see Chapter 8).[38]

From the beginning of the 1960s research in immunology had shown that antibody synthesis was the result of the cooperation of different cell types: macrophages captured and transformed antigens, whereas antibodies were produced by lymphocytes. Several groups showed that the information necessary for the synthesis of specific antibodies was transferred from the macrophage to the lymphocyte as an antigen-RNA complex.[39]

But it was in neurobiology, and in particular in the study of memory, that this pseudo-molecular approach apparently had its most striking successes.[40] At the beginning of the 1960s, studies on rats by Holger Hydén in Sweden (using microspectrophotometric techniques) and by Louis and Josefa Flexner in the United States, and work on the small worm *Planaria* by James McConnell, showed that the synthesis of macromolecules (RNA or proteins) was needed for memory formation and in particular for the acquisition of conditioned reflexes.

In itself, this was hardly surprising. Memory, like any biological process, requires the transcription of a certain number of genes and the synthesis of the corresponding proteins. But the authors of these studies went much further, suggesting that there was a direct molecular coding of memories and behaviors in these animals. Both McConnell, working on *Planaria,* and Allan Jacobson, working on the rat, claimed that it was possible to transfer *specific* behaviors from a trained animal to a naive one, in the form of molecules of RNA. These results, published by the most prestigious scientific journals and widely publicized in the media, provoked both excitement and major reservations. For example, both the possibility that *Planaria* could learn and the purity of the material used were questioned.[41]

In 1966, a number of researchers wrote a joint article in which they explained that they had been unable to reproduce the basic result—that is, the transfer of a conditioned behavior from a trained to an untrained rat by injecting RNA-enriched brain extracts.[42] At the end of the 1960s, George Ungar's results supported the possibility of such a transfer, but showed that the molecule responsible was not an RNA but an RNA-associated peptide. In the years that followed, Ungar generalized his initial observation to different kinds of acquired behavior—habituation to sounds, fear of the dark, and so on—purified the peptides responsible, determined their amino acid sequence, and argued that the results revealed the existence of a mnemonic code. The peptides extracted from the brains of animals that had been conditioned to prefer a color (Ungar called these proteins chromodiopsins) all had similar amino acid sequences, with a conserved part and a variable part.[43] According to Ungar, the conserved sequence corresponded to the preference behavior, whereas the variable sequence corresponded to the preferred color. Ungar seemed to be on the threshold of discovering the code for memory. But the quality of his work was increasingly criticized, as was his behavior. He published less and less frequently in specialized journals, reserving his announcements for the press or for popular science magazines.[44]

These molecular theories of memory gradually evaporated as the data proved either unreliable or nonspecific. In 1975 the discovery of enkephalins and endorphins (endogenous analogues of morphine), some of which have an effect on memory, provided another interpretation of the memory-transfer data. Ungar's peptides belonged to the same family of molecules as the endorphins and had a nonspecific effect on memory.[45]

In all three cases—the molecular bases of memory, embryonic induction, and the formation of antibodies—RNA molecules (or proteins in the case of Ungar's experiments) were seen as containers or transporters of information. They all integrated the new informational view of biology that characterized molecular biology. But the research was marked by a brutal reductionism, going directly from observed biological phenomena to macromolecules, ignoring what took place within organs or cells. In the experiments on antibody formation, as in the studies of embryonic induction, the observed effects implied that there was an interaction between cells. In the case of the acquisition of behavioral responses, it was known that different regions of the brain were involved in different behaviors, which therefore required the information to be transmitted through a neuronal network.

One explanation of this molecular fashion would be to frame events in terms of the politics of science. In the 1960s molecular biologists wanted to reduce as many biological phenomena as possible to the molecular level. As the years passed, however, molecular biology became stronger. Its struggle against the traditional biological disciplines became more subtle: molecular biologists infiltrated other disciplines, transformed them, and gradually took control. Once victory was attained, an armistice would be signed in which both sides agreed to respect the specificity of their respective approaches.

A complementary explanation, more psychological and conceptual than sociological, may also be valid. Many molecular biologists who had received little or no training in the traditional biological disciplines turned with a certain naiveté and ignorance to the investigation of complex biological phenomena. Their study of multicellular organisms led them to the unsurprising realization that life in fact has a high degree of organization. This in turn led to the development of a new paradigm that was not molecular nor cellular but something completely different.

Genetic Engineering

GENETIC ENGINEERING INVOLVES MANIPULATING, isolating, characterizing, and modifying genes as well as transferring them from one organism to another. The term now seems slightly old-fashioned and its use is waning as it is probably replaced by the less precise "biotechnology," but there is, as yet, no better phrase to describe this approach. It is not easy to describe the history of genetic engineering precisely because it is a technology and not a science. The history of science often consists of a detailed account of a series of experiments and the theories that enabled them to be understood. But the development of a new technology cannot be described as a linear succession of elementary steps. When a new technology appears, a set of individually unimportant elements become linked in a network and take on a new significance. Some of these elements are clearly defined technological advances that were developed in order to fulfill their function in the network; others are old discoveries that had not previously been exploited but which take on a new importance through their integration into the network. Finally, some elements in the network may be not discoveries at all but minor technical improvements that make a task possible by reducing the time and effort involved. For example, techniques for sequencing nucleic acids existed before 1970, but they were so slow and difficult that it was not until the development of rapid DNA sequencing techniques between 1975 and 1977 that the field of genetic engineering began to take off.[1]

The exact frontiers of a technological network are also difficult to define—genetic engineering benefited from discoveries and new techniques in other areas of biology, immunology, and physiology, such as the discovery of monoclonal antibodies or the development of in vitro fertilization. These techniques were incorporated into genetic engineering, enriched it, and modified it, while continuing to be a part of other technological networks.

The idea of deliberately changing the genetic makeup of an organism is as old as genetics itself. In 1927 this idea appeared to come a step closer when Hermann Muller, who had been Thomas Hunt Morgan's student, showed it was possible to induce mutations by X-rays. This method was limited, however, because the mutations it induced were just as random as spontaneous mutations: their frequency could be increased, but not their specificity. According to Oswald Avery, only in pneumococcal transformation was it possible to induce "by chemical means . . . predictable and specific changes which thereafter could be transmitted in series as hereditary characters."[2] The later discovery of transformation in other bacteria, together with conjugation and transduction, all reinforced the idea that the genetic makeup of an organism could be deliberately modified.

The term "genetic engineering" goes back further than experiments on genetic manipulation. It originally referred to any controlled and deliberate modification of the genetic constitution of an organism, either by classical genetic experiments—crossing organisms carrying particular genes—or by future techniques that would involve the manipulation of isolated genes and their introduction into another organism without crossing.

Molecular biologists soon realized that direct gene manipulation could have enormous practical applications. In 1958 Ed Tatum wrote,

> With a more complete understanding of the functioning and regulation of gene activity in development and differentiation, these processes may be more efficiently controlled and regulated, not only to exclude structural or metabolic errors in the developing organism but also to produce better organisms. . . . This might proceed in stages, from *in vitro* biosynthesis of better and more efficient enzymes to biosynthesis of the corresponding nucleic acid molecules, and to introduction of these molecules into the genome of organisms, either through injection, through introduction of viruses into germ cells, or through a process analogous to transformation.

Alternatively, it may be possible to reach the same goal by a process involving directed mutation.[3]

In 1969 Joshua Lederberg wrote the genetics entry in the *Encyclopaedia Britannica Yearbook* and predicted the extraordinary applications of the new genetics: viruses would be used to introduce new genes into the chromosomes of humans or plants, thus providing a treatment for human genetic diseases or a way of improving crops.[4] Viruses were widely seen as the vectors of choice for introducing new genes; oncogenic viruses were known to integrate themselves into the chromosomes of the infected cell and thus to transform its properties, like phage in bacteria (Chapter 18). But this power of transformation did not seem to be limited to viruses. Experiments on the direct transformation of the cells of multicellular organisms by foreign DNA had shown positive results as early as 1962, although they could not be replicated.[5] The isolation of purified fragments of DNA, and the total synthesis of a gene in 1970 (see Chapter 15), showed that the manipulation of genes was possible. Nothing, however, implied that molecular biology was on the threshold of a revolution even more thoroughgoing than that which had taken place twenty years earlier with the discovery of the double helix. The manipulation of genes still appeared to be a difficult objective, situated on a distant and ill-defined horizon.

HISTORIANS DO NOT AGREE on the origins of genetic engineering. Some see it as a natural development, largely anticipated during the years that preceded the first experiments.[6] Others have argued that the discovery, characterization, and use of restriction enzymes that can cut DNA at precise points played a decisive role. This seems to have been the opinion of the Nobel Committee, because in 1978 the first Nobel Prize for a discovery relating to genetic engineering was awarded to Werner Arber, Hamilton O. Smith, and Daniel Nathans. Arber had been the first to characterize what was termed restriction—that is, the ability of some bacteria to degrade (to cleave) any foreign DNA introduced into the bacterium, through the action of restriction enzymes.[7] Smith had been one of the first to purify and characterize the properties of these enzymes.[8] Nathans had used them to cleave DNA from the SV40 virus into fragments and thus make what would later be called a physical map of the viral genome, a prelude to the determination of the complete sequence of viral DNA.[9]

To my mind, the beginning of genetic engineering was marked by the experiment carried out at Stanford by David Jackson, Robert Symons, and Paul Berg and published in 1972 in the *Proceedings of the National Academy of Sciences*.[10] In this article, Jackson, Symons, and Berg describe how they obtained *in vivo* a hybrid molecule containing both the DNA of the SV40 oncogene virus and the DNA of an altered form of phage λ (λ dv*gal*) that included the *Escherichia coli* galactose operon.

To carry out this study, Berg first cut the parental DNA molecules by using a restriction enzyme, *Eco*R1, provided by Herbert Boyer's group at the University of California at San Francisco. The ends of the λ dv*gal* fragment and of SV40 had been altered by the action of another enzyme—terminal transferase—which had added repeated sequences of adenine and thymine, respectively. These repeated sequences would pair together, leading to the association of the DNA from phage λ and the DNA from SV40. The operation was terminated by the addition of DNA polymerase and of an enzyme—ligase—that could close the molecules of DNA, resulting in a circular molecule that was a hybrid of λ and SV40.

Strictly speaking, none of the steps used in Berg's experiment was original. Restriction enzymes had already been used to cleave SV40 DNA, and the effects of the ligase and terminal transferase were already known, thanks in particular to the work of Arthur Kornberg's group on ligase (see Chapter 19).

The idea of fusing two molecules of DNA using repeated sequences of adenine and thymine via the action of terminal transferase had previously been proposed by Joshua Lederberg in a research project submitted to the National Institutes of Health (NIH).[11] Finally, Peter Lobban, a student in the same department as Berg, had begun trying to associate in vitro two DNA molecules from phage P22.[12] Berg's article was original, however, because it presented an important result, a wide range of techniques, and a clear outline of future perspectives. It described the first in vitro genetic recombination between DNA molecules from different species.

Genetic recombination—the association of previously separate genes—is a natural process that occurs randomly in both bacteria and multicellular organisms. But it was thought to always take place between genes carried by organisms of the same species, or, in the most extreme case, between the genes from a virus and the genes of the infected cell. In Berg's experiment, bacterial genes were linked to those from a monkey virus. Furthermore,

Berg made simultaneous use of a large number of enzymes that were to form the basic tools of any experiment in genetic engineering: restriction enzymes to cleave the DNA and a ligase to close it up again, exonuclease to degrade it and DNA polymerase to repair it, and the terminal transferase used to make the ends of the DNA molecule "sticky."

Berg showed that the resultant DNA molecule could be integrated into the chromosomes of a mammalian cell. But because it also possessed DNA sequences from phage λ, it could replicate itself autonomously in bacteria. The hybrid molecule could thus be multiplied within these bacteria. In their article, Berg and his colleagues clearly described the dual use of genetic manipulation: the possibility of introducing genes into a given organism coupled with the use of bacteria to amplify hybrid molecules that had been created in vitro.

The article by Berg and his coworkers thus has the same founding value as Jim Watson and Francis Crick's first 1953 article. In it the authors united a series of elements that might have been dispersed over a large number of experiments and articles. They grouped together a project, a series of techniques, and a result that was important in and of itself. The article was a scientific work of art.[13]

To give a complete picture of the pioneering experiments in genetic engineering, three other results obtained around the same time need to be added.[14] First, in 1972, Janet Mertz and Ron Davis discovered that the restriction enzyme *Eco*R1 did not cleave DNA cleanly but left cohesive ends that could be stuck back together or associated with other DNA molecules that had been cleaved with the same restriction enzyme.[15] This result meant that it was possible to bypass terminal transferase, thus considerably simplifying in vitro recombination. Around the same time, the groups led by Herbert Boyer and Stanley Cohen used new bacterial vectors instead of phage λ.

Plasmids—circular DNA molecules—had been discovered in 1965. They carry genes that confer resistance to antibiotics and can autonomously replicate in a bacterial cell, where they can often be found in hundreds of copies.[16] The first recombinant plasmids, containing several resistance genes and / or DNA sequences from different species of bacteria, were made in 1973 and 1974. These plasmids could penetrate *E. coli* bacteria, where their genes could be expressed, even if they came from other bacterial species such as *Salmonella* or *Staphylococcus*.[17] Finally, in 1974 an experiment carried out by Cohen, Boyer, and their coworkers showed that DNA from the toad *Xenopus laevis,* once introduced and copied in a bacterium, was transcribed into ribonucleic acid (RNA). The possibility of getting bacteria to synthesize a protein from a multicellular organism became more real.[18]

MANY SCIENTISTS WERE WORRIED when Paul Berg announced his results prior to their publication in the *Proceedings of the National Academy of Sciences.* They feared that his research would lead to the development of *E. coli* bacteria carrying cancer-causing genes. These organisms might escape from the laboratory and spread their genes to the human population (which naturally carries *E. coli* bacteria). These fears were expressed publicly at a conference that took place in New Hampshire from June 11 to 16, 1973. Three months later, on September 21, 1973, Maxine Singer and Dieter Söll, who had chaired the New Hampshire conference, published a letter in *Science* in which they called for the creation of a committee to investigate the consequences of in vitro genetic manipulation.[19] In a letter sent to the *Proceedings of the National Academy of Sciences* and to *Science,* Paul Berg argued for at least a partial moratorium on such research and called for a conference that would define the conditions under which such experiments could be carried out.[20]

The conference took place from February 24 to 27, 1975, at Asilomar in California, and consisted primarily of a presentation of the experiments that had been carried out using the new technology. Only one day was devoted to their potential dangers. A report of these discussions, written by Paul Berg, David Baltimore, Sydney Brenner, Richard Roblin, and Maxine Singer, was published in *Science* on June 6, 1975.[21] The conference had tried to set out the risks associated with these new experimental techniques and the precautions that should be taken.

The conclusion was that work could begin again, but that strict physical and biological containment procedures would have to be adopted to limit the dangers. Physical containment involved limiting all contact of the potential pathogen with the experimenter or with the environment. This could range from wearing a lab coat and gloves during the experiment, through obeying the sterility rules enforced in microbiology laboratories, up to the use of extractor hoods or even isolated rooms subject to lower barometric pressure than the external environment and separated from it by an airlock. Biological containment involved carrying out the in vitro recombination on organisms that had been modified so that they could not survive outside the laboratory.

Recombinant genetic experiments were classified according to whether they involved DNA extracted from bacteria, from animal viruses, or from eukaryotic cells. The main hazard was associated with the last kind of experiment: it was thought that the genomes of eukaryotes contained inactive forms of potentially pathogenic viruses, in particular of tumor-causing viruses. Random manipulation of eukaryotic DNA would run the risk of provoking the incorporation of these viral sequences into bacteria, and their subsequent spread to the human population. In vitro genetic recombination with animal viruses appeared to be less dangerous, except when the viruses were potentially carcinogenic. In this case, genetic manipulation would be authorized only if the sequences responsible for the cancerous transformation had already been removed.

Experiments on bacterial DNA were by far the least dangerous. Indeed, the exchange of genetic information between bacteria is a natural phenomenon. The only risky experiments were those involving DNA from highly pathogenic bacteria or DNA coding for toxin genes. These experiments were completely banned.

A hazard level was associated with each type of experiment, with the recommended containment procedures adapted to the degree of risk. According to the report's overall approach, the proposed measures were provisional and were expected to evolve as a function of increasing knowledge and improvements in biological containment techniques.

The Asilomar recommendations were turned into precise rules, laid down for the United States by the NIH in January 1976. In Great Britain, the Williams report—published at the same time—set out a number of similar measures.[22] In France, a Délégation Générale à la Recherche Scientifique et Technique (DGRST) commission created in 1975 was charged with classifying projects and devising security rules. Given that there was no specifically French set of rules, the commission based its work on the recommendations made by the NIH and the Williams report.

A great media debate on recombinant DNA erupted, especially after the Asilomar conference, reaching a peak in 1976–1977.[23] There was in fact not one debate but several, all of which were characterized by a substantial degree of confusion. The possibilities raised by the new experiments on in vitro genetic recombination, and the potential risks linked to these new techniques, changed over time, at least as they were expressed in the opposing positions.

It was soon apparent that the main fear of the scientific community in 1971–1972 was that it would be possible to introduce oncogenes into *E. coli*. Berg's experiment had focused these fears because it had been performed with DNA from an oncogenic virus. These concerns were all the stronger because, between 1970 and 1975, most molecular biologists tended to accept viral models of cancer (see Chapter 18).

Although potentially dangerous, even tumor-causing, viruses had been studied for many years in a number of laboratories, such studies had never previously prompted any worries. With the first experiments in genetic engineering, a new group of molecular biologists, barely familiar with traditional biological disciplines or with the containment procedures used in laboratories that studied pathogenic organisms, began to study oncogenic viruses. The risk of an accident appeared much greater with these new experimenters, who were highly trained in the manipulation of DNA and restriction enzymes but were unfamiliar with the precautionary measures traditionally used.

The fear that these genetic manipulations would lead to the appearance of new bacterial or animal species that could reproduce and perhaps develop and establish themselves, to the detriment of existing species and the ecosystem, played only a minor role in these initial debates. Similarly, the fear that the manipulation of human genes would lead to a eugenic program aimed at the control and modification of the human genome was completely marginal.

After Asilomar, all that changed. The construction of new laboratories to carry out the most dangerous kinds of genetic experiments led to a series of confrontations. In the summer of 1976, when Harvard wanted to build such a laboratory, the mayor of Cambridge opposed the project, arguing that research should be carried out for the common good, not to get the Nobel Prize. After an enquiry that lasted several months, and after consulting experts, the town council overturned the mayor's decision and agreed to the construction of the laboratory.[24]

What had been a debate between experts exploded into the public domain. The arguments against recombinant DNA experiments became wide ranging—the creation of new pathogenic species and the unleashing of a lethal epidemic both appeared to be major dangers. In the longer term, recombinant DNA experiments ran the risk of encouraging mankind to control both itself and nature. It could also lead to the development of an exclusively genetic view of humanity and, as a consequence, to the development of eugenic policies.

The critics of recombinant DNA technology were part of a much larger current that was highly critical of science—as well as of its links with capitalism—and of the "scientific-military-industrial complex." The fears that these critics expressed seemed to be justified by the interest shown in the new technology by the main chemical and pharmaceutical companies. For the opponents of genetic engineering, this interest both revealed the existing relations between science and business, and heralded the development of even tighter links that would tend to deprive scientific research of what little independence and creativity it still had.

Throughout this period, the scientific arguments of the supporters of recombinant DNA technology remained more or less constant. They claimed that the potential for beneficial applications justified fundamental research, and they hoped that this technology would lead to the isolation and characterization of the genes of multicellular organisms and to the study of their organization and regulation. This new methodology was expected to show whether the mechanisms of cellular differentiation and intercellular communication involved new modes of genomic organization and regulation, or were merely minor variations of the mechanisms present in bacteria.[25]

The potential applications of the new techniques of genetic engineering proposed by its supporters were limited to a few examples that were regularly repeated throughout the long controversy. They claimed that genetic engineering would make it possible to reprogram bacterial cells so that they would synthesize clinically useful products from humans, such as insulin or interferon; to modify plants to resist insects and diseases, or to fix nitrogen from the air (thus avoiding the use of expensive nitrogen-rich fertilizers); and to correct genetic "mistakes" in humans.

A growing number of scientists also hoped that research would be able to be performed without hindrance or restrictive rules that would limit the number of future practical applications. In 1977 virtually the whole U.S. scientific community, led by the president of the Academy of Sciences, used these arguments as the basis of their opposition to research restrictions proposed by Senator Edward Kennedy of Massachusetts.

An unprecedented amount of experimental data was rapidly accumu-lated. The heat soon went out of the polemic, and biosecurity regulations, which originally had been extremely strict, were relaxed to previous levels. A number of more or less rational reasons lay behind these changes. The shift in attitude was not due to the development of more effective methods of biological containment or to proof that the procedures were sufficient to prevent the dissemination of any pathogenic genes that might be isolated during recombinant DNA experiments. Rather, it was a direct result of several conferences that took place between 1976 and 1978 to estimate the risks of dissemination. The historian of science Susan Wright has shown that the supporters of genetic engineering used these conferences to con-vince the scientific community and the general public that there was no danger.[26] The discussion was centered solely on the risk of an epidemic due to recombinant bacteria and not on the potential hazards for technicians or researchers manipulating recombinant DNA. Furthermore, the reports of the conferences did not always accurately reflect the debate—failing to mention, for example, that supplementary experiments had been commis-sioned or that the results were not yet available.

In fact, the risks associated with the transfer of a gene from a eukaryote into bacteria were reduced when it was realized that eukaryotic genes are generally split, interrupted by many long stretches of noncoding DNA, and that the excision of these sequences ("splicing") was necessary before a functional messenger RNA (mRNA) could be formed. Because bacteria lack the molecular machinery necessary for splicing, it would be impossible for the introduced gene to be expressed (Chapter 17).

What gradually convinced people of the safety of genetic engineering was the lack of accidents; in other words, people became convinced empiri-cally. As to the fear that genetic experimentation would lead to the modifi-cation of species in the future—to a clumsy attempt by humans to control their own evolution with the risk of a resurgence of eugenics—the early re-sults of genetic engineering and the ongoing technical improvements would hardly change matters, even if the practical consequences of the re-sults remained limited.

Early experiments showed that it was possible to carry out genetic re-combination in vitro and to introduce the product of this recombination into a bacterial cell. Before these experiments could open the road to the study of eukaryotic genes, techniques were needed to isolate these genes.

TWO COMPLEMENTARY APPROACHES WERE TRIED. The first, chosen by Tom Maniatis's group at Harvard, involved going from proteins to genes, which made it necessary to choose a cell or an organ in which a particular protein was strongly synthesized.[27] This was the case for the precursors of red blood cells, in which hemoglobin is synthesized in abundance. Messenger RNAs coding for hemoglobin are present at high levels in these cells, representing an important fraction of total mRNA. These molecules can be purified relatively simply (for example, by centrifugation), copied in vitro using reverse transcriptase (the enzyme discovered in retroviruses by Howard Martin Temin and David Baltimore)—thus producing "complementary" DNA (cDNA)—and then inserted into a plasmid or phage and amplified in bacteria.[28]

This approach had two weaknesses. First, the genes obtained by this method were not functional because they did not carry regulatory signals (in particular, promoter DNA sequences) upstream from the gene, which enable the RNA polymerase to recognize the gene and to copy it into mRNA. Second, this procedure could only be used to isolate genes coding for abundant proteins.

Another technique—cloning—was developed by David Hogness's group, which worked on *Drosophila* and wanted to isolate the genes that classic genetic analyses had shown were important in the development of the fruit fly (Chapter 22). The products of these genes, if they were known at all, were present in very small quantities. In this method, the whole genome is fragmented at random by restriction enzymes, and the resultant fragments are inserted into plasmids or phages, which are then incorporated into bacteria.[29] The result is a heterogeneous population of bacteria: each bacterium has a different phage or plasmid containing a different part of the genome. This population is called a genomic library. It is equally possible to create a cDNA library by copying all the mRNA molecules present in a cell into DNA, and then integrating these different cDNA molecules into plasmids, and then into bacteria.

Such libraries are useful only if those bacteria containing the plasmid or phage with the desired gene or stretch of cDNA can be isolated. The procedure is as follows: the bacteria that make up the library are separated by dilution and are then allowed to form isolated colonies on the surface of a petri dish. The colonies carrying the gene in question are then selected by molecular hybridization using a fragment of the DNA of the gene or of a similar gene. Thus, for example, in 1976 Maniatis's group first isolated a β-globin cDNA; two years later, they were able to isolate the gene from genomic DNA libraries.[30]

Molecular hybridization, developed by David Hogness, involves transferring the colonies onto a nitrocellulose filter then denaturing them—that is, separating the two strands of plasmid DNA contained in the colonies.[31] These filters are then incubated with a fragment of DNA (cDNA in Maniatis's experiment) called a probe, which has previously been denatured and radioactively labeled. This fragment of DNA will pair up ("hybridize") with the plasmid DNA that contains a DNA sequence identical or analogous to that of the probe. The filters are then washed, and the presence of colonies carrying DNA fragments homologous to the probe is revealed by the trace they leave on a photographic film. The experimenter then simply has to choose the colonies that produced the positive result on the film, grow them in a test tube, and isolate the recombinant plasmids. These plasmids, once cleaved by appropriate restriction enzymes, will release the desired DNA fragment.

The next step was the development of expression libraries. In such libraries the gene cloned in the phage (or plasmid) is expressed, and its protein products are produced within the bacteria. This means that it is possible to use an antibody to detect the presence of this protein in bacterial colonies. The technique is virtually the same as that for molecular hybridization: the bacterial colonies are transferred from the petri dish onto a nitrocellulose filter that is then incubated with an antibody rather than with a radioactive probe.

By comparing libraries either from organisms that do or do not possess a particular gene, or from cells or organs that do or do not express the protein coded for by the gene, it is possible to isolate genomic DNA or cDNA even though no fragment of DNA from the gene has previously been isolated, nor any antibodies been raised against the corresponding protein.[32]

Genetic engineering made its first fundamental step forward with the in vitro recombination of DNA molecules. The second step was cloning. But genetic engineering would not have developed so rapidly if new techniques and the improvement of existing methods had not simplified each step in the process. The following improvements—rapidly made by different research groups—had to be combined for genetic engineering to become an effective and efficient technology.

First, more and more restriction enzymes were purified and characterized, each of which recognized a specific sequence and thus cleaved DNA
into different size fragments. The commercialization of these enzymes,
which rapidly followed their discovery, made research much easier. Mapping a DNA fragment—that is, locating the different restriction sites—
became so easy that any student could do it. By 1978 approximately fifty
restriction enzymes and their cleavage sites were known.[33]

Next, the physical mapping of DNA molecules required a way of separating fragments following cleavage. Initial methods such as centrifugation
were rapidly replaced by electrophoresis on polyacrylamide gels, then on
agarose gels (agarose is a sugar polymer extracted from seaweed), with the
DNA fragments being revealed by the addition of a fluorescent dye, ethidium
bromide.[34] Once the fragments have been separated by electrophoresis,
those that carry a given DNA sequence have to be isolated. This again involves molecular hybridization after transfer of the gel content to a nitrocellulose sheet.[35]

A similar technique was developed to determine the size of RNA molecules corresponding to a given gene.[36] These two techniques were respectively named Southern and the Northern blots—"Southern" was named
after its developer, Edwin Southern, while "Northern" was coined as a joke.
(They were eventually joined by another joke term, the Western blot, or
protein immunoblot; various techniques have been called Eastern blots, but
in no case has the name stuck.) Finally, increasingly effective "vectors"—
phages or plasmids—were developed, containing several cleavage sites recognized by different restriction enzymes.

All these developments meant that fragments of DNA could be cloned
regardless of which restriction enzyme was used to cleave them. The
original plasmids generally contained a single antibiotic resistance gene,
whereas novel plasmids could contain several such genes, some of which
could be inactivated by the insertion of the DNA fragment into the plasmid.
As a result, bacteria that did not contain the plasmid would not develop in
the presence of antibiotics; bacteria containing the plasmid alone would
grow in the presence of any antibiotic, whereas those colonies containing
the plasmid with the insert would grow only in the presence of certain
antibiotics. This procedure made it possible to rapidly isolate those bacteria that have integrated the recombinant plasmid and to eliminate those
colonies that contain only the original plasmid. The majority of plasmids

still used today derive from a prototype pBR322 constructed in 1977 by the Mexican scientist Francisco Bolivar.[37] Many plasmids, adapted to various objectives (cloning, expression of the genes they carry), were later created from this plasmid and sold by various companies.

One of the limits of the first cloning experiments was that the plasmids or phages could incorporate only small DNA fragments (fewer than a few thousand bases). These vectors were not appropriate for cloning genes from eukaryotes, which are generally tens of thousands of base pairs in length. Cosmids—vectors that are derived from plasmids but that could be packaged in vitro into the heads of bacteriophage, resolved this difficulty.[38] With the development of research on the human genome, new methods for cloning, mapping, and separating large DNA fragments were developed (Chapter 27).

The first step in any research using genetic engineering is the isolation and amplification of the genes or purified DNA fragments. Once this stage has been completed, the DNA has to be characterized by making a physical map using restriction enzymes, and above all by determining the order of its nucleotides, its sequence. This final operation became much easier after the simultaneous development in 1977 of two competing sequencing techniques. One was created by the Americans Allan Maxam and Walter Gilbert,[39] the other by the British scientist Fred Sanger.[40]

Maxam and Gilbert's method used chemical reagents that cleave DNA at certain nucleotides, whereas Sanger used an enzyme-based method requiring DNA polymerase and specific inhibitors (see Chapter 19 for a fuller description of the development of Sanger's method). For both methods to be effective, polyacrylamide gel electrophoresis had to be improved. The development of thin gels and the use of new buffers substantially increased the resolution of electrophoresis, to the extent that DNA fragments differing by a single base could be distinguished. In one electrophoresis it became possible to determine the sequence of DNA fragments more than 300 base pairs in length.

Genetic engineering also involves all those techniques that enable researchers to alter the sequence of isolated DNA fragments in order to modify either the protein they code for or the signals that regulate their expression. Various experiments using chemical mutagens were initially carried out, but it was only in 1978 that Michael Smith developed a method of directed mutagenesis using synthetic oligonucleotides (short fragments of single-strand DNA).[41] At the same time new, simpler, and quicker protocols for synthesizing oligonucleotides were developed, and automatic synthesizer machines were soon commercialized.[42] Genetic engineering put the oligonucleotides produced by these methods to a number of uses, including inserting new restriction sites in the vectors and adding regulatory signals.

To study gene expression and function in eukaryotes, once the gene has been purified, characterized, and perhaps modified in vitro, it has to be re-integrated into cells. In the 1960s, experiments had shown that DNA could enter eukaryotic cells. Transfer was relatively inefficient, and the stability of the incorporated DNA was low. The use of animal viruses was not only potentially dangerous, as explained earlier, it also limited the size of the DNA fragments that could be incorporated. In 1973 Frank Graham and Alex van der Erb developed a new technique that substantially improved the transfer of DNA into cells or, to use the molecular biologists' jargon, the effectiveness of transfection.[43] Finally, in 1979, a method of cotransfection with a resistance gene was developed, showing that it was possible to isolate cells that had integrated exogenous DNA (this was analogous to the method developed several years earlier for bacteria).[44] This experiment opened the road to the development of stable cell lines with a modified genetic constitution.

These techniques were the fundamentals of genetic engineering. In 1982, following a practical course taught at Cold Spring Harbor Laboratory, a book called *Molecular Cloning: A Laboratory Manual* was published that described all these techniques in extremely simple terms.[45] Written by Tom Maniatis, Joe Sambrook, and Edward Fritsch, it soon became known simply by its first author's name and was found on the shelves of molecular biology laboratories all over the world. If there was a book of molecular biology, it was "Maniatis." The use of genetic engineering techniques became commonplace and marked the end of the developmental phase of the new technology. Despite (or perhaps because of) its popularity, this book was not well received by biochemists who considered that it abandoned the good experimental practices that they had painfully developed in the previous decades, replacing them by recipes that were blindly used by poorly trained "new" biologists. The disdain of the biochemists did little to dent the book's popularity.

THE MOST IMPORTANT initial results using the new techniques were the discovery of the discontinuous structure of genes in eukaryotes, the proof that genetic rearrangements exist in the cells of the immune system, and the demonstration that the genetic code is not, in fact, universal (see Chapter 17).

All these results threatened the dogmas of molecular biology. Some, such as the discovery of splicing, were extremely important, but others, such as the nonuniversality of the genetic code, were related only to certain species and ultimately posed no threat to our understanding of evolution, genetics, or molecular biology.[46] What contributed most to the transformation of molecular biology was the simple practice of genetic engineering: the isolation of genes and the decoding of their sequences. This practice led to the characterization of the genes involved in the cancerous transformation of cells (see Chapter 18) and enabled molecular biologists to isolate proteins that act as transcription factors (Chapter 20) and to characterize the master genes involved in the control of development (Chapter 22).

In 1976 the first cDNA molecules that were obtained corresponded to abundant RNA and proteins, such as β-globin, insulin, the rat growth hormone, and human placental hormone.[47] These results made it possible to characterize the corresponding genes.

The most sensational studies were not those that tried to isolate and characterize the genes of eukaryotes but those that sought to get bacteria to synthesize the products of these genes. These investigations were carried out in an atmosphere of frantic competition between laboratories on the West and East coasts of the United States, between the University of California at San Francisco (and the biotechnology company Genentech) and Harvard University.[48]

The genes studied were those for which the protein might have important medical applications, such as insulin and interferon. Two different strategies were employed: either the DNA corresponding to the protein to be produced by the bacteria was synthesized in vitro (thanks to the genetic code, which allows a corresponding nucleotide triplet to be assigned to each amino acid), or cDNA was obtained after mRNA had been isolated and then copied into DNA by reverse transcriptase. In both cases, eukaryotic genes were linked to a series of bacterial regulatory sequences that were well known from the work of François Jacob, Jacques Monod, and many others. Both approaches paid off; between the end of 1977 and the beginning of

1979 a series of experiments confirmed that it was possible to express proteins from eukaryotes in bacteria.

In the fall of 1977, Keiichi Itakura and Herbert Boyer of the University of California were able to get bacteria to synthesize human somatostatin, a low-molecular-weight hormone consisting of only thirteen amino acids.[49] The corresponding gene had been artificially synthesized and then fused to a fragment of the lactose operon. The resultant product was a fusion protein, a hybrid of β-galactosidase and somatostatin. A later chemical reaction released the somatostatin.

This might seem of limited importance—a small protein was expressed as a hybrid molecule. But this result showed that it was possible to express the protein of a eukaryote in bacteria, thus paving the way for subsequent major developments.

This first success was quickly followed by the expression of human insulin using the same approach.[50] (This development was itself preceded by the expression of insulin and of dihydrofolate reductase, an enzyme involved in nucleotide biosynthesis, from cloned cDNA.)[51] Finally, in 1979 and 1980, bacterial synthesis of growth hormone and then of biologically active interferon reinforced the idea that the new technology could have important medical implications.[52]

To this list should also be added experiments that led to a genetic "movement" in the opposite direction—that is, to the insertion into eukaryotes of genes that had been amplified in bacteria. As seen earlier, in 1973 a method for directly inserting DNA fragments into animal cells without using viruses had been developed and then improved by using selection genes (1979). In 1980 DNA was integrated into plant cells for the first time.[53] Because a plant can be reconstituted in vitro from a single cell, it thus became possible to obtain transgenic plants that incorporated new genetic information into their chromosomes.

The creation of transgenic animals followed shortly afterward. In 1980 Frank Ruddle injected mouse embryos a few hours old with foreign DNA that then integrated into their chromosomes. After several rounds of cell division in vitro, the embryos were implanted into a number of foster mothers, which, twenty days later, gave birth to a total of seventy-eight pups, two of which had integrated the foreign DNA into most of their cells.[54] Such transgenic mice were expected to be able to transmit the foreign DNA to their offspring, and this was later shown to be the case. This experiment had been made possible by the development, a few years earlier, of methods for manipulating and growing embryos in vitro, which were also used for the first in vitro fertilization in humans. These experiments ushered in the age of contemporary molecular biology.

Almost immediately, researchers began to apply this new technology to the early diagnosis of genetic disorders. By the 1960s, a certain number of genetic diseases or chromosomal disorders could be detected prenatally by culturing fetal cells present in the amniotic fluid and either looking for the products of the genes involved or observing the chromosomes—this was the principle of classic prenatal diagnosis. But if a given genetic disorder was not linked to a chromosome rearrangement that could be detected under the microscope, or it affected a gene that was expressed not in the cells to be found in the amniotic fluid but only in specialized fetal tissues, prenatal diagnosis was impossible.

With the new molecular technology, this difficulty disappeared. In 1976 Yuet Wai Kan of the University of California at San Francisco isolated DNA from normal subjects and from patients affected by a particularly serious form of anemia resulting from the loss of genes coding for α-globin.[55] He hybridized this DNA with a labeled cDNA corresponding to α-globin. By measuring the quantity of hybridized labeled cDNA, it was possible to distinguish normal subjects from patients with the disorder. Kan repeated the experiment with DNA extracted from amniotic cells and was able to show that the fetus concerned was not affected, which was confirmed at birth. A completely reliable diagnosis had been made without any blood being taken from the fetus—all the cells of an individual carry genes for α-globin, whether it is expressed or not. In the years that followed, researchers greatly improved this technique and applied it to many other genetic disorders.

WITH THE APPEARANCE of genetic engineering, molecular biology experienced a radical transformation that was further amplified by the development of the genome sequencing programs (Chapter 27). It has become a way of "reading" life.[56] The number of genes for which the sequence was known grew exponentially. From these sequences, it was possible to deduce the amino acid sequence of the proteins that were potentially coded by these genes. By contrast, the number of proteins for which the three-dimensional structure was known grew very slowly—even today that structure cannot be reliably predicted from the sequence and has to be determined directly. The molecular biologists of the 1980s and 1990s, like those of today, had to make the most of the one-dimensional information that was directly available and be satisfied with a linear reading of the book of life.

What could be done with such a text, written in a decipherable but un-known language, in which the three-dimensional meaning of the words re-mained hidden? Like an archaeologist who knows how to read a language without understanding its meaning, the molecular biologist looked for reg-ularities, repetitions, and resemblances. This approach led to the identifi-cation of families of proteins with similar structures, making it possible to guess the function of proteins that were known only through their corre-sponding gene. The description of such similarities thus had a heuristic value and raised the question of the ancestry of different proteins. It therefore led, quite naturally, to the study of evolution.

The fact that these similarities frequently concerned only one part of the protein implied that protein structure should be seen as composite, formed of several independent modules. A similar modular conception arose fol-lowing the study of DNA sequences that were found close to the gene and which scientists realized were necessary for the regulation of the gene's ac-tivity (Chapter 20).

In this period, molecular biology metamorphosed from a science of observation into a science of intervention and action. The book of life, even before it was understood from a structural point of view, could be altered.

The effects of the molecular modifications that were available at this time can be analyzed at two levels. At the molecular level the experimenter could induce subtle changes by using directed mutagenesis, replacing one amino acid with another. If the protein under study was well known and its three-dimensional structure had been determined, these substitutions made it possible to test different models of protein function. If the protein was an enzyme, it would be possible to study its catalytic mechanism (Chapter 21). If it was a regulatory protein, one could determine which amino acids interacted with the DNA sequence. The comparison of what a molecular biologist could do at the beginning of the 1970s with what it was possible to do with genetic engineering gives a measure of how much things developed in a few years. Before, a functional scenario for a protein could be proposed after the determination of the structure. But the model could be tested only indirectly and remained largely hypothetical. From the 1980s, molecular biologists could not only test models experimentally but also quantify the relative contribution of each amino acid or each bond to catalysis or binding.

At the level of the organism, the molecular biologist could now look for the effect of induced molecular modifications by introducing the modified form of the gene into an egg or cell, thus creating a transgenic organism. This made it possible to determine in vivo the function of each protein or DNA fragment. It also rapidly became possible to inactivate a gene or to alter its product by homologous recombination (the replacement of the normal copy by an altered copy) or to modify its level of expression—and all this in whichever organism was most appropriate to the topic (Chapter 24). With genetic engineering, biology became an experimental science at the molecular level. Even if the models remained relatively simple and overly mechanistic, the same distance separated the old and the new molecular biology as separates any science of observation from a science of action.

One way of highlighting the progress that was made is to return to the problem raised by Niels Bohr more than eighty years ago, in his "Light and Life" lecture.[57] Bohr pointed out that any molecular study of an organism required the destruction of its structures and the death of the organism and the cells that contain the molecules. Because of this problem, even the most efficient in vitro systems can always be criticized for not completely reproducing the properties of the living organism.

With the development of genetic engineering, this problem—the inherent limitation of any molecular study of biology—disappeared. The role of the molecules that form an organism could be studied without killing the organism. This was an essential development, but one that came late: few people still doubted that the molecular study of biology would contribute to understanding the elementary principles of the functioning of life.

Linguistic metaphors have often been used to describe the concepts and theories of molecular biology.[58] Molecular biology became integrated into the new informational vision of the world that became dominant at the end of the 1940s. These metaphors took on their full meaning in the context of genetic engineering. To decipher the language of life, molecular biologists use the same method that a child uses to learn to speak. They imitate the sentences of life, modify them, and see how organisms react to these new sentences. Through this dialogue, scientists learn the terms that are permitted and those that are not—the syntax of life. Sequencing the human genome and the genomes of other organisms has opened the book of life. It remains unclear, however, if it has truly enabled us to understand the language that is written there.

17

Split Genes and Splicing

THE EXISTENCE of the genetic code implied an alignment, a perfect co-linearity between genetic message and primary protein structure: this had been experimentally confirmed in bacteria, with all the precision that genetics could provide.[1] Despite the fact that eukaryotes are more complex than bacteria, most molecular biologists thought that in all organisms the same steps enabled the information contained in genes to be expressed as proteins. The startling discovery that in eukaryotic cells the coding DNA sequences are separated by noncoding sequences unfolded in a series of stages. The story of this discovery remains controversial. Many biologists consider that the contributions that were recognized by the 1993 Nobel Prize for this discovery are not the only, or even the most significant, ones.

During the Cold Spring Harbor Symposium of 1977, several groups—in particular those of Richard Roberts and Phillip Sharp—reported that in the adenovirus (a respiratory tract virus) the definitive messenger ribonucleic acid (mRNA) was a copy of noncontiguous DNA sequences; it was made up of a long fragment that coded for viral proteins, with, at one end, a series of small fragments that had been copied from distant parts of the DNA sequence.[2] By using restriction enzymes, gel electrophoresis separation of the DNA fragments, and hybridization, scientists were able to draw up a detailed map. Electron microscopy allowed them to directly visualize the hybridization complexes formed between mRNA and various parts of the DNA molecule. At Cold Spring Harbor Laboratory, the electron microscope observations were performed by the group of Louise Chow and Thomas Broker, who independently developed their own research project. Jim Watson, the head of Cold Spring Harbor, urged them to write a joint publication with the other group, led by Richard Roberts. The result was that Roberts got all the credit for the discovery.

After this initial discovery, it appeared that the fragmentation of the genetic message was limited to the extremity of the mRNA that did not code for proteins but which contains the leader sequence that initiates the translation of RNA. It was thought that this result would be limited to viruses and that an ad hoc evolutionary explanation would be all that was required to understand it. The need to compress genetic information into an extremely small volume had led viruses to adopt unusual solutions: in the ΦX174 bacteriophage it had recently been shown that the same DNA fragment could code for several completely different proteins by reading the nucleotides in different phases.[3]

The result obtained with adenovirus was confirmed by the analysis of another small animal virus, SV40, and then extended to the genes of eukaryotes.[4] By hybridizing mRNA to genomic DNA, several groups showed that protein-coding DNA sequences in eukaryotic genes were interrupted by noncoding regions; to use the nomenclature introduced by the American scientist Wally Gilbert, eukaryotic genes look like mosaics, consisting of a series of exons (DNA sequences that are expressed, and can therefore be subsequently found in mRNA) and introns (silent DNA sequences that are absent from the final mRNA and have no apparent function).[5] This result was confirmed for all subsequently studied eukaryotic genes, such as hemoglobin, immunoglobin, ovalbumin, and collagen.[6]

The number of introns varied but was always large; it could be as high as sixty for some genes, while the silent intronic sequences could represent up to 95 percent of the gene. In eukaryotic organisms the fragmentation of genes into introns and exons turned out not to be the exception but the rule. Apart from yeast, in which both the number and the size of introns seemed to be limited, all cells with a nucleus had mosaic genes. Even mitochondria—the "power plants" of eukaryotic cells, which have their own DNA molecule coding for ribosomal and transfer RNA and a few specific proteins—have split genes.[7]

Using radioactive isotopes, molecular biologists investigated how the cell processed DNA from a split gene. The gene was first copied into a long RNA molecule; this RNA, which was very short lived, was then converted into mRNA.[8] During this process, the RNA was spliced: the RNA sequences corresponding to the introns were eliminated, while the RNA fragments corresponding to the exons were lined up. This process of cleaving and ligation of RNA fragments took place in the cell nucleus. The only RNA molecules to be found in the cytoplasm were the definitive mRNA molecules, which were colinear with the eventual protein.

This discovery of split genes hit the world of molecular biology like a bombshell. It was described at the time as a mini-revolution.[9] Although this is undoubtedly true, in retrospect it should have been less of a surprise because a number of results had hinted that the mechanisms that led from DNA to proteins were much more complex in eukaryotes than in bacteria. For example, as early as 1971 it was known that eukaryotic mRNA molecules were modified at the 3' (right-hand) extremity by the addition of a long stretch of adenylic acids.[10] Molecular biologists had begun to characterize the various kinds of RNA molecules present in the nucleus—known by the catch-all term heterogeneous nuclear RNA (HnRNA)—and shown that they were much larger than mRNA molecules, while only a fraction of them (10 percent) were the precursors of mRNAs. Other modifications, also shared by mRNAs and HnRNAs, occurred at the 5' extremity—with, in particular, the addition of a methylated guanosine.[11] Results from Georgy Georgiev's group in Moscow had led to the hypothesis that mRNA molecules corresponded to the 3' part of HnRNA and were thus produced by the breakdown of the HnRNA molecule.[12]

The fact that these scientists did not discover the fragmentation of genes can perhaps be explained by the difficulty of their experiments, which were carried out before the development of genetic engineering. The lack of a mechanism explaining how fragments of DNA that were far apart in the genome could produce RNA sequences close together in the same mRNA molecule perhaps led some experimenters to ignore—or perhaps exclude—some observations. The French biologist Pierre Chambon realized that before the 1977 Cold Spring Harbor Symposium he already had experimental data suggesting that the ovalbumin gene was split, but he had not fully understood them.[13]

THE TRUE EXPLANATION appeared too outlandish even to be considered. Indeed, the existence of interrupted genes was completely unexpected.

Studies of bacteria had suggested that the genetic message was clear, well punctuated, and economically written. In fact, it turned out to be long and complex. Nevertheless, scientists were enthusiastic about the discovery of splicing: they had finally found a clear difference between bacteria and eukaryotes. From a neo-Darwinian point of view, the evolutionary conservation or creation of a process as aberrant as gene splitting suggested it played an essential role in the cell—for example, by regulating gene expression.

Several models were proposed along these lines. Control of splicing could stop the expression of genes that had been transcribed by mistake, playing the role of a final check. But in most models the process was given a far more significant regulatory role. Some researchers took the regulatory mechanisms of transcription that had been discovered in bacteria and applied them to eukaryotic cells, so specific proteins—inhibiting or activating—were thought to fix onto the HnRNA and to modulate splicing. Other models involved the simultaneous splicing of several different genes and sought to explain the coordinated changes of expression observed during cell differentiation and embryogenesis—a new path of differentiation was thought to be the result of a new splicing enzyme.[14] These models adopted the spirit, if not the letter, of Roy Britten and Eric Davidson's model of the genetic control of development (see Chapter 22).[15]

Even more daring hypotheses were put forward. While studying introns contained in yeast mitochondrial DNA, the French geneticist Piotr Slonimski (who had been Boris Ephrussi's student) noticed that if splicing did not take place normally, a hybrid protein could be formed from the genetic information in the intron, together with information from the neighboring exon. Given that genetic experiments had shown that changes to the intron could alter splicing, Slonimski suggested that this hybrid protein, which he called maturase or m-protein, participated in the splicing of its own mRNA. Slonimski's model thus stated that a gene contained the information required for its own splicing. This model allowed for spontaneous autoregulation because the maturase, by acting and by splicing the HnRNA, eliminated the RNA that produced it.[16]

Slonimski adapted this model—first developed for mitochondrial genes—to nuclear genes and both enriched and extended it.[17] The study of the amino acid composition of what were called theoretical nuclear maturases

suggested that these proteins were to be found at the nuclear membrane
and thus coordinated the splicing of HnRNA with the passage of RNA
through the pores of the nucleus. If there were only a limited number of
pores, the different RNA maturases might have to compete for a place near
one of the pores. This competition would be won by those maturases that
had been synthesized first. Slonimski's model suggested that cellular regulation might depend on the initial state of the cell. If this cell were a fertilized
egg—the embryo at the beginning of its development—this would explain
the existence of nongenetic hereditary constraints that affect development
but are not the result of the action of embryonic genes. An analogous model
was used by another French molecular biologist, Antoine Danchin, to explain cancerous transformation and aging in cells.[18]

Thus, in the short space of time between 1977 and 1980, and on the basis
of a proven fact—the existence of split genes—molecular biologists developed a series of models that gave this phenomenon a fundamental regulatory role. Rather than simply revealing the imagination of experimenters,
these models also expressed their frustration at the lack of major discoveries
that could explain the mechanisms of cell differentiation and embryonic
development.

Unfortunately, experimental results did not confirm these hopes. None
of the proposed models turned out to be false, but their biological significance is in fact extremely limited. For example, Slonimski's model applies
only to a few mitochondrial introns. The regulation of splicing does indeed
take place during certain steps of cell differentiation, but this is a qualitative regulation (the cell controls production of two different mRNAs from
a single genomic fragment) rather than a quantitative regulation (the quantitative level of expression remains essentially that of the transcription of
DNA into RNA). This regulation of splicing plays only a secondary, minor
role in controlling gene expression. Wherever it has been observed, it corresponds to a principle of economy that enables the cell to make two slightly
different proteins from a single gene.

Molecular biologists eventually abandoned the idea that an explanation of the complexity of multicellular organisms could be found in the
fact that their genes were split. Molecular biology was thus complicated
but not transformed by the discovery of exons and introns. It was the biochemical approach to splicing that produced the biggest surprises: a series
of different mechanisms was discovered, requiring a complex molecular

machinery and implying that the three-dimensional structure of RNA played an important role.

In some ribosomal RNAs, splicing turned out to be spontaneous, requiring no external protein, being catalyzed simply by the RNA molecule.[19] In itself, this result might seem unimportant because it was limited to the splicing of a few types of RNA. In fact, it was extremely interesting; together with data on ribonuclease P, it showed that nucleic acids could catalyze a chemical reaction, thus behaving like an enzyme.[20] Up until this point, molecular biology had strictly separated the functions of biological macromolecules: nucleic acids were involved in stocking and transferring genetic information, whereas proteins had structural and catalytic functions (see Chapter 25 for the consequence of these discoveries for the place of RNA, and for scenarios describing the origin of life). The discovery of split genes forced molecular biologists to pay more attention to the mechanisms of evolution.[21] It was easy to explain the origin of point mutations where one amino acid is replaced by another—Jim Watson and Francis Crick had proposed a model in 1953 (see Chapter 11)—but it was much more difficult to explain how amino acids could be inserted into a protein. The existence of introns could account for this. A point mutation at the junction of an exon and an intron that removed the distinction between the two sequences would add amino acids to the middle of the protein.[22]

A number of observations confirmed that, over the course of evolution, there had indeed been insertions or deletions of amino acids at intron–exon splice junctions. Furthermore, in most cases these junctions appeared to correspond to those parts of the protein chain that were on the outside of the macromolecule.[23] Changes to these junctions—which would not change the overall structure of the protein—could lead to the creation of new catalytic and regulatory sites that come into contact with molecules in the cell, or even on the outside of the body.[24]

On the other hand, differential splicing allowed the evolution of a new form of protein without leading to the disappearance of the old form.[25] Researchers thus discovered a mechanism that created new forms without altering previous function. Susumu Ohno had suggested that the duplication of genes had a similar function.[26]

Some molecular biologists hypothesized that an analogous evolutionary process could have associated fragments of genes—exons coding for protein substructures. The complex proteins currently found would thus be the result of the assembly of several different elementary protein domains by recombination of the exons coding for them.[27] Evolution could use this extremely powerful erector set to tinker with great effect.[28]

Once again, however, seductive hypotheses were only partly verified. Some complex proteins, such as immunoglobulins, are the result of the repeated duplication of a single exon. But although the recombination of exons can explain the appearance of some proteins, such as those forming the extracellular matrix or eukaryotic transcription factors, it is by no means a universal rule. Many exons do not correspond to precise structural domains, and the position of introns may differ in two structurally similar proteins.[29]

The theory of exon shuffling does not seem to be as general as was first hoped. Evolution has not always used the same bricks to build complex protein structures. In fact it has probably reinvented the same basic elements several times over. The appearance of introns certainly made the recombination and duplication of genes easier, but it did not profoundly change the speed of evolution.

Molecular biologists initially thought that introns and exons were specific to eukaryotic cells. The existence of introns in mitochondrial genes, which were thought by some to be of prokaryotic origin, suggested that this was not the case. In addition, the exon shuffling model suggested that introns must have existed in all organisms during the first phases of evolution. The role of exon shuffling was probably more fundamental at these early stages than later during evolution. From this point of view, today's prokaryotic bacteria could be considered the most highly evolved organisms, having eliminated all the junk DNA to be found in introns.[30]

It is particularly striking that the discovery of split genes was rapidly integrated into evolutionary models. There was a paradox here: many molecular biologists present themselves as the most ardent supporters of the evolutionary synthesis, yet their enthusiasm for new evolutionary mechanisms implies that this theory, at least in its canonical form, cannot explain the evolution of multicellular organisms (see Chapter 23 for further discussion of this paradox). Furthermore, the new mechanisms, such as exon shuffling, are not faithful to a rather simplistic version of Darwinism because they provide no immediate advantage to the organisms that possess them. The maintenance of introns appears at best to be a neutral process, merely tolerated by evolution.[31]

IN THE LATE 1970S AND 1980S, four more discoveries questioned the dogmas of classical molecular biology. In 1977 Claude Jacq, working in the laboratory of George Brownlee at the Medical Research Council in Cambridge, England, discovered the existence of pseudogenes—stretches of DNA that were structurally similar to genes but were apparently silent. They were initially considered to be evolutionary residues, the inactive product of gene duplication. Three years later, however, Jacq and his colleagues discovered that some pseudogenes had lost their introns, as though they had been formed from mRNA by the action of reverse transcriptase.[32] Ten years after the development of Howard Temin's proto-oncogene model (see Chapter 15), there was a new threat that the genome could integrate information from the cytoplasm, raising with it the specter of Lamarckism. (Chapter 23 describes how the menace evaporated.)

Another surprise came from the study of the mitochondrial genome. The comparison of mitochondrial and nuclear DNA and protein sequences revealed that the meaning of several codons in the genetic code was not the same in mitochondria as in the canonical code worked out for *E. coli* and eukaryotic nuclear genomes.[33] For example, the UGA codon normally stops protein synthesis, but in mitochondria it leads to the incorporation of the amino acid tryptophan. This deviation from the universal genetic code is, however, restricted to a few codons, to some mitochondria, and to the nuclear genome of some unicellular organisms such as *Tetrahymena* and *Paramecium*.[34]

The exploration of other unicellular organisms—trypanosomes—revealed the existence of a new mechanism called editing, which alters the transfer of information between DNA and proteins. RNA molecules, produced by the faithful transcription of the gene, are not functional.[35] They lack some nucleotides, generally uridines, which are added after transcription. The number of missing nucleotides varies between one and several hundred. There are several different versions of editing, some of which can be observed in multicellular organisms. But the phenomenon is most important in the mitochondria of unicellular eukaryotes, where it has been most closely studied. Editing is linked to splicing—small RNA molecules ("guides") act as nucleotide donors.[36] Editing again confirmed the catalytic properties of RNA, but, despite its strangeness, its quantitative significance remained limited. (Note that, despite some similarities of vocabulary, RNA editing is unrelated to genome editing, the powerful new technique that enables scientists to precisely and efficiently modify genomes, created by the discovery of CRISPR—see Chapter 24.)

A final discovery resolved, at last, the mystery of antibody synthesis.[37] As we saw earlier, Linus Pauling thought he had solved this problem by hypothesizing that the antigen played an active role in antibody synthesis. But this model had not been supported experimentally, and above all it had suffered because it did not fit with the dogmas of molecular biology. Francis Crick, in his famous 1957 lecture "On Protein Synthesis," accepted that the synthesis of both antibodies and adaptive enzymes was an exception to a general rule, a rare case where the action of genes was insufficient to shape proteins.[38] Following François Jacob and Jacques Monod's work, the synthesis of adaptive enzymes was seen to agree with the dominant view of molecular biology. The formation of antibodies ought similarly to comply with the new conceptual framework. Even before this, in 1959 Joshua Lederberg proposed the following hypotheses regarding antibody formation in an article in *Science*.[39] The structure of antibodies depends solely on their amino acid sequence; this sequence, which is different for different antibodies, is genetically determined; and finally, because there cannot be as many genes as there are antibody molecules, the genes responsible for antibody synthesis must have mutated, and mutated differently, during the formation of the cells that produce antibodies.

Lederberg's theory has to be understood in the context of the theory of antibody formation proposed by Niels Jerne in 1955, and the clonal theory proposed by David Talmage and the Australian scientist Frank MacFarlane Burnet in 1957.[40] Niels Jerne became convinced that the instructive model proposed by Pauling was wrong, as shown by his own observation of "natural antibodies" that recognize molecular structures with which organisms had never been in contact. Jerne hypothesized that, quite spontaneously, there existed a high diversity of antibodies with different specificities. The complex formed by an antibody and a new antigen was ingested into the cell and reproduced through the action of RNA. For Jerne, as for Max Delbrück who communicated his article to the U.S. Academy of Sciences, only a Darwinian mechanism combining spontaneous variation and natural selection was able to explain the capacity of the immune system to react so efficiently against any invader. However, Jerne's model did not explain the origin of antibody diversity, nor the precise mechanism of antibody replication.

Burnet was not only an immunologist—after training as a virologist he had become interested in the biology of cancer. He was convinced that both tumor development and viral epidemics were due to the growth of one or more clones—that is, to the existence of a few cells or viruses that could multiply extremely rapidly. Burnet transposed these ideas to antibody synthesis: each antibody was synthesized by one or more cellular clones that multiplied when they encountered an antigen with an affinity for the antibody they synthesized. These clones were preprogrammed to make the appropriate antibody. To explain how the cell could start to multiply in the presence of the antigen, Burnet suggested that the cell might carry its antibody on its surface. Contact with the antigen would modify the cell membrane and initiate cell replication. Antibodies cannot be made against the components of the organism because the corresponding clones are eliminated during development.

Burnet's theory was rapidly accepted by the scientific community, and in particular by the molecular biologists. However, Pauling's theory was not abandoned because of convincing experimental data—the rare sequences of antibody molecules known in 1959 were not sufficient to reject the hypothesis that all antibodies were formed from the same polypeptide in differently folded forms, while the reality of the natural antibodies proposed by Jerne remained unclear. A year earlier, Lederberg had carried out an experiment that suggested that each cell in the immune system made antibodies that could recognize only one kind of antigen.[41] The fact that this was an isolated result, together with the technical difficulties involved in carrying out the experiment, meant that its impact was limited. Lederberg himself recognized that Pauling's theory had become unacceptable, not in its detail, but rather in its spirit. The new theory was "more closely analogous to current conceptions of genically controlled specificity of other proteins."[42] The theory had to be selective and not instructive, faithful to the neo-Darwinian spirit of molecular biology.[43]

In fact, Burnet and Lederberg had merely pushed the problem of diversity back to genes and DNA. Their proposed solution—the existence of somatic mutations in immunoglobulin genes—remained unproved until 1976, when it was corrected following the direct analysis of the genes responsible for making antibodies. Somatic mutations were indeed involved, but they were associated with a highly complex process involving the rearrangement of gene fragments. The study of antibody molecules showed that they contained two polypeptide chains, each of which was formed of variable and constant regions. In 1976 Nobumichi Hozumi and Susumu Tonegawa showed that the DNA sequences coding for the variable and constant regions were far apart in embryonic cells, but close together in antibody-producing cells: the genes coding for antibody molecules were rearranged during development.[44]

Subsequent studies revealed the complexity of these rearrangements, which involve several different genetic regions. This genomic rearrangement during development turned out to be limited to the cells of the immune system, taking place in genes coding for immunoglobulins in B lymphocytes and in genes coding for receptors in T lymphocytes.[45] Jacob and Monod's view that the genome was stable during development (see Chapter 14) was thus not seriously altered. The movement of fragments of DNA in the genome is of limited importance and does not play a major role in either the regulation or the evolution of multicellular organisms. But these rearrangements do play a fundamental role in unicellular organisms such as ciliates.

18

The Discovery of Oncogenes

BEGINNING IN 1975, a series of experiments showed that cancer, whatever its direct cause, is the result of the activation or inactivation by mutation of a highly conserved set of genes called oncogenes.[1] These genes are variously involved in the control of cell division, cell survival, and DNA repair. Some of their products form part of the regulatory network that relays signals from outside the cells to the nucleus and enables cells to adapt their rate of division to the needs of the organism. These discoveries form what can be called the oncogene paradigm. The slow and complex development of this view was possible only because of the tools of genetic engineering. The rise of the oncogene paradigm shows how molecular biology became integrated into other biological disciplines to form a new biology.

Many molecular biologists turned to the study of small tumor-causing animal viruses in the hope of finding both a model system for studying animal cells and a tool for understanding the molecular bases of cancerous transformation. The most widely studied were the ribonucleic acid (RNA) viruses, called type-C retroviruses because of their appearance under the electron microscope. The RNA that forms them is recopied into a double strand DNA molecule through the action of reverse transcriptase, and this DNA integrates itself into the genome in an inactive form called a provirus. Some of these viruses cause tumors, but others do not.

Around 1975, an influential model of the origin of cancer was proposed by two American scientists, Robert Huebner and George Todaro, which suggested that retroviruses played a key role in the process. Huebner and Todaro argued that such viruses existed in all genomes: silent and not expressed, they were nevertheless transmitted from generation to generation. Under the effect of mutagens or by the action of other DNA or RNA viruses, these silent proviruses could be activated and become oncogenic. According to this model, all cancers, be they spontaneous, induced by chemical agents, or induced by viruses, are the result of the activation of proviruses.[2] Another model, called the protovirus model, had been proposed by Howard Temin, the discoverer of reverse transcriptase. This model was not very different from that of Huebner and Todaro, and also implicated proviruses in the development of cancers. But for Temin, proviruses (which he called protoviruses) were not silent tumor-inducing structures that passively waited in the genome to be activated, but were instead part of the normal cellular genome, involved in the transfer of information from DNA to RNA and vice versa (see Chapter 15).[3]

In 1973 Edward Scolnick showed that a sarcoma-inducing tumor virus could be produced by the recombination of genetic material from various sources. He suggested that the development of a type-C transforming virus could involve the capture by these viruses of oncogenic information provided by the cell.[4]

These different theories about the origin of cancer were largely based on data from bacterial genetics obtained ten years earlier. In 1960 François Jacob had suggested that studies of lysogeny and bacterial gene regulation might provide models for understanding cancer.[5] Huebner and Todaro's provirus model was the adaptation of the prophage concept to tumor viruses. The possibility that a virus could take over and transduce genes near its chromosomal insertion site had been proved for bacteriophage; Scolnick merely extended this idea to tumor viruses.[6]

Furthermore, it had been shown that the integration of a prophage in the genome of a bacterium could disturb the functioning of neighboring genes. Several research projects on tumor viruses tried to characterize the genomic virus-insertion sites, on the grounds that cancer might also be the result of the alteration of the functioning of normal genes in the cell after the integration of a tumor virus.

These models were not the only ones that were focused on explaining the development of cancer. In a lecture given in 1974, Temin reviewed five different models that explained the origin of cancer, including the simplest, according to which a cancer-causing gene would appear by the simple mutation of a normal cellular gene.[7] Surprisingly, these different models were not in competition and were not even mutually exclusive.

THIS FERMENT of ideas and results shows that the very concept of cellular oncogenes—normal genes that can become transforming genes responsible for cancer after mutation or transduction by a virus—had already been proposed and even accepted by the biological community before the discovery of the first cellular oncogene, *sarc,* which was soon renamed *src.* The experiments that led to this discovery were described in two articles published in the *Journal of Molecular Biology* (1975) and *Nature* (1976).[8]

The experimental system used was the virus that causes Rous sarcoma—the first tumor virus to have been discovered, in 1911, in the chicken. In the early 1970s a series of studies, in particular by Hidesaburo Hanafusa, had shown that in this disease a single gene was responsible for the malignant transformation of infected cells. The French scientist Dominique Stehelin and the Americans Harold Varmus and Michael Bishop isolated a complementary DNA probe corresponding to the viral gene responsible for cancerous transformation. The hybridization techniques used for the subsequent steps in the study were not novel, having already been used by a number of other groups with the same objective. Stehelin and his colleagues showed that normal chicken cells carried copies of a gene that was very similar to the transforming gene of the Rous sarcoma virus. Using differential hybridization, they compared the structure of this gene in different birds and showed that it had varied over the course of evolution in a way similar to other avian genes.

In their conclusion, the authors suggested that the cellular gene they had discovered must play a role in the regulation of cell growth and in development. The results of this important experiment were well received by other scientists but were not considered to be revolutionary. Even its authors interpreted them in the context of the existing models of cancer. According to Huebner and Todaro's model, the *src* cellular gene might come from a type-C provirus in which the other viral sequences had evolved to the extent that they could no longer hybridize with the rest of the Rous sarcoma virus. The result could also be interpreted in the light of results previously obtained by Scolnick, in which case the *src* gene was a normal cellular gene that had been transduced by the Rous sarcoma virus.

Stehelin and his colleagues did not generalize their results to other vertebrate species; they merely claimed that birds carried sequences similar to the *src* transforming gene. Strikingly, no similar sequence was found in mammals. The fact that this negative result was at first readily accepted may seem surprising; in today's oncogene paradigm the products of oncogenes have fundamental roles in all cells, and because of this ubiquitous function they have been highly conserved over evolution. It was two years later, in September 1978, that Deborah Spector and her colleagues showed that there was a gene homologous to *src* in both mammals and fish.[9] Subsequent publications and historical descriptions of the discovery of oncogenes do not indicate that there was a two-year delay before the oncogene discovered by Stehelin and his colleagues was accepted as a ubiquitous component of the cells of eukaryotes.[10]

The importance of the discovery of the *src* oncogene was also hidden by the disappointing results of the study of the expression of this cellular oncogene: it was expressed at a low but identical level in all avian cells, be they normal or cancerous, and its expression did not change over the course of development.[11] These two characteristics were not those that were expected of a transforming, tumor-inducing gene. As shown in Chapter 14, cancer was considered to be a process of deregulation: the expected result was therefore that *src* would be strongly expressed in tumor cells.[12] Furthermore, malignant transformation is often accompanied by the reexpression of genes that are active during development. Indeed, tumor cells, like embryonic cells, are able to multiply extremely rapidly. Researchers therefore expected that oncogenes would also be strongly expressed in the embryo. The properties of the *src* gene thus appeared to be very strange.

In the years that followed, these results for *src* were extended to other retrovirus transforming genes. But this generalization was insufficient to lead to the birth of a paradigm of cellular oncogenes. Instead, this new vision was established between 1981 and 1984 as the result of a set of simultaneous discoveries showing that the same genes were involved in "spontaneous" cancers and in those induced by chemical agents or by viruses, and that oncogene-coded proteins were involved in the control of normal cell growth.

A new experimental approach developed by the American scientists Robert Weinberg and Geoffrey Cooper, known as the transfection assay, played an important part in the establishment of this new view. These experiments were the natural consequence of the work carried out on RNA and DNA tumor viruses. They were initially developed to answer the following question: was it possible, after the integration of a virus into the genome as a provirus, to recover the virus (by isolation and fragmentation of genomic DNA) in an infectious form? The DNA of cells that had been transformed by a virus were cleaved by restriction enzymes and the resultant fragments were added to normal, untransformed cells in the presence of calcium, which enabled the fragments to be incorporated into the cells and integrated into the genome. Many colonies of transformed cells appeared among the transfected cells, confirming that the provirus had conserved its transforming properties.

Researchers now wondered if, using the same approach, normal cells could be transformed with DNA extracted from cells that had been transformed by chemical agents, or with the DNA from spontaneous human tumors.[13] Previous studies had shown that known tumor-inducing chemical agents were also mutagenic after modification within the organism.[14] This suggested that the transformation of cells by chemical agents was the result of mutations induced in one or probably several normal genes. In the long debate between supporters and opponents of the genetic origin of cancer, these experiments tipped the balance in favor of the former; in this case, "genetic" means that cancer results from the modification of the genetic material of some cells within the organism, not that these mutations will be transmitted to the progeny or that they were inherited from the parents.

Cooper and Weinberg's results were unambiguous: in a significant fraction of chemically transformed cells or tumors, it was possible to isolate DNA fragments that were able to transform normal cells (fibroblasts) after transfection.[15] But DNA extracted from normal cells was nontransforming.[16] Thanks to the new tools of genetic engineering, it was not long before these transforming DNA fragments were isolated (cloned) and characterized (in particular sequenced). Unexpectedly, these transforming sequences turned out to be identical to a number of cellular genes that had previously been characterized by their homology with genes from transforming retroviruses (the *ras* oncogenes).[17] Less than a year later, scientists characterized the mutation that led to transformation in DNA extracted from a bladder cancer. This mutation, which had vital consequences for cell physiology, was the result of a minor alteration in the coding part of a *ras* gene: a change in a single nucleotide led to the replacement of the twelfth amino acid in the protein by another amino acid.[18]

Other transfection experiments revealed modifications in different cellular oncogenes. Two results were particularly unexpected. The same genes were found to be involved in very different forms of cancer, and it was discovered that an oncogene mutation may not involve the promoter—and thus not change gene expression—but may instead alter gene structure. It was soon shown that during tumor formation, the same oncogenes could be activated by three other molecular mechanisms, including chromosomal translocation.[19]

The simultaneous proof that, in different tumors, using different mechanisms, the same small group of genes was activated and became transforming showed that these genes were key actors in this transformation. But on its own this result would not have been enough to give rise to the oncogene paradigm; rather, it was merely a first step in the establishment of this paradigm. The second step involved a series of discoveries about oncogene function that explained their central role in transformation and the development of cancer.

These discoveries were largely a matter of chance, as a brief chronology shows. In 1983, scientists sequenced platelet-derived growth factor (PDGF), a growth factor (a small protein that can stimulate cell division) present in blood platelets. Computer comparison of this sequence with known protein and nucleotide sequences revealed that it was similar to that of a transforming gene present in the simian sarcoma virus.[20] This suggested that the viral oncogene was a slightly altered version of one of the normal genes coding for PDGF—a small change in a growth factor or its expression was thus enough to induce cellular transformation and cancer. The following year, an oncogene, erb-B, carried by a retrovirus that causes avian erythroblastosis, was sequenced and was found to be virtually identical to part of the sequence of the receptor protein for another growth factor, epidermal growth factor (EGF).[21] This link between oncogenes and growth-factor receptors was not completely unexpected; in 1980 several groups had shown that some growth-factor receptors and oncogenes had the same enzymatic activity and induced similar modifications in a number of cellular proteins. In the same year, the sequence of a G protein was determined for the first time. G proteins play an essential role in the transmission of information from extracellular receptors to the inside of the cell; they turned out to be similar to products of *ras* oncogenes.[22] Finally, other oncogenes, such as *myc* and *fos,* also play a role in the regulatory pathways that control cell division. Adding growth factors induces the transitory expression of these oncogenes and leads to cell division.[23]

The development of this functional aspect of the oncogene paradigm was supported by the discovery that the structure of cellular oncogenes had been highly conserved during evolution. In 1983 genes were discovered in yeast that were partially related—both in functional and in structural terms—to *ras* genes.[24] From the neo-Darwinian point of view, which was shared by most molecular biologists, this result was entirely expected for cellular components that play a key role in growth and in cell division.

At the same time, scientists took a decisive step forward in their understanding of the regulatory networks controlling cell division and cell differentiation. In addition to cyclic adenosine monophosphate (cAMP), an already characterized second messenger (a small molecule that relays an external regulatory signal inside the cell), new second messengers were discovered, and the mechanisms leading to their production were determined. The characterization of the enzyme protein kinase C, which is activated by these second messengers and intervenes in signal transduction, took place at the same time as the discovery of the role of cellular oncogenes in these regulatory pathways.[25]

The third step that led to the acceptance of the oncogene paradigm was anticipated by François Cuzin of France and Robert Kamen of the United Kingdom.[26] Brilliantly realized by the U.S. scientist Robert Weinberg and his coworkers, there was proof that cancer develops through the cooperation of different oncogenes.[27] The discovery that a series of mutations in different oncogenes was required before a cancer could develop helped physicians accept the new paradigm. It had been widely held in medical circles that cancer could not be the result of a single mutation; the frequency of point mutations was much higher than the frequency of cancer. Furthermore, cancer rates increase with age, and the disease takes a long time to develop (for example, even after a strong dose of radiation, the development of cancer takes several years), suggesting that several steps are necessary for tumor formation.

The discovery of cooperation between oncogenes led to their provisional classification into two groups—immortalizing and transforming oncogenes.[28] The activation of the former—such as *myc* or *fos*—led cultured cells to divide indefinitely. The products of these genes act in the nucleus and directly control gene transcription and cell division. The transformation of the normal form of the oncogene into an altered, transforming version is the result of the overexpression of the oncogene. Together with immortalizing oncogenes, transforming oncogenes such as *ras* are responsible for cellular transformation, for the acquisition of cancerous properties. They code for products that are present in the cell membrane or cytoplasm. A structural modification of these proteins is required for the cells to go from a normal to a transformed state. Since its introduction, this classification of oncogenes has repeatedly changed and become more complex (as we will discuss later).

Many facts have been omitted from this brief history of the emergence of the oncogene paradigm. For example, recessive oncogenes (also known as anti-oncogenes or tumor-suppressor genes), first described in 1971, have not been mentioned.[29] Transformation occurs only when both copies of the gene are altered (inactivated). The first member of this new family of tumor-causing genes to be characterized was the retinoblastoma gene in 1986.[30] The discovery of tumor suppressors suggested how DNA tumor viruses might function: the products of these viruses inhibited the action of tumor suppressors. Nevertheless, this discovery did not lead to the development of a new paradigm. The products of tumor suppressors, like those of oncogenes, are involved in the same regulatory networks.

Taken together, these findings provided a rational explanation of the development of cancer. In accepting the oncogene paradigm, most biologists agreed that studying this small group of genes was the best way to understand cancerous transformation. Alterations in these genes and their functioning were considered to be the cause of cancer. The oncogene paradigm is thus both a community of objects (cellular oncogenes and their products) and a community of methods that permits the characterization of the structure and function of these oncogenes and their products.

How can the virtually unanimous acceptance of this paradigm be understood, given that, for more than a century, the field had seen so many controversies? Several practical reasons explain its acceptance. The new paradigm made it possible, at least theoretically, to develop a series of diagnostic and prognostic tests. Furthermore, it opened an important area of biomedical research to molecular biologists, and it enabled virologists to use the skills they had acquired in their long and often fruitless search for cancer viruses.[31] Through the structural characterization of the proteins encoded by these oncogenes it also opened the door to the development of new drugs specifically targeted against them.

There were also two more fundamental reasons for this widespread and rapid adoption of the new view. The first was that many of the discoveries that led to the establishment of the paradigm were in fact completely unexpected. Computer-assisted sequence comparison played a key heuristic role in the development of the paradigm's functional aspects. Such apparently serendipitous discoveries are particularly significant, at least for biologists, because they were facts that had been produced independently of any theory or model. This heuristic value of sequence comparisons is characteristic of the new form of molecular biology that was ushered in by genetic engineering (see Chapter 16).

To understand the second reason the paradigm was accepted so easily, one must remember that the word "cancer" does not mean the same thing in different biological disciplines. The observations and facts that a new theory would have to explain were not the same for different specialists. For physicians, a theory of cancer would have to explain its gradual development and its multifactorial origin; it would also have to include the well-established correlation between the mutagenic and the oncogenic effects of a variety of chemical compounds or physical treatments. For biochemists, a model of cancer would have to explain the many biochemical and structural changes that take place in cancerous cells. For cell biologists, cancer was a disease that disturbed intercellular communication. Furthermore, many cancers appeared to be linked to the existence of chromosomal break points, followed by translocation between chromosomes of the resultant fragments—something that could be seen under the microscope.[32] For virologists, most human cancers probably had a viral origin. Even if this were not the case, the mechanisms of nonviral oncogenesis had to be identical to those of virally induced cancers. Finally, for molecular biologists, the isolation and characterization of the genes involved in cancer were natural twin objectives.

Interestingly, but probably not surprisingly, the oncogene paradigm led to the unification of these different conceptions of cancer, through the alliance of the concepts of molecular biology with those of other biological disciplines. The changes that take place in tumor cells are indeed biochemical, but they are linked to changes in a small number of genes. Very few cancers are caused by viruses, but viruses change cells by the same mechanisms—the alteration of oncogenes—as radiation or chemical mutagens. Cancerous changes directly affect cells and not the whole organism, but the cells are altered in their ability to communicate with the external milieu and with other cells. Finally, several oncogenes have to be modified to obtain cell transformation, while chromosomal translocations are directly linked to a change in the expression of oncogenes.

Even opposing visions of cancer were reconciled in this new paradigm: regulatory and structural changes in genes are required for full cell transformation while changes affecting whole chromosomes or isolated genes can lead to either the activation or the modification of cellular oncogenes. Most biologists thought that the site of tumor virus integration into the genome played a key role in transformation. (As previously explained, this idea had its origin in studies of bacterial genetics from the 1960s, and contradicted results that showed that some tumor viruses themselves had transforming genes.) These two different visions were united in the new paradigm. The activation of cellular oncogenes could be the result of their incorporation (and, perhaps, modification) within a retrovirus, or of the integration, close to these genes, of a retrovirus that had just activated their expression (see Chapter 20).

It is striking that in the ten years that separated the discovery of the *src* oncogene from the final acceptance of the paradigm, no single system, no particular gene, played a dominant role. Although the discovery of *src* was extremely important, for a long time its precise mechanism of action remained unknown. Similarly, the discovery of changes in the *ras* oncogene in human tumors had a substantial psychological impact, but it took nearly ten years for scientists to begin to understand the role of *ras* in cancerous transformation.

The development of the oncogene paradigm also reveals two highly specific aspects of modern biological research. First, the technology available to the biologist is increasingly sophisticated: isolating a gene and characterizing its modifications have become routine. The heuristic value of the simple comparison of DNA sequences has been added to this technological power. It took only a few years for the main oncogenes (or tumor-suppressor genes) to be characterized and their function inferred. Second, the history of oncogenes underlines the limits of molecular knowledge. The greater the number of oncogenes characterized and the better understood their modifications and functions, the less biologists were able to distinguish the fundamental from the secondary, the driver from the passenger mutations. As a result, they found it difficult to put some order into the ever-expanding number of oncogenes, and the two initial categories of oncogenes were converted into eight hallmarks.[33]

Targeted drugs, which the knowledge of oncogenes made possible to design, inhibited enzymatic activity or monoclonal antibodies. These were shown to be efficient, but they were not the magic bullets that had been expected. They turned out to be active only against certain types of tumors, and cancer cells that are resistant to these drugs rapidly appear. The simplicity of the 1980s paradigm has been diluted and lost in the subsequent accumulation of data.

This contradictory effect has been reinforced by the complete description of the mutations occurring in the different cellular clones present in a tumor at different stages of its development, enabled by the new, fast, and cheap techniques of DNA and RNA sequencing developed at the beginning of the twenty-first century (Chapter 27). This is analogous to the difficulty encountered in other domains of biology—for example, in the study of development (Chapter 22). It raises two important questions: is it possible to go beyond today's molecular systematics, and does this require the development of a new logic of life, a postgenomic logic based on a science of complexity (Chapter 28)?

19

From DNA Polymerase to the Amplification of DNA

IN 1983 Kary Mullis developed a technique for amplifying DNA called the polymerase chain reaction (PCR).[1] PCR can amplify virtually any DNA fragment, even if it is present in only trace amounts, thus allowing it to be characterized. It can aid forensic medicine by characterizing DNA molecules present in biological samples such as hair, traces of blood, and so on. It can even permit the detection and characterization of the rare DNA molecules that persist in animal or human remains that are tens or even hundreds of thousands of years old (for its use in studies of human evolution, see Chapter 23). PCR can also be used in genetic diagnosis on the basis of a single cell, such as, for example, a cell of a very young embryo prior to implantation in the uterus. Finally, it permits the early detection of bacterial or viral infections.[2]

Despite the fact that PCR appeared several years after the main techniques of genetic engineering, it shares several essential characteristics with them. Although it does not, in and of itself, open new experimental horizons—it merely amplifies DNA, which can also be done by other, more time-consuming methods such as cloning—in practice it has made possible a number of experiments that were previously impossible.

In reality, PCR is a postmature development; there is no reason it should not have been developed at any point after the early 1960s. The principle of PCR is based on the properties of DNA polymerase, which was isolated and characterized by Arthur Kornberg in 1955. The history of PCR goes back to the path that opened the way to the development of molecular biology. In particular, it involves a series of studies that provided biochemical confirmation of the double helix model of DNA proposed by Jim Watson and Francis Crick in 1953. It highlights once again the difficult problem of the relationship of molecular biology and biochemistry. It also emphasizes the important degree of technological continuity between the molecular biological studies of the 1940s and the work of the 1980s.

When scientists discovered that an enzyme could copy molecules of DNA in vitro, they removed one of the weaknesses of Watson and Crick's 1953 model of DNA replication—it relied upon the remarkable properties of entirely hypothetical enzymes. Furthermore, the discovery of DNA polymerase also confirmed a number of characteristics of the double helix model. It is only briefly described by the historians of molecular biology. Horace Judson gives it only a few lines (a transcription of Crick's reaction on hearing Kornberg's initial results at a 1956 conference on the chemical bases of heredity, held at John Hopkins University).[3] In *A Century of DNA*, Franklin Portugal and Jack Cohen provide only a brief and cautious description.[4]

This discretion contrasts sharply with the official recognition of Kornberg's work at the time. In 1959, together with Severo Ochoa, he was awarded the Nobel Prize for his work on enzymes that form polynucleotides. The Nobel Prize committee's response to these discoveries was remarkable both for its rapidity and for its perspicacity. Kornberg's work on the purification and characterization of DNA polymerase began only in 1955; the first results were presented in 1956, and the first articles were published in 1958.

Kornberg's scientific career was typical of a physician who decided to turn to research during World War II.[5] His first studies were on nutrition and vitamins—in 1945, supervised by Ochoa, Kornberg helped purify two enzymes that are essential for the cell's energy metabolism. Kornberg had a short stay in Gerty and Carl Cori's laboratory at Saint Louis—called "the Mecca of enzymology" because the laboratory studied very complex enzymes involved in the synthesis of sugar polymers.[6] Kornberg returned to Washington, where he studied an enzyme that synthesized one of the essential coenzymes to be found in the cell, nicotinamide adenine dinucleotide (NAD). This work made him familiar with nucleotide metabolism; from

1953 to 1955 he characterized several enzymes involved in nucleotide biosynthesis.

From Kornberg's autobiographical writings it is difficult to understand what led him, in 1954, to begin studying the enzymes that make DNA and RNA. Was it a natural prolongation of his previous studies? Did he want to follow the same path as other eminent biochemists, such as the Coris, who, after working on the enzymes involved in sugar metabolism, then studied enzymes that polymerize sugars into long polysaccharide chains? Or, despite Kornberg's protestations to the contrary, was he influenced by molecular biology and the impact of Watson and Crick's 1953 model?[7]

In 1954 Kornberg used a radioactively labeled nucleotide, thymidine, in his attempt to characterize and purify two enzymes involved in the synthesis of ribonucleic acid (RNA) and of DNA. As the source of the enzymes Kornberg chose an extract of *Escherichia coli* bacteria because, as he said, "the rapidly growing *E. coli* cell had become a favored object of biochemical and genetic studies."[8] Initial results showed that a small but reproducible fraction of the thymidine was incorporated into a polymer that was probably DNA.

These promising initial studies were interrupted by the announcement that Severo Ochoa and Marianne Grunberg-Manago had discovered an enzyme—polynucleotide phosphorylase—that could polymerize long chains of RNA in the absence of any matrix.[9] Kornberg, together with Uriel Littauer, purified this enzyme from *E. coli* bacteria.[10] This enzyme later turned out to be extremely useful for synthesizing RNA molecules with different nucleotide compositions, used in experiments aimed at deciphering the genetic code (Chapter 12). But Kornberg and his colleagues soon realized that this enzyme incorporated nucleotides randomly, and thus could not play any role in intracellular information transfer. They returned to their work on DNA polymerase and succeeded in purifying the enzyme that they then used to show that all four deoxyribonucleotides were necessary for in vitro DNA synthesis. The DNA template was also indispensable, and all synthesis stopped once an enzyme that digested DNA, DNase, was added. By contrast, adding RNase, an enzyme that digested RNA, had no effect, showing that the replication of DNA did not require a passage through an RNA intermediary. (This result is true for DNA polymerase, but it is not valid for the whole process of chromosome replication where RNA primers can be required for the subsequent synthesis of DNA.)

With even purer enzyme fractions, Kornberg was able to replicate DNA twentyfold in vitro: at the end of synthesis, the parental DNA represented only 5 percent of the DNA present in the tube.[11] Kornberg found that the best substrate for replication was a single strand of DNA, which agreed with the replication model initially proposed by Watson and Crick (see Chapter 11). Furthermore, in the newly synthesized DNA, the percentage of A was always equal to that of T, and the percentage of G was always equal to that of C. The base composition of the newly synthesized DNA was identical to that of the DNA in the test tube at the beginning of the experiment.[12]

All these results showed that the DNA present at the beginning served not only as a primer for DNA synthesis (like enzymes that need a primer made of a few links of sugar to make polysaccharides), but also as a template that was faithfully copied by DNA polymerase. To determine the nature of the nucleotides situated close to the incorporated radioactive nucleotides, Kornberg carried out a sophisticated labeling experiment with radioactive nucleotides, followed by digestion with highly specific nucleases. This experiment showed that the newly synthesized strand was oriented in the opposite direction from the strand that had served as the matrix, a result that again agreed with Watson and Crick's model.[13]

Kornberg's experiments were remarkably elegant. They also represented a considerable amount of work: in order to have the necessary molecular tools, Kornberg had been obliged to purify several enzymes involved in nucleotide metabolism. Above all, these experiments showed that the extraordinary replicatory ability of DNA was based on the properties of a single enzyme, DNA polymerase, which could choose with a high level of accuracy among the nucleotides present in the environment to pair up the bases, as suggested by Watson and Crick.

However, nothing in Kornberg's work proved that the DNA synthesized in vitro had the same biological properties as the parental DNA. The quick decision of the Nobel Committee to award him a prize was undoubtedly a result of the biochemists' determination to affirm their participation in the final stages of the race for the secret of life that was represented by the development of molecular biology. If the Nobel Prize was awarded for Kornberg's confirmation of Watson and Crick's model, it would have been more logical to give the prize to them first, and then to Kornberg! Watson and Crick received the Nobel Prize only in 1962, whereas Alfred Hershey,

Salvador Luria, and Max Delbrück had to wait until 1969 before the contribution of the phage group was officially recognized.

The fact that Ochoa was a cowinner of the Nobel Prize for his discovery of polynucleotide phosphorylase—an enzyme whose precise physiological role remains poorly understood but which is probably involved in the breakdown of RNA—shows that above all the gesture was intended to reward those whose successful research had shown that intracellular transformations of informational molecules were also part of enzymology, and thus of biochemistry.

Molecular biology and biochemistry should be seen as having complementary, not opposing, roles: the molecular biologist deciphered the main pathways of information transfer, whereas the biochemist dealt with the details of the molecular machinery. These mere "details" required enzymes that carried out tasks that were astonishing and somewhat magical.[14] This division of labor provided molecular biologists with a confirmation and a concretization of the mechanisms that they had hypothesized; for the biochemists, it confirmed that everything in the cell was biochemical, from intermediate metabolism to gene replication.

Nevertheless, Kornberg's studies lacked a final experimental confirmation: they needed to show that DNA synthesized in vitro was indeed active. But the "activity" of DNA, the proof of its informational role, was not easy to detect in vitro. The best experimental system for proving this was transformation (see Chapter 3): it should have been possible to amplify the transforming factor using Kornberg's purified enzyme. However, as seen earlier, the transformation experiments pioneered by Avery were difficult to control and quantify. As a result, Kornberg's experiments aimed at amplifying transforming DNA by the action of DNA polymerase produced negative results.[15]

It was only in 1967 that Kornberg was able to amplify in vitro the DNA of a phage, ΦX174.[16] This experiment worked only because of the addition of another enzyme, DNA ligase, which closed the newly synthesized molecules. This experiment, which the scientists involved modestly described as their twenty-third and twenty-fourth contributions to the enzymatic synthesis of DNA, was presented by journalists as the creation of life in a test tube, and as the most important scientific discovery of 1967.[17] This gap between the reactions of journalists (and the public) and the attitude of scientists shows the change that had taken place in the minds of biologists. The definition of life had been altered: for biologists, replicating a virus in vitro is not the equivalent of creating life. Life does not lie in molecules—it is to be found in the complexity of the systems under study.

This late but enthusiastic reception for Kornberg's work explains the disappointment that followed the discovery, in 1969, that his purified enzyme was not in fact involved in DNA replication in *E. coli*.[18] The media circus that followed, which lasted more than two years, can be explained—like the enthusiastic welcome for the in vitro replication of phage ΦX174—by the long period in which molecular biology was, as the French phrase has it, "crossing the desert" (1965–1972).[19] Although molecular biology had precise concepts, its tools were still inadequate. The fact that few breathtaking discoveries were made during these years led to an overinterpretation of data and to a continual questioning of the dogmas of the young science under the slightest experimental pretext. The debate that accompanied the discovery of reverse transcriptase was a good example of this situation (Chapter 15).

Later experiments confirmed that Kornberg's purified enzyme was not essential for DNA replication, but instead intervened in DNA repair. But the enzymes involved in in vivo DNA replication have a structure that is close to Kornberg's DNA polymerase, and, most importantly, they function according to the same principles. Above all, the discovery that Kornberg's DNA polymerase had only a minor physiological role did not call into question the fact that enzymes are able to replicate DNA, as set out in Watson and Crick's model.

The polymerization activity of Kornberg's enzyme was used to incorporate labeled nucleotides into an unlabeled molecule and thus to make radioactive probes that would permit the detection of homologous RNA or DNA molecules by hybridization on gels or in libraries. Kornberg's enzyme is all the more useful because, in addition to its polymerizing activity, it possesses a degrading activity that enables it to eliminate mismatched nucleotides in vivo and to replace nonradioactive nucleotides in vitro with their labeled equivalents.[20]

There is also another use for Kornberg's DNA polymerase. In 1977 Frederick Sanger showed that it could be used to determine the sequence of DNA molecules. (PCR is an offshoot of this sequencing technique.) Like many discoveries, this was the fruit of a meeting between an idea and pure chance.[21] Sanger's idea was to follow the incorporation of nucleotides into the newly synthesized DNA molecule as the polymerase proceeded along the DNA strand. Chance played its part in the design of the experiment. Because Sanger and his colleagues wanted to incorporate as much radioactivity as possible into the synthesized DNA, they used three unlabeled nucleotides; the fourth—X—was labeled but not diluted with unlabeled X and therefore was added at a very low concentration. They found that this low concentration sometimes made the enzyme stop and detach at a point where it should have incorporated an X nucleotide, producing incomplete strands of varying length. The researchers could separate these strands using electrophoresis on acrylamide gels, and directly determine the relative positions of the different X nucleotides. The experiment was then carried out with the three other nucleotides, and the four reactions were run simultaneously on a gel to derive the complete sequence. The principle of the sequencing technique had been found: stop the elongation of the DNA strand at a given nucleotide and then determine the length of the different fragments thus obtained.

Sanger subsequently made a number of improvements to this method. Its canonical form is as follows. The DNA fragment to be sequenced is cloned into an M13 phage, which can exist as a single strand of DNA.[22] Recombinant single-strand phages are isolated, and an oligonucleotide complementary to the phage DNA, and which hybridizes close to the insertion site of the fragment to be sequenced, is added. This oligonucleotide is simply a primer for Kornberg's DNA polymerase, which cannot start synthesizing DNA *ex nihilo* but can only elongate previously existing strands. The four nucleotides are then added, together with a small quantity of an analogue of one of the nucleotides—a dideoxynucleotide—which, when it is randomly inserted into the sequence in place of the proper nucleotide, immediately halts DNA synthesis. The experiment is performed with four different analogs, each corresponding to one of the nucleotides, and the synthesized strands are analyzed by electrophoresis.

There is another method for sequencing DNA, named after its inventors, Allan Maxam and Walter Gilbert (Chapter 16).[23] This method uses chemical reagents to cleave DNA at precise positions. Both methods are equally simple, but Sanger's method was subsequently adopted for automatic DNA sequencing.

This example again underlines the important role that enzymes play in molecular biological research and as tools in genetic engineering. Researchers use enzymes because of their remarkable specificity. Because DNA polymerase faithfully copies DNA strands, it can be used to determine the sequence. But previous studies by Kornberg's group had also shown that this enzyme could be tricked into incorporating analogs, such as dideoxynucleotides, into the DNA strand in the place of normal nucleotides. A precise understanding of DNA polymerase was a key prerequisite for its use.

KARY MULLIS HAS PROVIDED a detailed account of his discovery of PCR.[24] The idea came to him one Friday evening in April 1983, when he was driving on the hilly roads of California on the way to the chalet where he was to spend the weekend. Mullis had received a doctorate in biochemistry, and in 1979 he had been recruited by the Cetus biotechnology company to make oligonucleotides for use as probes (see Chapter 16). However, the arrival on the market of oligonucleotide synthesizers had freed researchers like Mullis for other projects. While he was driving that evening, he was thinking of how to develop a technique that, on the basis of an extremely small biological sample, would be able to determine the identity of a nucleotide at a precise position in the DNA molecule. The project was interesting because many genetic diseases are produced by the substitution of a single nucleotide at a precise position in a gene. Such a technique would mean that it would be possible to make a genetic diagnosis on the basis of a very small biological sample.

Mullis's idea was to use Sanger's sequencing technique. The first step would be to synthesize an oligonucleotide immediately next to the nucleotide that was to be determined. The two strands of DNA would be separated by heating, and the oligonucleotide would be hybridized with the complementary strand. This oligonucleotide would function as a primer for the DNA polymerase. The four radioactively labeled dideoxynucleotides would then be added separately, but the only one that would be incorporated would correspond to the one at the position of interest.

The idea was a good one, but it suffered from the fact that oligonucleotide fixation is not always specific. Mullis decided that it would be possible to confirm the result by synthesizing a second oligonucleotide corresponding to the sequence immediately downstream of the nucleotide, on the complementary DNA strand. This oligonucleotide would hybridize with the DNA strand and would be elongated by incorporation of a dideoxynucleotide that was complementary to the dideoxynucleotide incorporated in the first experiment. The two experiments should thus lead to complementary results.

There was one further difficulty: the DNA sample might contain free nucleotides that could be incorporated instead of the dideoxynucleotide. Mullis's idea was to do the experiment in two stages: first, he would add the oligonucleotides but not the dideoxynucleotides. The DNA polymerase would use the nucleotides present in the milieu to elongate the oligonucleotides. Once the reaction was complete (and the free nucleotides used up), all that would be necessary would be to heat the mixture to separate the two strands of DNA from the oligonucleotides of varying lengths to which they had hybridized, and then rehybridize the DNA with new oligonucleotides, this time adding the dideoxynucleotides.

But what would happen if the oligonucleotides had been sufficiently elongated to hybridize with the second oligonucleotide? Mullis immediately had the answer: the result would be the specific amplification of the DNA sequence between the two oligonucleotides. Familiar with computer programming and with the loops that are often used in programs, Mullis quickly understood that by repeating the elementary steps—hybridization, synthesis, and heating—he could amplify the DNA sequence between two oligonucleotides. Furthermore, Mullis realized that the amplified DNA fragment would have a precise size and be bounded at each end by the oligonucleotides that had been added as primers. Nothing prevented the oligonucleotides from being very far apart, and thus from amplifying very large DNA fragments.

Mullis thought that the idea was too simple for someone else not to have thought of it first. But when he asked his colleagues, none of them had ever heard of anything like it. They could see no reason it would not work, but none of them was particularly enthusiastic, either.

Preparing the experiment took several months. Using Kornberg's original articles, Mullis had to determine the optimal concentrations for the reagents, the size of the oligonucleotides, and the composition of the medium. After this long preparation, the experiment was an immediate success. The first publication announcing the development of the technique dealt with the prenatal diagnosis of sickle cell anemia, showing the practical, applied aspect of the discovery (Mullis's name was lost in the middle of a long list of authors).[25]

A number of improvements have been made to the initial protocol. Most importantly, Kornberg's enzyme has been replaced by a DNA polymerase extracted from *Thermus aquaticus* (Taq), a bacterium that lives in hot springs.[26] This enzyme is not denatured by the temperatures required to separate the DNA strands after the elongation phase. It was thus no longer necessary to add the polymerase at the beginning of each elongation phase. This led to the development of automatic machines that could be programmed to reach the required temperatures for oligonucleotide hybridization, elongation, and the denaturing of the synthesized strands. Furthermore, the whole set of operations—hybridization and elongation—could be carried out at a higher temperature, thus limiting the risk of nonspecific hybridization of the oligonucleotides and increasing amplification efficiency.[27]

THE DISCOVERY—OR PERHAPS MORE PRECISELY, the invention—of PCR raises a number of points relating to both the internal functioning of science in general and the specificity of molecular biology in particular. Without doubt, it was postmature.[28] All the necessary tools for its realization had existed in the 1960s. Indeed, in the 1960s Joshua Lederberg and Kornberg had discussed the possibility of obtaining large quantities of DNA using DNA polymerase.[29] Har Gobind Khorana and his colleagues had gone even further, suggesting that replication could be carried out using short complementary oligonucleotides from each of the two strands of DNA.[30] At the end of their article, they outlined what became the three phases of PCR—hybridization with oligonucleotides, elongation by polymerase, and denaturation of the synthesized molecules—and the idea of repeating the process many times over. But there was a fundamental difference between this proposal and Mullis's idea. Khorana's aim was to copy an already well-characterized DNA molecule in vitro. Mullis's goal was to amplify a molecule of DNA sufficiently to be able to characterize it. This fundamental difference in the aim of the experiment completely altered the significance of the procedure.

Strangely enough, Mullis's idea initially was met with a polite but unenthusiastic response. In fact, the discovery of PCR was really a discovery only because it enabled DNA to be amplified. Many factors—nonspecific oligonucleotide fixation, unexpected stoppage of the DNA polymerase, errors in nucleotide incorporation—could have made the technique ineffective, and indeed the first results obtained with invention were relatively modest. It was the use of Taq polymerase that made PCR sufficiently simple and effective to democratize the PCR and to allow "the practice of molecular biology without a permit."[31] Mullis's colleagues, who were prudent about the possibility of developing the technique, were not blind but were realistic. How many apparently revolutionary discoveries have ended up in the trash can of history?

It is also important to note that PCR is perfectly representative of molecular biology and could virtually symbolize the discipline. It is a simple technique that uses biological properties—in this case, enzymes that replicate DNA—as research tools.

Two other characteristics make the discovery of PCR emblematic. The name—polymerase chain reaction—was chosen because of its reference to nuclear chain reactions.[32] In both its spirit and the person of some of its founders, molecular biology is the descendant of the physics of the 1930s and 1940s. It is also the sister of computing: the PCR protocol, with its loops of repetitive operations, is analogous to the methods of programming. PCR shares with these methods the simplicity of a series of elementary steps. Its potential, like that of a computer, is simply the product of the monotonous repetition of these steps.

The principle of PCR is so simple that, when Mullis was awarded the Nobel Prize for Chemistry in 1993, a previous winner remarked that it was a mere technical trick, without the intellectual richness that should be expected from a Nobel Prize–winning study. Nevertheless, PCR, more than any other technique, has changed the work of molecular biologists. And, after all, if it was simply a trick, why had no one discovered it before?[33]

PART FOUR

BEYOND MOLECULAR BIOLOGY?

IT IS MORE THAN SIXTY YEARS since scientists first described the structure of DNA and of a handful of proteins. Forty years have passed since the development of genetic engineering provided molecular biologists with the tools they needed to isolate and characterize the macromolecules involved in complex biological processes, in both prokaryotes and eukaryotes.

These key developments have led to a massive accumulation of data. It is not possible to present all these subsequent discoveries at the same level of detail as those described in the first three parts of this book—a history of any one of the major developments that took place during this period could easily be as long as the three previous sections. One option would be to end the book at this point, and to direct the reader to various detailed historical accounts that have been published on these more recent discoveries. But this would run the risk of implying that there was some kind of qualitative break between the history of molecular biology covered in the first parts of the book and these more recent developments, as has indeed been suggested by some of the more audacious authors of recent discoveries. And it would equally fail to emphasize the very real continuities that exist between the two periods.

The question addressed in the following chapters is very simple: does the explanatory framework created by molecular biologists between 1940 and 1980 still apply today? This final part will therefore look first at the results of the interactions between molecular biology and a number of other disciplines—biochemistry (in particular the study of proteins), developmental biology, evolutionary biology, and, in the case of gene therapy, medicine. It will then explore the challenge posed by the discovery of a series of phenomena that were unsuspected by classical molecular biology, in particular the importance of regulatory RNAs and of epigenetic modifications. The most significant changes that took place at the turn of the century were the completion of the first genome sequencing programs, followed by the emergence of the new disciplines of systems biology and synthetic biology, both of which claim to put forward a radically new approach to living phenomena. This raises the question of whether they have sounded the death knell of molecular biology.

This final part opens with an introductory chapter that emphasizes that most research in molecular biology, which was made possible by genetic engineering, involved precise and straightforward descriptions of biological phenomena that were previously barely understood.

A final chapter focuses on the images, metaphors, and representations that are used in molecular biology. Early representations, such as those of the operon model, helped give molecular biology its seductive power. By exploring these ways of portraying scientific discoveries and concepts we can gauge the extent to which the transformations that have taken place over the past five decades are indeed truly radical, or whether what appear to be major changes and novel developments in fact conceal hidden continuities.

20

The Molecularization of Biology and Medicine

THERE IS NO SHORTAGE of examples to illustrate the molecularization of biology and medicine that occurred from the early 1980s onward. Among the most striking were the precise descriptions of the action of hormones and growth factors (already touched on in Chapter 18), of cell division, and of the many mechanisms of DNA repair.[1] This chapter focuses on three issues: transcription mechanisms in eukaryotes, the molecular characterization of neurodegenerative diseases, and the description of apoptosis (programmed cell death). Initially, results in the first two areas were considered to be a challenge to the dogmas of molecular biology, while the third was viewed as a radically new cellular phenomenon that was foreign to the reductionist program of molecular biology.

▌ Mechanisms of Transcription in Eukaryotes

During the 1960s and 1970s, our understanding of transcription in prokaryotes became increasingly precise, but progress in the description of eukaryotic transcription lagged behind. The technical and conceptual difficulties produced by the complexity of eukaryotic cells and the relatively large amount of genetic material they contain compared to prokaryotes could only be overcome by using the tools of genetic engineering. The discovery of ribonucleic acid (RNA) splicing led many biologists to suspect that transcription and its regulation were probably also entirely different in eukaryotes and prokaryotes. This idea was supported by all those who believed that the models of molecular biology, developed in bacteria, could be of only limited use in explaining how multicellular organisms function (see Chapter 22).

Despite these arguments, when the first eukaryotic genes were cloned in the late 1970s, a research program began that took as its starting point the contemporary understanding of prokaryotic transcription regulation. The first step was to construct an experimental system that could measure the expression level of a given gene. In the absence of an effective in vitro eukaryotic transcription system, two methods were used. The first of these involved the transfection of the gene into eukaryotic cells, followed by the measurement of the amount of protein produced from the transfected gene after a brief lapse of time (generally forty-eight hours)—an experiment that harked back to the transformation of *Pneumococcus* by Oswald Avery forty years earlier. Taking as their starting point the results in bacteria that showed that the regulatory sequences were located upstream of the genes, researchers began to replace the coding part of the gene with a reporter gene encoding an enzyme whose activity was easy to measure. Another possibility was to use microinjection to introduce a gene into a giant *Xenopus* oocyte, which produced a high level of expression in a single cell.

Once expression of the introduced gene had been obtained, researchers gradually trimmed the upstream sequences of the gene with exonucleases, or cleaved the DNA with restriction enzymes, thus eliminating fragments of the DNA sequence from a role in transcription. Despite the obvious elegance of this approach, the picture provided by the early results, published in 1979–1980, was far from clear. Studying the sea urchin *H2a* histone gene, Max Birnstiel and his collaborators made two puzzling observations. First, the deletion of a sequence located 30 base pairs upstream of the transcription initiation site, now called the TATA box (and analogous to a sequence found at a similar position in prokaryotes), reduced the expression of the gene but did not abolish it as expected. Similar results were obtained with the SV40 promoter sequence.[2] Second, other sequences located far upstream either activated or inhibited the expression of the downstream gene.

In 1981, Walter Schaffner and his collaborators made a discovery that had a massive impact on the study of transcriptional regulation in eukaryotes. They studied a short sequence, 72 base pairs long, that was present in the oncogenic virus SV40, which had previously been shown to be required for a high level of replication and transcription.[3] They found that when this sequence was inserted upstream of any gene it dramatically enhanced the rate of transcription of that gene.[4] Schaffner's group described three new characteristics of these sequences, which they termed enhancers and which differed in a number of ways from the regulatory sequences that had been described in bacteria. Not only did the enhancers remain active when positioned at a great distance from the transcription initiation site (many thousands of base pairs away), but—more puzzling still—their orientation had no significance. Furthermore, they remained active when they were placed downstream of the transcription initiation site, even when they were positioned in an intron.

These findings were soon confirmed by other teams, using a similar sequence present in the closely related polyoma virus. Similar sequences were also found in the long terminal repeats of retroviruses. This discovery explained some enigmatic results reported in 1981, in which some oncogenic viruses that lacked any transforming sequences nevertheless transformed the cells that they infected by inserting their retrotranscribed genome in close proximity to a host proto-oncogene.[5] The initial explanation had been that the retrovirus provided a strong promoter to the proto-oncogene (Chapter 18). A better explanation was provided by the presence of enhancers in the retrovirus because these did not require a precise positioning of the retrovirus for efficient activation of gene expression to occur.

The existence of enhancers was soon demonstrated in eukaryotic genes. A striking example was provided by the enhancers present in the introns of immunoglobulin genes, which explained the high rate of production of immunoglobulins by antibody-forming cells.[6] The activating power of enhancers was also rapidly confirmed by the creation of transgenic animals in which enhancers had been positioned in proximity to different genes, and by the study of gene expression in these organisms.

Three mechanisms were proposed to explain the action of enhancers. The first posited that they were the entry points of the eukaryotic RNA polymerases that had been so painstakingly purified in previous years. Considering the size of the eukaryotic genome, it seemed highly unlikely that these enzymes could depend on random encounters with their promoters in order to initiate gene transcription. It was therefore proposed that there were entry points for the RNA polymerases on the genome, from which point they slid down the DNA sequence until they met promoter sequences—as had been shown for the *Escherichia coli* RNA polymerase.[7]

The second hypothesis flowed from the new interest in chromatin that had grown since the 1976 discovery that its structure contained areas where DNA was more susceptible to the action of endonucleases such as DNAase I. These domains of increased nuclease sensitivity were associated with a high level of gene expression. The enhancer sequences of the SV40 and polyoma viruses were shown to be sensitive to the action of nucleases.[8] It was thought they might correspond to areas where chromatin is reorganized in order to favor transcription initiation. It had been known since 1974 that the basic element of chromatin was the nucleosome, an assembly of eight histone proteins (Chapter 26). Because the formation of nucleosomes induces a positive supercoiling of DNA, the raison d'être of enhancers might be to decrease DNA supercoiling in order to favor DNA unwinding and transcription initiation.[9]

The third explanation for enhancer activity was based on Ulrich Laemmli's observations that chromatin is organized in loops, and that these loops are the active domains of transcription.[10] Enhancers might contribute to the formation of these loops and to their attachment to the nuclear matrix.

These three explanations were not mutually exclusive, and some researchers combined them in increasingly baroque models. Strikingly, the actual mechanism of action shown by enhancers—transcription factors binding to DNA—was not suggested, or at least not immediately.

Although their precise mechanisms of action remained unclear, enhancers were immediately converted into a research tool. Dissecting the 5' upstream sequences of a given gene required a certain level of gene expression that many cloned genes did not reach. Adding enhancer sequences was a means of stimulating gene expression to the point where dissection of the promoter became possible.

Although viral enhancers were effective in many cell types, in most cells the enhancers were cell specific. The three mechanisms described previously could not easily account for this. An alternative hypothesis rapidly followed, in which enhancers were the binding sites for proteins that activated transcription.[11] This hypothesis was supported by the characterization of the glucocorticoid receptor as well as of the heat shock transcription factor, which is involved in the response to proteotoxic stress.[12] Both these factors were able to activate transcription by binding at sites that were distant from the promoters.

The characterization of the transcription factors that bind to the sequences present in enhancers (or to promoter sequences) took many years. New electrophoretic techniques, such as footprinting and gel shift assays, were designed to characterize the sequences that were recognized by these transcription factors. Once the transcription factors had been partially purified and their binding sequences had been determined, the factors were purified to homogeneity by affinity chromatography, then crystallized in the presence of their DNA targets, enabling the structure of the bound complexes to be analyzed by X-ray diffraction.

At the beginning of the 1990s the structures of the first transcription factors were described, along with the characterization of the components of the transcriptional machinery.[13] Specific DNA interaction motifs were described—zinc fingers, helix-loop-helix—the existence of which enabled whole families of transcription factors to be cloned. This work occurred in parallel to the development of eukaryotic in vitro transcription systems, which proved particularly difficult owing to the complexity of the transcriptional machinery in eukaryotes (involving a high number of factors), unlike the relative simplicity of bacterial systems. These in vitro techniques made it possible to confirm the activator role of the transcription factors that had previously been isolated.

As a result of all this work, the mystery of eukaryotic transcription was explained.[14] Enhancers were shown to be DNA transcription factor binding sites, which could be far distant from the gene they affected. By forming loops of DNA, the transcription factors that were bound to the enhancers interacted with the transcription factors and the transcription machinery close to the promoter. The formation of such loops through the interaction of two kinds of repressor molecule had previously been observed in bacteria.[15]

There are indeed differences between eukaryotic and prokaryotic transcription. Most eukaryotic transcription factors are activators, many different factors control a eukaryotic gene, and in eukaryotes chromatin structure plays an important role. But the principles of gene regulation are the same in the two branches of life: by interacting directly or indirectly with RNA polymerase, proteins control the binding of this enzyme to the promoter, and hence the initiation of transcription.

The characterization of enhancers turned out not to be a challenge for molecular biology, but rather a confirmation and an extension of the results obtained in bacteria. It was an additional blow to the idea that it is possible to give a precise structural and functional definition of a gene. Enhancers are distant from the genes they control and are often shared by different genes. We cannot exclude the enhancer from the gene, but we cannot include it, either. But who cares about the definition of a gene? Definitely many philosophers of biology and maybe some geneticists too, but very few molecular biologists. What interests them is to provide a good explanation for the phenomena they observe, as was the case with enhancers.

Researchers are still arguing about how the activity of enhancers is controlled by the organization of the nucleus in different domains.[16] But the action of enhancers in transcription is understood, and their fundamental role in the control of gene expression during development has been amply confirmed.

▌ Neurodegenerative Diseases

Neurodegenerative diseases are among the best examples of how molecular descriptions can shed light on the mechanisms of the diseases that in most cases are not hereditary, which had previously seemed mysterious. Even in diseases that are not so mysterious, such as diabetes, the molecular approach has enriched our understanding and made it possible to distinguish different forms that were previously undetected.

This section focuses on neurodegenerative diseases that can be detected by postmortem examination for the presence of protein aggregates in the brain: Alzheimer's disease, Parkinson's disease, and Huntington's disease, together with Creutzfeldt-Jakob disease (CJD) and other prion diseases including kuru and bovine spongiform encephalopathy (BSE or "mad cow disease").

Advances in understanding the etiology of these diseases were very slow. Alois Alzheimer first described the disease that was eventually given his name in 1906, but the structural characteristics of the aggregates he identified were not understood until the 1960s, when electron microscopy revealed the presence of extracellular amyloid plaques and intracellular neurofibrillary tangles in the brains of people who had died from this disease.[17] Accurate hypotheses about the pathophysiological mechanisms of Alzheimer's were first proposed in the 1980s and were gradually confirmed in the 1990s by the new molecular technologies. Molecular biology was not defeated by the challenge of these atypical diseases; instead it demonstrated its explanatory power. Nevertheless, many questions still remain unanswered.

Before exploring the transformation in our understanding of Alzheimer's disease, we must first trace the development of ideas about prion diseases, because these diseases represented a model for other neurodegenerative conditions—indeed, it is now accepted that they all share common mechanisms.

The first prion disease to be identified was scrapie, which was noticed in sheep and goats as early as the eighteenth century. There is no evidence that the disease has ever been transmitted from sheep to humans. In 1936 experimental proof of transmission between sheep was obtained when the French veterinarians Jean Cuillé and Paul Louis Chelle injected a spinal cord extract from a diseased sheep into the eye of a healthy one; after a long incubation period, the previously healthy sheep developed the symptoms of scrapie.[18]

The proteinaceous infectious particles, as they were known in the early 1980s—later to be shortened to "prions," that were the agents of this disease became the object of intense scientific interest in the mid-1960s. Various studies suggested that this particle did not contain any nucleic acid and was not a virus: it was too small, and it was not sensitive to physical, chemical, or biochemical agents that inactivated nucleic acids.[19]

Given that molecular biologists, led by André Lwoff, had clearly distinguished the different types of microorganisms and had refined the concept of a virus to our modern understanding, these observations were puzzling.[20] In 1967, J. S. Griffith proposed three mechanisms to explain how a pathogenic agent might be devoid of nucleic acid.[21] The prion could be an antibody with a structure identical to the antigen; it might induce the expression of a gene encoding a protein identical to it; or it might be a dimer, and promote the polymerization of a monomer with a primary structure identical to the structure of its subunits. The last hypothesis is similar to our current view of prion diseases, and, like the second, it was inspired by the models developed by the French school of molecular biology.

In the 1960s and 1970s, the U.S. physician D. Carleton Gajdusek described kuru, a neurodegenerative disease endemic to certain tribes of

Papua New Guinea. Gajdusek attributed its incidence to the local tradition of funerary cannibalism, in which families ate the bodies of dead relatives, including the brains. He further showed that the disease was similar to scrapie in sheep, and suggested that other neurodegenerative diseases in humans, and in particular Alzheimer's disease, might be caused by what he called slow viruses.[22] In 1976, Gajdusek was awarded the Nobel Prize in Physiology or Medicine for this work.

In 1982, Stanley Prusiner published an article that was to become famous, in which he coined the term "prion."[23] He used a simple and reproducible experimental system to study these prions: hamsters that had been injected with extracts from scrapie-infected sheep brains developed a disease that could then be transmitted from hamster to hamster by intracerebral injection of infected brain material. The short incubation period enabled Prusiner to partially purify the prion, which he described as a protein of 27–30 kilodaltons.[24]

How prions reproduced remained unclear. Prusiner favored the most heretical hypothesis—heretical in terms of the central dogma of molecular biology, that is. He suggested that prions could reproduce their protein structure (that is, their amino acid sequence) within infected cells.[25] After being awarded the Lasker prize along with Oswald Avery's colleague Maclyn McCarty in 1994, Prusiner compared his experiment with Avery's on the chemical nature of the transforming principle in *Pneumococcus,* performed half a century earlier.[26] Although there were indeed parallels between the two experimental protocols, the impact of their results on biology was dramatically different. Prusiner's characterization of prions turned out to be a mere epiphenomenon compared to the earth-shaking discovery of the genetic role of DNA.

Studying aggregated proteins is not easy. Nevertheless, it proved possible to obtain antibodies against the partially purified prions, and those antibodies could be used to screen libraries of complementary DNA (cDNA) to isolate first cDNA clones of the prion protein, then genomic clones.[27] To everyone's surprise, these clones revealed that the protein was expressed in normal cells and was not overexpressed in infected animals. Furthermore, the amino acid sequences of the normal prion and of the infectious form were identical. The hypothesis that the pathogenic form of the prion was produced by post-translational modifications of the normal protein sequence also had to be abandoned. The only explanation that remained was that the infectious prion consisted of an altered, pathogenic, conformation of the normal protein. A link between the prion gene and a gene controlling the incubation time of the disease was an additional argument for the prion protein playing a central role in the disease.[28]

Mutations in the prion gene were later described and were shown to be responsible for rare genetic diseases. The first was discovered in a familial form of dementia, Gerstmann-Sträusler disease, and others were subsequently described in Creutzfeldt-Jakob disease (CJD).[29] The hypothesis that was proposed was that these mutations favored the formation of the pathogenic form of the prion protein. They also reinforced the connection between the protein encoded by the prion gene and neurodegenerative diseases.

The characterization of the mechanism of Alzheimer's disease followed a parallel course. A peptide was isolated from the amyloid plaques of patients who had died from the disease, and its sequence was determined. From this sequence, an oligonucleotide was synthesized that made it possible to isolate cDNA and genomic clones. The peptide present in amyloid plaques was shown to be derived by proteolysis from a membrane protein called amyloid precursor protein (APP).[30] The gene encoding this protein was localized on chromosome 21, which coincided with the observation that patients suffering from Down syndrome, or trisomy 21, often suffer from premature forms of Alzheimer's disease.[31]

It was expected that mutations in this gene would be found that corresponded to early onset familial forms of Alzheimer's disease, but the initial results were disappointing. No mutations were found in the *APP* gene of Alzheimer's sufferers, whereas *APP* mutations were detected in patients with another brain disease, Dutch type hereditary cerebral hemorrhage, in which aggregates of the amyloid peptide had also been observed.[32] In 1992, another gene linked with a familial form of Alzheimer's was identified.[33] Mutations in this gene are present in populations of German origin living in Russia and were characterized in 1995.[34] The protein produced by this gene, presenilin, is a component of one of the proteases that cleave APP to generate the β-amyloid peptide. Soon a second presenilin gene was isolated that was also mutated in familial forms of Alzheimer's. Meanwhile, mutations in the *APP* gene were finally described in some familial cases of Alzheimer's.[35] In various ways all these different mutations favored the

formation of the peptide and its consequent aggregation. A coherent scenario for the origin of the disease had finally emerged.

Significant developments in our understanding of neurodegenerative
diseases came in the late 1980s and 1990s. In 1996 the British government
admitted that there was a probably a link between BSE in cattle and the
growing number of cases of a variant form of CJD in humans. Amid the
major crisis that this triggered in Great Britain, increased funding was
made available for research into prion diseases. Experiments using transgenic animals confirmed earlier hypotheses about the mechanisms that
produce this category of diseases. First, the different strains of hamster
prions were reproduced in mice carrying a hamster prion gene, while the
disruption of the prion gene prevented the development of the disease in
mice.[36] The introduction into mice of the mutated forms of the genes present in humans suffering from CJD, Alzheimer's, or other neurodegenerative
diseases led (in most cases) to the development of pathologies that were
similar to the human diseases.

The mechanism of transconformation / aggregation was also explored
through the study of prion-like phenomena that were discovered in yeast.[37]
The existence of different strains of prions was explained by the existence
of different pathological conformations. A general mechanism of aggregation that could explain all these pathologies was proposed, which consisted
of two steps: the slow formation of "seeds," followed by the rapid elongation of fibrils.[38] In vitro reproduction of the pathogenic form of the prion
was finally obtained in 2004.[39]

In the early years of the century, two major breakthroughs occurred that
gave the field an even greater coherence. The first was the 2005 description
of a common structural motif—a highly stable β-sheet structure—that was
at the origin of the fibrils observed in these diseases.[40] The second was the
demonstration that the difference between infectious prion diseases and the
noninfectious neurodegenerative diseases (Alzheimer's, Parkinson's, and
Huntington's diseases) is not due to differences in the mechanisms at the
origin of the diseases. The protein aggregates present in all these conditions
can propagate from one cell to another in infected organisms.[41] Instead, the
differences reflect the unusual stability of the prions outside the organism,
during their transmission from one individual to another.

Despite the hopes of some biologists (and of many philosophers of biology), the gradual unveiling of the mechanisms at the origin of these neuro-

degenerative diseases did not challenge the central dogma of molecular biology.[42] What was forbidden, according to Francis Crick, was the possibility of a protein transferring its sequence to another protein (or to a nucleic acid). The central dogma did not forbid a protein from transferring its conformation to another protein—indeed, this was implicit in the allosteric model of protein transconformation proposed by Jacques Monod, Jeffries Wyman, and Jean-Pierre Changeux in 1965.[43] The new vision of allostery that emerged in the early 2000s, according to which multiple different conformations of a given amino acid sequence could coexist (see Chapter 21), could accommodate the existence of different strains of prions with greater ease than the classic allosteric model, in which proteins had access to only a limited number of conformations. The heterodox models proposed by Prusiner in 1982 were rapidly discarded.

Today, it is clear that the formation of protein aggregates is closely related to the development of these neurodegenerative diseases, but the pathophysiological process is still not fully understood.[44] In the case of Alzheimer's disease, the relation between the two types of aggregates—the intracellular neurofibrillary tangles and the extracellular amyloid plaques—remains unclear.

Although the relationship between the large aggregates visible under the microscope and the pathological symptoms was at the origin of the identification of the disease, paradoxically we still do not understand the nature of that link. Some patients who show obvious symptoms of the disease have no plaques in their brains, while others who show no detectable signs of disease may have many. Nor do we understand the link between the formation of aggregates and the death of neurons, which is at the heart of the disease's symptoms. The role of the inflammatory response in the progression of the disease is also unclear, although it is increasingly believed to play a crucial part. Although a major susceptibility gene has been identified that encodes apolipoprotein E, its precise link to the disease and to the formation of aggregates has not yet been explained. Many studies have suggested that toxic agents present in the environment may be involved in the development of neurodegenerative diseases, but with the exception of one type of molecule in Parkinson's disease, this has yet to be proved.

The greatest disappointment of all is that the fantastic advances in our understanding of Alzheimer's disease have not yet led to any major therapeutic progress. Precise descriptions of its mechanisms have opened up many avenues for new therapeutic approaches, but so far these have been explored in vain. Some researchers have even cast doubt on the value of our current explanations. The problem probably lies not in any erroneous representation of the early molecular events involved in the diseases, but rather in the distance—both in the number of steps and in time—between these early events and the complex development of the disease.

▌ Molecular Mechanisms of Apoptosis

A good example of a successful cohabitation of molecular biology with another discipline is cell biology. No other discipline—apart from genetics—could have felt more threatened by the development of molecular biology. After all, molecular biologists thought that intracellular structures were produced simply by assembling macromolecules.

But the 1980s were indisputably the golden age of cell biology. This renewal of the discipline, despite such a difficult context, can perhaps be partly explained by the development of very efficient but simple methods for studying the cell. One of these—immunofluorescence, which combined the specificity of antibodies with the easy detection of fluorescent molecules—has revealed the architecture of the cell and the presence of a cytoskeleton. Because it is simple to use, immunofluorescence does not require any great skill, unlike the electron microscope.

Nevertheless, new techniques such as this would not have saved cell biology had they not revealed the unexpected wealth and range of intracellular traffic. The transport of proteins between the cell surface and the various intracellular organelles is carried out by a set of vesicles. In their amino acid sequences, proteins carry signals that enable them to be taken up by these different transport systems. The decisive first step in revealing this system involved the hypothesis of a signal peptide put forward by Günter Blobel and David Sabatini.[45] Next came the deciphering of the protein secretory pathway from the mid-1970s. The work was done using a combination of biochemical studies and genetic analyses performed on yeast.

Cell biologists naturally carried on this work until all the proteins and enzymes involved in this intracellular traffic had been isolated, characterized, cloned, and sequenced. But intracellular events are not solely explained at the molecular level but also in terms of compartments and vesicles. The cellular and molecular approaches are complementary.

Another excellent example of this perfect dovetailing of molecular and cellular descriptions can be found in studies on cell death. The existence of cell death in normal organisms, specifically during their development and in the absence of any pathological process, was first described in the nineteenth century and was regularly noted in the first part of the twentieth century.[46] But it was only in 1964 that Richard Lockshin and Carroll Williams clearly distinguished this form of programmed cell death (PCD), which they observed in insect development, from accidental death (necrosis) resulting from the absence of nutrients or of oxygen.[47]

In 1972, John Kerr and his collaborators described a particular form of PCD, characterized by the fragmentation of the nucleus and the rapid elimination of dead cells by phagocytosis. They called this process *apoptosis,* in reference to the mechanism of cell death responsible for leaves falling in autumn.[48] In 1980, an additional characteristic of this particular type of cell death was discovered: it involved the action of an endonuclease that cut between nucleosomes and led to fragmented DNA, which generated a characteristic ladder profile when subject to electrophoresis in agarose gels.[49] This fragmentation of DNA became the characteristic hallmark of apoptosis and was detected by the development of a simple technique known as TUNEL (a striking but inaccurate acronym for terminal deoxynucleotidyl transferase–mediated digoxigenin-deoxyuridine nick-end labeling). The characterization of apoptosis was a direct consequence of the development of the tools of genetic engineering (Chapter 16). Apoptosis was soon observed in various developmental systems and in illnesses such as cancer and neurodegenerative diseases.

In the absence of any clues as to the molecular mechanisms that might operate during this process, apoptosis remained a mystery. The breakthrough came from the small worm *Caenorhabditis elegans,* chosen by Sydney Brenner as a model for the study of development in the 1960s (Chapter 22).[50] Brenner embarked on the genetic study of this organism, and his collaborator, John Sulston, gave a complete description of its cell lineage in the early 1980s. To understand the "logic of development" and the role of genes in cell fate decisions, mutations affecting cell lineage were isolated.[51] Some of these mutations were heterochronic—that is, they brought forward or delayed cell fate decisions.[52] These genes were isolated, and in the 1990s the developmental role of microRNAs encoded by these genes was revealed (see Chapter 25). Another project involved the characterization the genetic mechanisms controlling the formation of one particular organ, the worm's vulva.

Perhaps the most significant result of this first description of the complete cell lineage of an organism was the precise description of PCD in *C. elegans*. During the development of the adult worm (it contains 959 cells), it was shown that 131 cells die, including 105 neurons. In *C. elegans* the pattern of PCD is perfectly reproducible (which is not the case for many organisms). This phenomenon was a huge help for the project that Robert Horvitz developed at the beginning of the 1980s, which was to characterize the genes involved in PCD in *C. elegans* through an extensive program of mutagenesis.[53]

The first mutations to be isolated, *ced-1* and *ced-2* (*cell death-1* and *-2*, respectively), altered the phagocytosis of dead cells, an event that occurs late in apoptosis.[54] Horvitz was hoping to find mutations that acted earlier, controlling the decision for cell death.

Such early-acting mutations began to be isolated in 1983. The first was *ced-3,* which allowed extra cells to survive, followed a year later by *ced-4,* which had similar effects.[55] Thanks to the precise genetic maps of *C. elegans* that had been constructed previously, the two genes were easily cloned and sequenced. No similar sequences were found in gene banks, however, providing no clues as to the functions of these genes.

The story of "death genes" has much in common with that of oncogenes (Chapter 18). The answers came a couple of years later, when, following a routine interrogation of the gene banks, it was discovered that the sequence of *ced-3* was similar to that of a recently identified cysteine protease that was involved in the maturation of a lymphokine, interleukin-1.[56] The first member of the large family of caspases (cysteine proteases involved in cell death) to be characterized was *ced-3*.[57] Shortly afterward, *ced-4* was shown to be an activator of *ced-3*. Meanwhile, the overexpression of another mutated gene, *ced-9*, was found to permit the survival of extra neurons. In this case, a mammalian homolog was immediately found, in the form of *bcl-2*.[58] This gene was well known as an oncogene, favoring the survival of cancer cells. These initial analyses of the molecular mechanisms of apoptosis were rapidly complemented by a description of the cell-signaling pathways that convey death signals from "death receptors" to the caspase machinery, and of the caspase targets—this turned out to explain the characteristic features of death by apoptosis.

The molecular descriptions made possible by the *C. elegans* system gave apoptosis its full identity and established its reputation. It was now possible to imagine how cell death could be stimulated or inhibited in pathological conditions.

The existence of a program of cell death revived philosophical speculations about the complex relationship between life and death, which were revealed to be both antagonistic and complementary. This program was shown to be inscribed in the genome, at a time when the human genome project was expected to reveal "who we are." The characterization of proto-oncogenes had previously been presented as the discovery of "enemies within."[59] The discovery of cell death mechanisms in our genome was an even greater shock—it showed the presence of "death within."

For Horvitz, the characterization of the mechanisms of apoptosis revealed a "principle of biological universality," and highlighted the importance of new animal models that were well adapted to particular research projects. The same conclusions flowed from the study of the mechanisms of transcription in eukaryotes, and from the analysis of the molecular mechanisms behind the development of neurodegenerative diseases. The heuristic power of the new molecular biology, which flowed from the interchange between organisms and data banks, had brought about a profound transformation of the work of biologists.

Thirty years have now passed. Apoptosis occupies its full place in biological explanations, but it has lost some of its apparently extraordinary features. Apoptosis is just one specific form of cell death among many others, and these various forms share some of its mechanisms. Genes encoding components of the apoptotic machinery are no longer dramatically described as death genes: they turn out to have other functions in the organism. They are simply genes that have been recruited during evolution to be involved in apoptotic pathways. Like developmental genes before them (see Chapter 22), death genes are no longer seen as a well-defined and separate category of genes.

Although Horvitz's research project of describing the genes involved in apoptosis has proved highly productive, it has not yet fully attained its initial objective. The exact mechanisms that lead to a decision of cell death remain unexplained. We still need a systemic view of the different pathways involved in this decision. Contingent variations in cell networks and their complex dynamics appear to be crucial actors in this decision, but their role is not yet fully understood (see Chapter 28).

21

Protein Structure

THE WAY THAT macromolecular structures, and in particular proteins, are characterized has undergone a series of major changes since the 1960s. The methods involved have been developed by experts in a range of fields— X-ray crystallography, nuclear magnetic resonance (NMR), high-resolution electron microscopy, and molecular dynamics—and many molecular biologists do not fully understand the principles and practices hidden behind these esoteric terms. Nevertheless, these multidisciplinary approaches have their place in a history of molecular biology, for a number of reasons.

The first is historical. When the structures of transfer ribonucleic acids (tRNAs), ribosomes, and DNA and RNA polymerases were finally determined, they were seen as the culmination of molecular studies that had begun in the 1950s and 1960s. For example, the determination of the crystallographic structure of the complex between DNA and the λ phage repressor was seen as a final confirmation of the operon model.[1]

The precise identification of the molecular structures of key cellular components also enabled scientists to find evidence for some of the hypotheses of the 1950s that lay at the heart of the dogmas of molecular biology. The first such hypothesis was Francis Crick's 1957 suggestion that protein folding is a spontaneous process. This implied that the native state of a protein corresponds to what is called its free energy minimum—or, if this minimum cannot be reached for kinetic reasons, to a secondary free energy minimum—and that there was a spontaneous route to this minimum that could be reached on a physiological timescale. A second hypothesis, which might appear contradictory to the previous one, was that the same amino acid sequence could lead to different structures (this was later extended to nucleic acids). Various models of how this could occur, such as the allosteric model and the induced-fit model, were initially developed to explain how enzymes function (Chapter 14). Any result that made possible a better description and explanation of these transconformations contributed to solidifying the structure of molecular biology.

In fact, virtually any result from structural biology reinforced the central conviction that flowed from the molecular vision: any biological phenomenon can ultimately be reduced to physicochemical explanations. There was a second assumption, less clearly stated but evident in molecular biological models: macromolecular explanations will be mechanistic, and the world of biological phenomena is deterministic. (See Chapter 28 for some recent criticisms of the last statement.) The principles that guided the study of proteins for more than half a century underpinned the development of molecular biology.

▌ From Protein Structures to Macromolecular Machines and Drug Design

When molecular biology triumphed at the beginning of the 1960s, only three macromolecular structures were known: DNA (although the double helix still remained a model), myoglobin, and hemoglobin.

The exponential growth in the determination of macromolecular structures was the result of technological improvements at all steps in the process. X-ray sources became increasingly powerful, in particular with the use of synchrotron radiation, while the collection of X-ray data was automated. The path from diffraction data to three-dimensional representations of electron density and structures benefited from increases in computing power and the development of appropriate software. In 1953, Jim Watson and Francis Crick had to use slide-rules, primitive mechanical calculators, and metal templates to come up with their model. Those days were long gone.

There were also advances in the preparation of biological samples, such as the development of automated protocols to find the right conditions for protein crystallization. The most significant step forward came with the ability to produce large amounts of pure proteins, which were sometimes modified to facilitate crystallization or to solve what was known as the phase problem (a loss of information that occurs when measurements are made) by replacing natural amino acids with analogs bearing heavy atoms. In the 1980s these techniques of genetic engineering swept through science laboratories, ushering in a golden age of structural biology.

In the 1950s, it was not clear that the structures revealed by X-ray crystallography would offer an understanding of how proteins worked. Many scientists suspected that the rigid structures obtained through X-ray diffraction, constrained as they were by the crystallization process, were different from the active, native structures. These doubts were soon dispelled by a series of structural descriptions, including lysozyme, ribonuclease, chymotrypsin, and carboxypeptidase A, each of which led to the development of elaborate hypotheses to explain the catalytic power of these enzymes.[2]

By the end of the 1960s the X-ray diffraction study of protein crystals had become the fundamental heuristic tool that it remains today. The way that structural biology advanced was not related to the size of the molecules that were being studied. Despite their relatively large size, the structure of some viruses was completed relatively early, in the late 1970s, thanks to the existence of repeated subunits and of symmetries within the structure.[3] Nevertheless, one major event was the characterization of the huge structure of the ribosome in 2000, which revealed the precise role of ribosomal RNA in protein synthesis (Chapter 25). This was marked by the 2009 Nobel Prize in Chemistry awarded to Venkatraman Ramakrishnan, Thomas Steitz, and Ada Yonath.[4]

There was a slow shift in the functional categories of the proteins that were characterized during this period. In the 1960s and 1970s, most of the work was done on enzymes and soluble proteins. This was followed by DNA-protein complexes and, from the mid-1980s, by descriptions of membrane proteins. Uncovering the structure of DNA-protein complexes provided important insights into the mechanisms of transcription, which were actively studied in the 1980s (Chapter 20). They also led to the discovery of exotic new protein motifs, such as the helix-loop-helix, the leucine zipper, and the zinc-finger. At the same time, the isolation and characterization of developmental genes revealed the functions and conserved nature of these genes, particularly through the characterization of these motifs (Chapter 22).

The first structural descriptions of membrane proteins—including a bacterial photosynthetic center, the histocompatibility antigens, bacteriorhodopsin, a proton pump, and the potassium channel—had an even bigger impact.[5] Each of these structures represented a major step toward understanding functions that had previously remained mysterious, such as the precise role of the histocompatibility complex. These studies also represented technical tours de force because membrane proteins are reluctant to form crystals after their extraction from the lipid membrane bilayer. New technologies also emerged from the study of membrane proteins, such as electron diffraction for bacteriorhodopsin, a first step toward the simultaneous use of X-ray diffraction and electron microscopy to describe the largest structures.

However, early studies did not reveal any simple rules for protein structures. Every protein, with the exception of the evolutionary-related myoglobin and hemoglobin, appeared to be different. But by the mid-1970s scientists identified protein motifs, resulting from the association of elements of secondary structure in a precise order, which emerged as a way of grouping proteins in structural families and of classifying them. In 1981, Jane Richardson proposed a new way of representing proteins that made these motifs clearly visible, and it was rapidly and widely adopted (see Chapter 29). This new representation made it possible to describe the conformational changes that affect protein structure. The clearly visible motifs were stable structural elements—rigid parts of the protein machine—that moved during functioning by leverage, rotation around hinges, and so on.

In the late 1990s, the concept of the macromolecular machine became widespread (macromolecules had previously been only sporadically described as machines). The trigger for this shift was the characterization of ATPase by X-ray diffraction in 1994 and Paul Boyer's hypothesis that adenosine triphosphate (ATP) synthesis was generated by the rotation of one part of the protein around another, just as a rotor moves inside a stator.[6] This suggestion was rapidly confirmed by direct observation, in which one part of ATPase was coupled to a molecule labeled with a fluorescent probe while the other part of ATPase was attached to a microscope slide.[7] ATPase—the most important cellular enzyme involved in the production of the form of energy used by cells—worked like a three-stroke motor.

This result was welcomed by the scientific community, and in 1998 the journal *Cell* devoted a special issue to "macromolecular machines."[8] However, while "machine" was an apt description of proteasomes and chaperones, the term was less appropriate when it came to describing spliceosomes or nucleocytoplasmic transport.[9]

The interest in such nanomachines was prompted by the surprising discovery about ATPase, and it also was supported by the development of new technologies that made it possible to study the movements of single molecules and the forces that they exerted.[10] These new technologies formed part of the rapidly developing field of nanotechnology. Throughout the twentieth century, biologists had been studying the mechanisms of muscle contraction; in the final decade it became possible to study the movement of single myosin molecules.[11] Molecular motors (dynein and kinesin) that permitted the transport of proteins and vesicles along microtubules were also intensively studied. Many researchers also characterized RNA and DNA polymerases not only to measure the forces exerted during gene replication and transcription but also to describe the precise mechanisms of these processes.[12]

The fashion for nanomachines waxed and waned. The new technologies revealed machines that were not perfectly deterministic but instead were stochastic; proteins were found to be in permanent balance between different conformations. In the early years of the twenty-first century, these findings would drastically change the vision of macromolecules (as discussed later).

Advances in structural descriptions also led to the synthesis of targeted drugs that minimized the side effects of chemotherapies. Anticancer drugs that target the macromolecules (receptors, kinases) involved in oncogenic transformation are now commonplace, but the first striking evidence for the effectiveness of this strategy came with the development of drugs to target human immunodeficiency virus (HIV), in particular protease-inhibitor-based combination therapy.

These new tools to attack HIV were developed with extraordinary rapidity. The virus was isolated in 1983 and its sequence determined in 1985.[13] The sequence revealed the nature of the proteins encoded by the virus, which made it possible to synthesize them in vitro. The first viral protein to be targeted was reverse transcriptase. Azidothymidine (AZT), a known inhibitor of the activity of this enzyme, was soon shown to be effective against the virus.[14] A second potential target was viral protease, which is required to cleave the polyprotein that is the precursor of the different viral proteins. Site-directed mutagenesis showed that the HIV protease was an aspartate protease; this was confirmed by the effectiveness of pepstatin A, a well-known inhibitor of this family of proteases.[15] That same year, it was found that the HIV protease could be inhibited by peptides with sequences resembling those recognized by the viral protease but with different amino acids at the cleavage site.[16] In 1989, the structure of a complex of HIV protease and a peptide inhibitor was determined.[17] This opened the way to a wave of precisely designed synthetic HIV protease inhibitors and to their incorporation in tri-therapy for HIV.[18] In parallel with the development of these new targeted drugs, the production of monoclonal antibodies able to specifically bind and inhibit some cellular components was the second facet of this pharmacological revolution.

Protein Engineering, Protein Stability, and Protein Folding

Site-directed mutagenesis—the possibility of replacing one amino acid by another in a protein—offered a tool for determining the function of the different amino acids present in the active site. What could be understood could, in principle, be enhanced. Projects soon emerged aimed at modifying the substrate specificity of enzymes or at increasing their stability—a requirement for their use in biotechnology. Complementary studies to explain the origin of protein stability were developed with equal speed.[19] Throughout this period, scientists were not focused on the mechanisms underlying protein folding.

A good starting point for understanding the ensemble of interwoven studies addressing these issues is the work performed by Brian Matthews and his team at the University of Oregon on the bacteriophage T4 lysozyme. The initial impetus for this work came from the unexpected observation that proteins extracted from thermophilic organisms had no particular structural characteristics, nor any specific amino acid composition; they were homologous (showing a high degree of similarity) to proteins extracted from closely related nonthermophilic organisms. Explaining the stability of these molecules at high temperatures therefore required a precise study of the stabilizing bonds involved in their structure. Researchers working on phage were very familiar with the lysozyme, which is synthesized by the bacteriophage and helps it to break out of the host; isolating mutants that affected lysozyme activity was relatively straightforward.

Matthews's team identified a mutant in which the enzyme was destabilized; the mutant enzyme was crystallized and its structure compared to that of the wild-type enzyme by use of the difference Fourier map technique, which was quicker than determining the mutant structure from scratch. They found that the mutant was missing a single arginine, which had been replaced by a histidine.[20] However, this result did not explain why the mutant protein was destabilized, which led to more than a decade of attempts to explain the origin of protein stability.

Techniques gradually changed as the study of spontaneous mutations was slowly replaced by site-directed mutagenesis. Among the factors explored during this period were the roles of hydrophobic interactions, disulfide bonds, salt bridges, and hydrogen bonds, as well as the ability of acidic or basic amino acids localized at either terminus of a protein helix to stabilize the dipole generated by the formation of the helix.[21]

Some of the most striking findings were that hydrophobic interactions played a major role in protein stabilization, that disulfide bonds played a stabilizing role, and that minor changes which did not alter the global conformation of the protein could nevertheless have a considerable effect on its stability. This confirmed the existence of "structural invariants in protein folding," a term coined by Cyrus Chothia in 1975 on the basis of the fifteen crystallographic protein structures that were then available.[22] These invariants included evidence that the interior structure of proteins is hydrophobic and closely packed. Above all, Chothia's insight emphasized the power of X-ray diffraction studies.

Alan Fersht played a key role in demonstrating the opportunities offered by protein engineering and wrote the first article in the first issue of the journal *Protein Engineering,* which appeared in 1986.[23] His first contribution to the field focused on the common editing mechanisms used by DNA polymerase and aminoacyl-tRNA synthetases, which are involved in error correction during DNA replication and protein synthesis. Fersht then turned to tyrosyl-tRNA synthetase (or more precisely a long *N*-terminal fragment of it) found in *Bacillus stearothermophilus,* describing the first reaction catalyzed by this enzyme, which involves the formation of a tyrosyl-adenylate intermediate (the second step is the binding of tyrosine to tRNA). By using site-directed mutagenesis to replace the various amino acids present in the active site, Fersht showed that the activity of this enzyme was a perfect example of catalysis by stabilization of the transition state of the reaction, a hypothesis put forward by Linus Pauling decades previously.[24]

Fersht's work was not limited to catalytic mechanisms. He also demonstrated the importance of the electric charges of amino acids present on the surface of proteins and helped describe the nature of the amino acid residues and bonds that stabilize the structure of proteins.[25] The work on stabilization led him to study a new model, barnase, a ribonuclease isolated from *Bacillus amyloliquefaciens,* which was small enough for its three-dimensional structure to be solved by NMR.

In the space of a few years after 1982, the work of these researchers and others led to a clear understanding of the origins of protein stability and of the catalytic power of enzymes. This in turn led to attempts to design more stable enzymes with new specificities, capable of producing new types of reaction.

An indication of these new powers acquired by biologists was the ability to transform one enzyme into another—turning lactate dehydrogenase into malate dehydrogenase by a single mutation. But this experiment also highlighted the limits of this approach: mutation could produce radical changes, but not always in a predictable manner.[26] One of the difficulties protein engineers encountered was that many mutations had an unexpected destabilizing effect.[27] This became particularly clear when a random mutagenesis strategy was added to systematic site-directed mutagenesis.[28] It turned out that nature was subject to the same constraints: the spontaneous appearance of enzymes that conferred antibiotic resistance to bacteria was constrained by "a trade-off between stability and activity."[29]

As a consequence, the language and concepts of evolutionary biology crept into the fields of protein studies and protein engineering. Alan Fersht's team played a pioneering role in describing the conflicting evolutionary trends of protein stabilization and increased catalytic efficiency.[30] This approach typified the functional synthesis that slowly emerged from the meeting of the modern evolutionary synthesis with functional and molecular biology.[31] It had huge consequences, not only for evolutionary biology but also for our understanding of the origin of human diseases that are caused by destabilizing mutations leading to protein degradation or protein aggregation.[32]

This period was also characterized by an increased interest in the evolutionary history of proteins. A series of studies showed that the ancestors of modern proteins were probably more "promiscuous" (less specific). The idea was not new, but it now became widely accepted.[33]

A major, but ultimately limited, breakthrough was the development of catalytic antibodies (or abzymes, from antibody-enzymes) that would bind to the components of a chemical reaction and also catalyze the reaction. Their discovery flowed from Pauling's model of enzymatic catalysis and from the studies of tyrosyl-tRNA synthetase that confirmed Pauling's conception. Proteins that bind with high affinity to the transition state of a chemical reaction—or to a stable analog of this transition state—would catalyze this reaction. Antibodies could be raised against any molecule, any analog of a transition state, so in principle it became possible to make antibodies that could catalyze any reaction. All that was needed was the ability to design a stable analog of the transition state. The production of unlimited amounts of pure antibodies, a requirement for the characterization and utilization of these catalytic antibodies, became possible following the development of monoclonal antibodies by Georges J. F. Köhler and César Milstein in 1975.[34]

In 1986 Richard Lerner at the Scripps Clinic in California catalyzed the hydrolysis of an ester, using antibodies that recognized analogs that mimicked the tetrahedral transition state of the reaction.[35] Antibodies that could catalyze more complex reactions—bimolecular reactions, cyclization, or rearrangements—were soon obtained.[36]

But this approach rapidly reached its limits. The discovery of identical, convergent antibody structures suggested that there were a limited number of possible forms. The diversity of natural antibodies was limited to variation in three short loops of the variable domains of immunoglobulins. Even more problematic was the low catalytic efficiency of these antibodies, which was far below that of natural enzymes. Although it was known that the catalytic antibodies were selected for their ability to recognize the transition state analogs and not to catalyze the reaction, and that analogs often only imperfectly mimicked the transition state, this did not fully explain their low efficiency.[37] The addition of random mutations led to a degree of progress, but the main problem lay in the rigid structure of antibodies, which was incompatible with the radical rearrangements of protein structure that were associated with enzymatic catalysis.

From Rigid Proteins to Ensembles of Conformations in Equilibrium

It was well known that proteins could adopt different conformations. The allosteric model developed by Jacques Monod, Jeffries Wyman, and Jean-Pierre Changeux suggested that proteins could exist in two different conformations, and the induced-fit model explained how a protein could alter its conformation to adapt to that of its ligand.[38] However, in the allosteric model this flexibility was limited to the specific group of oligomeric proteins (formed from multiple identical subunits) and to just two rigid states. According to the induced-fit model, protein flexibility was not strictly a property of the protein itself, but rather of its interaction with a ligand. At the beginning of the century a new vision emerged in which protein flexibility found its proper place, a consequence of the evolutionary history of proteins. All this was the result of new approaches, new technologies, and new ways of thinking about protein flexibility.

The first change occurred in the field of protein folding. In the late 1960s, Cyrus Levinthal challenged the hypothesis that polypeptide chains spontaneously reach their energy minimum by a random process; he showed that the time necessary for this to occur was incompatible with physiological timescales.[39] The solution lay in the existence of folding pathways, and many experimental studies provided insight into the nature of these pathways. The formation of disulfide bonds was a good reporter of protein structure and was widely used for this reason.[40] These studies revealed that secondary structures formed rapidly, together with an initial hydrophobic collapse; there were many alternative protein folding pathways, and there was a late rate-limiting step of transconformation between what was poetically called a "molten globule"[41] and the native, more rigid, protein conformation. This last point was confirmed by protein engineering, which also shed light on the precise nature of this last transconformation.[42]

The discovery of chaperones in 1986 was initially interpreted by some as a blow to Francis Crick's 1957 sequence hypothesis that the structure of proteins was entirely encoded in the genetic sequence of DNA. The obvious function of chaperones was to play a part in the folding process of other proteins, to drive them along particular folding pathways. However, this was not true—chaperones were soon found to have a less glamorous function:

the prevention and elimination of "improper bonds" that might otherwise prevent the formation of native structures.[43]

All these studies contributed marginally to the major change that took place at the beginning of the 1990s: the emergence of a new model drawn from statistical physics, known as the energy landscape, which was supported by simulations and was based on the metaphor of the folding funnel.[44] The different denatured forms of a protein, located all around the top of the funnel, would progressively fold while sliding along its irregular sides. This new model was able to account for previous observations and was rapidly accepted.

The new vision of protein folding quickly and logically led to a new vision of protein structure. A major driver of this shift was the discovery that some native proteins were "intrinsically disordered," reaching a fully folded state only after interaction with their substrate, for example.[45] Early on, a number of researchers argued that proteins were an ensemble of structures in dynamic equilibrium, but this view was not widely accepted.[46] Part of the reason for this was because computational approaches to revealing protein dynamics by describing the movements around the bonds that form them were restricted by the limited computing power of the time. For decades, the direct measurement of these movements by NMR was similarly limited by the power of the instruments used in data acquisition and of the software that was required to process the information.

This situation began to change in the late 1990s. Advances in NMR made it possible to demonstrate the existence of an ensemble of conformational states and to reveal the dynamics of interconversion between these different states.[47] At the same time, computational structural biology was now no longer limited to determining rapid, local movements within proteins, but could reveal the slowest, global, conformational transitions involved in catalysis.[48]

As a result, a new vision of enzymatic catalysis emerged. The catalytic pathway was seen as simply the succession of different populations of conformations accessible to the enzyme.[49] The transition from one catalytic step to another could be described as a reequilibration between these different populations. Induced-fit effects were not excluded from this vision, but the internal dynamics of the protein structure were the major players.

This new approach also embraced the conformational changes associated with allosteric regulation and fitted perfectly with Monod, Wyman, and

Changeux's intuition that the different conformations were preexisting. But it was realized that the phenomenon was not limited to oligomeric proteins—every protein existed in different populations of conformations, and these were not simply limited to two, as in the classic allosteric model.[50]

Perhaps the most significant result showed that catalysis and allosteric regulation did not obey different mechanisms, but instead emerged via a common route, opened up by the dynamic characteristics of the protein.[51] The result was a new and unified vision of protein structure, function, and regulation in which evolution played a major role, gradually sculpting the network of interactions between amino acids that generate the complex dynamics of proteins.

This new vision was not limited to proteins. The behavior of RNA and DNA could also be explained by the existence of an ensemble of different populations of conformations in equilibrium.[52]

There is not the space to describe all the protein-related discoveries made in this period, such as quantum effects in catalysis or the discovery of the ability of some proteins to splice themselves.[53] The ability of prions to spread their particular conformation to other molecules of the same nature has already been described in Chapter 20. The most significant conclusion to emerge from all these studies was that the importance of proteins has not been usurped by genes or by RNA, and that the description of the ways in which they fold and exercise their functions remains central to biology today. Understanding this process requires joint work by structural chemists and evolutionary biologists.

22

The Rise of Developmental Biology

FROM THE EARLY 1970S, the term "embryology" was gradually replaced by "developmental biology." Behind this shift lay the growing influence of molecular biology, as seen through the gradual introduction of its techniques and its explanations, which were based on genes and molecular mechanisms, into the study of embryology. This chapter is not a history of developmental biology, but rather traces the ways in which our understanding of embryological development was transformed by the new molecular vision.

▌ From Bacteria and Bacteriophage to Complex Organisms

In the 1960s, some of the pioneers of molecular biology turned their efforts to the study of multicellular organisms. Seymour Benzer began his work on the genes controlling behavior in *Drosophila*.[1] Meanwhile, Sydney Brenner adopted a new animal model—the nematode *Caenorhabditis elegans.* Over subsequent decades Brenner's group and many other researchers collaborated in a complete description of what became known as "the worm" in terms of gene mutations affecting development and behavior, and indeed the whole cell lineage. They also used electron microscopy to provide a complete description of the worm's nervous system (see also Chapter 20).[2] Similarly, in the early 1970s François Jacob turned to the study of mouse development.[3] George Streisinger chose the zebrafish, *Danio rerio,* publishing his first results after nearly ten years of silence.[4] Gunther Stent studied the leech, while other molecular biologists began to study the simplest organisms that showed cellular differentiation, such as sporulating bacteria or aggregating slime molds. Others again, such as Mark Ptashne, continued to dissect the complex mechanisms controlling the alternative fates of bacteriophages after bacterial infection, in the belief that this could still offer insights into the processes involved in multicellular organisms.[5] Boldest of them all was Francis Crick, who eventually turned to the neuro-biological study of consciousness in humans. Some of the model organisms, such as the zebrafish or the worm, were radically new whereas others were simply "turned molecular," with new systems finding their place through the use of natural or artificial genetic tools that facilitated the isolation of mutants.[6]

These different strategic choices reflected two shared convictions on the part of molecular biologists. The first was that the main principles permitting organisms to use information stored in their genetic material were now understood. The second, flowing from this, was that the study of multicellular organisms and of the nervous system was the new frontier of biological knowledge.

The operon model explained how the expression of genes could be controlled by other genes and by signals coming from the environment, through the binding of regulatory proteins to DNA and the control of transcription initiation. But however valuable the operon was as a model, it did not prove that identical mechanisms operated in multicellular organisms. As Jean Brachet pointed out, any attempt to explain development through the study of organisms that do not develop (such as *Escherichia coli*) was profoundly paradoxical.[7] New molecular mechanisms for the control of gene expression probably existed in eukaryotic organisms and were waiting to be discovered.

The central issue was the extent to which these new mechanisms were like those in microorganisms. If they were significantly different, there was a danger that they would be impossible to find in simple models of cell differentiation such as sporulation. But the direct investigation of mammalian (mouse) development was a challenge, partly because of the complexity of the organism and of its development, and partly because of the lack of the appropriate molecular tools to study the processes that were involved.

Other questions were still unanswered. How seriously should scientists take the metaphor of a genetic program of development? Or to put it another way, how logical was development? It is interesting to compare the answers of two pioneer molecular biologists, Sydney Brenner and François Jacob, in the late 1960s and early 1970s. Brenner spent over a year studying computer programming in order to gain insight into development. His systematic efforts toward a complete description of nematode development were in line with his conviction that there was a logic of development that could be revealed. He suggested, following Crick and Stephen Jay Gould, that heterochronic genes controlling the rhythm of development might play an essential role—a hypothesis that turned out to be only partially correct.[8]

François Jacob, one of the fathers of the modern notion of the genetic program, adopted a much more empirical approach to the study of early development in mice. He focused on the control of intercellular interactions by studying the role that membrane proteins might play in the embryo. At the time, it was not possible to identify the regulatory genes controlling development, but there is nothing to indicate that Jacob saw this as a major obstacle. The work carried out in his laboratory flowed from the key hypotheses that had guided the work of experimental biologists since the beginning of the twentieth century, in particular the importance of cell–cell interactions and of embryonic induction, whereby one group of cells directs the development of another group.

In 1975, Dorothea Bennett proposed a model for early mammalian development, according a major role in the control of this process to a pseudoallelic complex called the T-complex.[9] This gene complex had been studied by mouse geneticists since the 1920s, and mutations in it had been shown to interrupt development at early stages. Bennett proposed that this complex, which was mistakenly considered at that time to be evolutionarily related to the major histocompatibility complex, encoded membrane proteins involved in cell–cell interactions. These interactions led to a modification of the pattern of gene expression in these cells, and in particular to the expression of other genes of the T-complex, which in turn led to new types of intercellular interactions. This mechanism was hypothesized to drive the early steps of mammalian development.

This model was supported by experiments carried out in collaboration with Jacob's laboratory. Antibodies were raised against a membrane protein or set of proteins called the F9 antigen that were expressed in undifferentiated embryonal carcinoma (EC) cells—derived from tumors of the gonads called teratocarcinomas, and similar to embryonic stem cells. The F9 antigen was also expressed in the early embryo, but not in EC cells after differentiation or at later stages of embryonic development. With this antiserum, it was discovered that the level of the F9 antigen was reduced in one mutant of the T-complex that blocked the compaction of the embryo at the morula stage, an essential step in early differentiation. Jacob did not take part in the elaboration of the model, but he supported it enthusiastically, and it reinforced his conviction that he had found the right experimental system to uncover the mechanisms of early mammalian development.

Sadly, this was not the case. The immunological findings from the T-complex mutant turned out to be artefacts. The T-complex did not exist; it was an illusion created by an absence of recombination due to a chromosomal inversion. The model evaporated, and with it the hope that the program of development was directly inscribed in the genome.[10]

Another possibility open to molecular biologists was to short-circuit the complexity of multicellular organisms by working on cell systems that retained a limited degree of differentiation ex vivo. The embryonal carcinoma cell lines derived from teratocarcinomas were one example, but other systems, such as the differentiation of precursors of muscle cells or blood cells, were also exploited at this time.

One technique was particularly promising. Cell hybridization, invented in 1960 by Georges Barski, Serge Sorieul, and Francine Cornefert at the Institut Gustave Roussy at Villejuif near Paris, made it possible to combine genetic and biochemical studies.[11] Two years later, Boris Ephrussi and Sorieul observed that hybrid cells exhibited a gradual reduction in the number of chromosomes, and Henry Harris then obtained hybrids between cells of different species.[12] Ephrussi found that after hybridization between a differentiated cell and a nondifferentiated cell, the latter had an inhibitory effect, leading to the disappearance of the specialized functions of the differentiated cells in the hybrid.[13] Intriguingly, the differentiated characters could reappear through a loss of chromosomes in the hybrid, suggesting that the undifferentiated cellular state was a repressed state, and that differentiation corresponded to a derepression.[14] This had already been suggested by Jacob and Jacques Monod in their discussions of the implication of the operon model. This observation opened up a way of characterizing the chromosomes and genes encoding repressors involved in the control of differentiation.

However, hybrid cells slowly lost their significance in the 1970s and 1980s as new experimental strategies gave researchers direct access to the genes controlling development.[15] These cell lines also fell out of favor owing to their instability and the complexity of the results they produced, which led to increasingly convoluted experiments and interpretations. In hindsight, it seems likely that the puzzling observations made with these hybrid cells were partly due to epigenetic modifications that would have been difficult to control, even if they had been recognized.

▌ Eric Davidson and the Rise of Molecular Embryology

Eric Davidson played a major role in the transformation of embryology, not as a molecular biologist who took up embryology but as an embryologist who adopted the tools and concepts of molecular biology. His aim was to cast a new light on development, without abandoning the methods of classical embryology. In this respect it is highly significant that in 1973 he chose the sea urchin as his preferred experimental system, and he remained faithful to it for over forty years. This animal had played a fundamental role in the early years of experimental embryology, in particular through Theodor Boveri's famous experiments at the Marine Zoological Station in Naples in the closing years of the nineteenth century.

Eric Davidson's book *Gene Activity in Early Development,* published in 1968, had a huge influence, especially on American embryologists.[16] Davidson placed the control of gene expression at the heart of the new embryology, describing the molecular techniques available to embryologists and highlighting the key phenomena that still needed explaining.

The molecular techniques he described were still in their infancy: the use of drugs affecting specific steps in the transfer of genetic of information from DNA to protein; radioactive labeling of newly synthesized DNA, ribonucleic acids (RNAs), and proteins; and molecular hybridization, which made it possible to distinguish different populations of RNA and DNA sequences. Each of these techniques had problems, and it was difficult to interpret the results they produced.[17]

The main result of the work done with these techniques was the demonstration that the earliest phases of development were not dependent on gene transcription: information was prestored in RNAs in the urchin egg, which were translated only after fertilization. This result was seen as a blow to the oversimplified models of molecular biology and a boost to the traditional embryological view that development was mainly controlled by the cytoplasm, not by the nucleus.

Later phases of development were correlated with important shifts in the populations of RNAs (and therefore in transcription) that were revealed by labeling and molecular hybridization. As we have seen, these experiments also revealed the existence in the nucleus of high-molecular-weight RNAs, called heterogeneous nuclear RNAs (HnRNA), only a small fraction of which were converted into messenger RNA (mRNA) (Chapter 17). Molecular hybridization studies also showed that there were many large families of repeated sequences in eukaryotic genomes.[18]

For Davidson, and for embryologists in general, these puzzling results showed that the control of development was fundamentally different from the control of physiological processes in microorganisms. Explaining these processes would therefore require the discovery of new, specific mechanisms.

In 1969, Davidson and Roy Britten (who had been the first to demonstrate the wide-scale presence of repeated sequences in the genome) proposed a model of gene regulation in multicellular organisms that was compatible with all these observations. They suggested that, in response to sensor genes, integrator genes produced activator RNAs that interacted with receptor genes to induce the expression of a battery of producer genes.[19] In a second article published two years later, Britten and Davidson proposed that mutations in integrator and / or receptor genes formed key steps in the evolution of life (Chapter 23).[20]

Britten and Davidson's model was welcomed by embryologists such as Conrad Waddington, who criticized those molecular biologists who had recently turned to embryology for reducing development to differentiation, when it is in fact the consequence of differentiation and morphogenesis. In addition, the embryologists maintained the fundamental distinction between determination and differentiation (determination being the cellular decision that precedes differentiation), which had been established by experimental embryologists in the first part of the twentieth century and had been totally neglected by molecular biologists.

The most striking phenomenon in the years between the triumph of molecular biology in the early 1960s and the isolation of the genes controlling development in the early 1980s was the range of mechanisms that were pro- posed to explain the control of differentiation and development. For ex- ample, amplification of genes coding for ribosomal RNA was found in am- phibians, and later amplification of chorion protein genes was described in *Drosophila*.[21] As a result, the amplification of genetic sequences was con- sidered to be a mechanism potentially involved in development. We have already seen how RNA splicing and its control were thought to play a major role in differentiation and development (Chapter 17). It was argued that the transfer of mRNAs from the nucleus to the cytoplasm and the inhibition of mRNA translation by associated proteins were important factors in the con- trol of gene expression. Researchers were also interested in the potential role of both RNA stability and protein stability. Protein modification by phosphorylation played a major controlling role in multicellular organisms but was presumed to be absent in bacteria.

This emphasis on hypothetical multilevel control of gene expression in differentiation and development was a consequence of the lack of solid data regarding the mechanisms of transcription in eukaryotes in the period be- tween 1960 and 1980 (Chapter 20).[22]

The Roots of Evo-Devo

The roots of evolutionary developmental biology, or Evo-Devo as it is known, can clearly be seen in the work of Jacob and Monod, going back beyond the operon model to their earlier distinction between structural and regulatory genes, the latter controlling the expression of the former.[23] A mu- tation in a regulatory gene was expected to have a more dramatic effect than a mutation in a structural gene. This created a distinction between the evolutionary role of the different genes (and mutations), which went against the spirit of Modern Synthesis, but was along similar lines to Richard Gold- schmidt's iconoclastic writings on micro- and macromutations.[24]

Monod and Jacob did not address this issue directly, although in a cryptic publication for the Pontifical Academy of Sciences they showed an aware- ness of the potentially different effects of mutations affecting the two cat- egories of genes.[25]

Allan Wilson immediately grasped the significance of this distinction. He spent years characterizing the effect of mutations in regulatory genes in bacteria, while simultaneously beginning to study mammalian and human evolution.[26] In 1975, he published a famous paper with Mary-Claire King that compared protein sequences between humans and chimpanzees, using these data to calculate the small genetic distance between them.[27] This paper is still cited today for the blow that it dealt to the assumption that humans were very different to other animals. Interestingly, this area of study, now called phylogenetics, was predicted by Francis Crick in his 1957 central dogma lecture.[28]

The real message of King and Wilson's paper was that the small genetic distance between humans and chimpanzees suggested that the huge observable differences between them must be the result of variations in a limited number of genetic sequences controlling development, and not in variations of the kind of protein-encoding "housekeeping" genes that had so far been studied. These developmental genetic sequences remained unknown, and so were not included in Wilson and King's study.

Wilson was also convinced that the genetic variations associated with human evolution, as with any other process of major morphological transformation, were not simple point mutations altering a single base at a time, but rather were genetic events of a different nature, such as chromosomal translocations.[29] His conception of evolution became more and more heterodox, and closer and closer to the ideas of Goldschmidt.

Humans were clearly not the right system in which to study the genetic mechanisms of development. *Drosophila* was far more promising. Antonio Garcia-Bellido and Gines Morata took the first major steps in revealing the genetic control of development, through their work on fly wing formation. They concluded that a decisive step was the formation of tissue compartments in which all the cells had the same developmental fate, and that this process was controlled by selector genes.[30] Influenced by the work of Jacob and Monod, they viewed selector genes as a specific class of regulatory genes operating in development. This model was well received and was publicized by Crick and Peter Lawrence.[31]

The idea that development was guided by master control genes was shared by many biologists. The evolutionary biologist Stephen Jay Gould suggested that some genes might play a major role by controlling the rate of development.[32] Previous experiments and models contributed to the emergence of the notion of the developmental gene, and to the elaboration of what William Baker described in 1978 as "a genetic framework for *Drosophila* development."[33] This new vision was proposed by Rudolf Raff and

Thomas Kaufman in their highly influential book *Embryos, Genes, and Evolution,* published in 1983.[34]

All these efforts would have been in vain had *Drosophila* molecular geneticists not decided to clone these important developmental genes using the new tools of genetic engineering. As early as 1973, David Hogness employed these tools to study the molecular mechanisms controlling the development of *Drosophila*.[35] In 1952–1954, Hogness worked with Monod at the Pasteur Institute in Paris, and more recently he had been impressed by the Britten–Davidson model. Hogness used a wide range of cutting-edge techniques to pursue his aim, creating banks of DNA fragments of the *Drosophila* genome. He employed in situ hybridization to precisely localize DNA fragments on the insect's giant chromosomes. Hogness was also the father of chromosomal walking (later known as positional cloning); this painstaking technique enabled researchers to clone a gene that was in close proximity to another gene that had already been cloned and characterized (see Chapter 27 for the use of this technique in the isolation of the genes involved in various diseases). The density of *Drosophila* genetic maps meant that the fly was well suited to chromosome walking.[36] In 1978–1979, Hogness spent a year's sabbatical in Walter Gehring's laboratory in Basel, and as a result the many techniques he had developed spread rapidly through the global *Drosophila* community.

Gehring also played an important part in the molecularization of developmental genetics.[37] In his younger days he had studied under Ernst Hadorn, a specialist in *Drosophila* development who had done extensive work on the transdifferentiation of imaginal discs in *Drosophila*. Imaginal discs are the groups of cells present in the larva that contribute to the formation of adult tissues. Hadorn had noticed that these cells were able to abruptly reorient their fate, giving rise to other cell types. This phenomenon of transdetermination obeyed rules that had been studied by theoretical biologists who were convinced that this system would provide clues to the control of development.[38] Although this did not prove to be the case, the phenomenon reinforced the idea of a tight genetic control of cell fate in development.

In 1978, Christiane Nüsslein-Volhard and Eric Wieschaus of the European Molecular Biology Laboratory (EMBL) in Heidelberg used saturation mutagenesis to begin the identification and classification of the *Drosophila* genes involved in the early steps of development, and to describe their precise roles in this process.[39] Firmly situated in the genetic framework of development that had recently been elaborated, this work was a milestone in the characterization of developmental genes and their functions. However, the EMBL viewed this project as peripheral to its principal objectives of developing complex technologies to study macromolecular structures. The laboratory did not appreciate that the rapid spread of gene-cloning technologies had rendered traditional approaches in molecular biology partially obsolete, and that new avenues were now open to the molecularization of all aspects of biology.

The final step in the emergence of Evo-Devo was the cloning of homeotic genes and their relatives, and their characterization by sequencing. Homeotic mutations had first been characterized in the late nineteenth century, and their extraordinary phenotypes—such as the presence of legs instead of antennae on the head of the *antennapedia* mutant of *Drosophila*—had captured the imagination of biologists. *Drosophila* homeotic genes were grouped in two different pseudoallelic complexes and were difficult to study genetically. Despite their striking phenotypes, the exploration of the underlying genotype had attracted few geneticists, the most famous exception being Ed Lewis.

The first developmental gene to be cloned and sequenced was not a homeotic gene but another type of selector gene, called *fushi tarazu* (*Drosophila* genes have often been given quixotic or esoteric names). Gene sequencing of homeotic genes rapidly revealed the existence of a shared sequence of 180 nucleotides, corresponding to a protein structural motif called the homeobox.[40] This is a DNA-binding domain that was shown to be evolutionarily conserved.[41] This conservation rapidly led to the cloning of homeobox-containing genes in *Xenopus* and in the mouse.[42] Comparison of these genes revealed that although the homeobox was the most conserved sequence, the homeotic genes themselves were basically the same in each of these very distantly related species, which last had a common ancestor over 500 million years ago.

These results from 1984 were doubly important. They confirmed that selector genes were regulatory genes, following the definition provided by Jacob and Monod in 1959, and they showed that developmental genes had been conserved in evolution. This was a complete surprise because the development of insects and mammals, for instance, seemed to follow very different paths and to obey different principles. Some years earlier, in 1978, Ed Lewis had published an important article summarizing the state of knowledge on the *bithorax* complex. Nothing in the article suggested that the mechanisms he described were involved in the development of organisms as different from insects as mammals.[43] In 1977, when Jacob argued that evolution did not design but instead "tinkered," he strikingly did not suggest that tinkering was involved in the genetic mechanisms of development, proposing instead that new forms of development required the acquisition of new genetic material.[44] He later acknowledged that the process of tinkering was more extensive than he had initially imagined.[45]

1984 to the Present Day: A New Molecular Vision of Development

The discovery of the homeobox, and the presence in other organisms of genes homologous to those involved in the development of *Drosophila,* led to an explosion of interest in the area. The pathway for discovering developmental genes had dramatically changed. Researchers now began in *Drosophila* and then looked for homologs in other species.

The most striking results, which were initially also the most ambiguous, were obtained with the homeotic genes, the homologs of the *bithorax* and *antennapedia* complexes of *Drosophila.* In most organisms, these two complexes were found to be grouped in one unique complex that was duplicated twice in the evolution of mammals, generating the four (incomplete) complexes found today in mice and humans. The order of the genes in the complex was also found to be conserved, as were the relationship between the position of the genes in the complex and the timing and localization of their expression in the embryo.[46] The simplest explanation was that these genes had the same functions in all organisms. Given that in insects these genes seemed tightly linked to the presence of segmentation, it was hypothesized that the homologous genes were also responsible for the partial signs of segmentation visible in mammals.

This supposed identity of the mechanisms of segmentation in insects and mammals did not hold true, however. Homeotic genes were shown to exist in nonsegmented animals such as *C. elegans.*[47] In mammals these genes have also been recruited for the formation of limbs. Their role has gradually been revised so that they are now seen as producing positional clues or providing a "code." Evolution has tinkered with them, giving them new functions, exactly as Jacob suggested for nondevelopmental characteristics.

The conservation of the organization of these gene complexes in the genome remains a puzzle. From the 1930s onward, many studies had been carried out on pseudoallelic complexes—gene complexes generated by gene duplication—with the aim of revealing the relationship between the structural organization of the genome and the functions of the genes within it. In most cases, as with the *T*-complex, these bold attempts met with little success.[48] It has gradually been accepted that the maintenance of these gene complexes is a consequence of the conservation of the regulatory elements (enhancers) that control gene expression, and of the organization of the chromatin in what are called topologically associated domains (TADs).[49] For developmental genes in general, conservation of their localization in the genome is the exception rather than the rule.

In parallel, the other developmental genes cataloged by Nüsslein-Volhard and Wieschaus, as well as other *Drosophila* genes known to be involved in development such as *notch,* were also shown to be present in a range of organisms, often with an increased copy number in mammals.[50] The general conclusion was that a small ensemble of multigene families controlled animal development.[51]

These genes were gradually organized in pathways and networks, while their expression pattern in development was revealed by in situ hybridization or by immunofluorescence. The functional role of these genes was tested in amphibians by the injection of antisense RNA or oligonucleotides, and after 1990 by the knockout strategy in mice (Chapter 24).

The characterization of developmental genes through their homology with genes in *Drosophila* (and occasionally in *C. elegans* or the zebrafish) was not the only path that was followed. Important genes involved in differentiation were also isolated through the study of ex vivo differentiating cell systems—the most spectacular experiment of this type being the isolation in 1987 by Harold Weintraub's group of *MyoD,* a master control gene of muscular differentiation.[52] Weintraub took as his starting point the observation that one cell line of embryonic fibroblasts was able to differentiate into myoblasts (dividing precursors of muscle cells) when treated with 5-azacytidine, which can be incorporated into DNA as an analog of one of the four bases, cytosine, but which unlike cytosine cannot be methylated (see Chapter 26). This suggested that cell differentiation into myoblasts involved the demethylation of a gene's DNA, and that this demethylation in some way activated the gene. Through careful experimentation a clone called *MyoD* was isolated, which induced the differentiation of fibroblasts into myoblasts. The *MyoD* gene coded for a transcription factor, and one of its functions was to directly stimulate the expression of genes coding for muscle proteins. *MyoD* is also a member of a family of genes, and its DNA binding motif is present in many other developmental genes. Along with *pax-6,* it is one of the best examples of the power of these master control genes.

It was immediately recognized that the overlap in function and sequence of the different members of a family of developmental genes—a form of partial redundancy—could potentially play a protective, buffering role. This also explained why the first knockout experiments of these master control genes in mice often produced disappointing results—only when two (or more) members of the same gene family were inactivated did important developmental effects appear. A good example of this was the simultaneous inactivation of *MyoD* and *Myf5* (another member of the *MyoD* family), which prevented the formation of skeletal muscles, whereas mutations in each of the genes alone had limited effects.[53]

Many of these genes were transcription factors, activating the expression of other genes in a tissue-specific manner. But other developmental genes encode signaling molecules, membrane receptors, and other components of the signaling pathways.

Characterization of the upstream sequences of genes that are differentially expressed during development also led to the identification of the elements that controlled their expression (promoters and enhancers), and of the transcription factors binding to them. This often led to an interaction—either by chance or by design—with studies directly targeted at developmental genes. The traditional cellular description of development was thus gradually enriched by a description of the developmental genes that are successively expressed in cells. The distinction between cell populations was increasingly considered to depend on their pattern of gene expression, rather than on morphology.

The results obtained by Walter Gehring's team on the *eyeless* gene in *Drosophila* in 1995 also showed the power of developmental genes. The sequencing of the *Drosophila eyeless* gene revealed it to be homologous to the *small eye* or *Pax-6* gene in mice, which, when mutated, affects the formation of the eye.[54] This gave Gehring the bold idea of testing the consequences of overexpressing this mouse gene in *Drosophila*. The remarkable result was the same as that previously observed with the overexpression of the *Drosophila* gene: the formation of ectopic *Drosophila* eyes on the insect's legs, wings, and other areas.[55] Flies and mice have been separated for over half a billion years, and their eyes have radically different morphologies; yet shared, conserved genetic motifs tell each form of body to produce the appropriate type of eye.

At the beginning of the 1990s, the same techniques were applied to plants, and the teams of Hans Sommer, Enrico Coen, and Elliott Meyerowitz isolated and characterized the genes involved in particular in the development of flowers.[56] They showed that these genes encoded transcription factors with a DNA-binding motif similar to the homeobox, but not homologous to it. These experiments involved transposon mutagenesis, which made possible the rapid isolation of the mutated genes. As early as 1991, a genetic model of flower development was proposed.[57] A few years earlier, *Arabidopsis thaliana* had been adopted as a model for studies of plant genetics, which facilitated the rapid expansion of genetic investigations of plant development.[58]

At first these molecular studies seemed to support established models of development. In the past, embryologists such as Charles Manning Child, Paul Weiss, and even pioneer computational biologists such as Alan Turing, had proposed the existence of gradients of morphogens that guided development. In 1971, Lewis Wolpert proposed a general model for the generation of positional information.[59] Also, Christiane Nüsslein-Volhard published a wonderful demonstration of the in vivo existence and function of such morphogens, using the example of the bicoid protein in the early *Drosophila* embryo.[60]

But the marriage was not perfect. Distinctions between differentiation and determination, much loved by embryologists, found no support at the molecular level—for example, the *MyoD* master control gene is active in both determination and differentiation. And despite their intuitive attractiveness, only a few examples of morphogenetic gradients have been described at the molecular level, such as capentaplegic (dpp) and Sonic Hedgehog (Shh).

One famous example, known as the organizer, a structure of the early embryo able to organize its development, offers a good illustration of these difficulties. Attempts to chemically characterize the organizer were made in the 1930s (Chapter 8), but this surprisingly led to a dead end. In 1991, it was discovered that the *Xenopus* homeobox gene *goosecoid* (which shares its DNA-binding specificity with the *Drosophila* gene *bicoid*) was expressed in the dorsal lip of the blastopore (the precise localization of the organizer). Injecting *goosecoid* mRNA mimicked the results obtained by Hans Spemann and Hilde Mangold's transplantation experiments in 1924.[61] Similar results were obtained in mice.[62]

Unfortunately, this simple model was rapidly obscured by the involvement of many other genes in this process, which function as a system. In fact, embryologists had already shown that the phenomenon was less simple than had initially been thought in 1924, and understood that the mechanisms in mammals and amphibians were not identical.

No one today would dispute the importance of these master control genes in development. Their discovery opened up a new era in developmental biology.[63] What has probably changed since those early observations is the power attributed to each gene. Instead of focusing on individual genes, researchers now recognize the significance of the system or network formed by the ensemble of these genes. Genes act within cells, giving them the specific properties that allow those cells to participate in the process of morphogenesis. In addition, developmental genes are not members of a select club: many of them remain functional at the adult stage. To the extent that scientists still argue about the functions of these genes, that debate now focuses on their role in evolution. That is the subject of the next chapter.

23

Molecular Biology and Evolution

DURING THE YEARS when the molecular vision of life was emerging, it might have seemed to the outside observer that the relationship between molecular biology and evolutionary biology was plain sailing. Both the evolutionary biologists who adopted the Modern Synthesis and the new band of molecular biologists were convinced that all organisms shared common mechanisms that could explain their properties and their evolution. Darwinism was central to the thinking of Max Delbrück, one of the leaders of the phage group which was so active in the development of molecular biology, while Theodosius Dobzhansky, one of the founders of the Modern Synthesis, paid considerable attention to the early results of molecular biology, highlighting Oswald Avery's discovery even before it was published.

Perhaps surprisingly, the fundamental evolutionary process of variation and selection played a huge part in molecular biologists' day-to-day experiments on bacteria and bacteriophages; indeed, one of the earliest and most significant contributions of Delbrück and Salvador Luria was to show that the raw material of evolution by natural selection—mutations—arose spontaneously in bacteria, irrespective of the environmental conditions.[1] The chief strategy employed by the bacterial and phage geneticists was to explore a particular biological mechanism by creating or isolating mutants that affected the process being studied. This approach was also to prove extremely influential later in the century as researchers turned their attention to more complex, multicellular model systems. The pioneer molecular biologists had no doubt that they would be able to obtain the mutants they needed, whatever the genetic properties they were seeking. Implicitly, they fully acknowledged the creative power of the process of variation and selection, a central aspect of the Modern Synthesis.

In reality, relations between molecular biologists and evolutionary biologists were not quite so amicable. There were repeated disputes, sometimes prompted by professional jealousies as well as conceptual differences. The first signs of disagreement appeared at the beginning of the 1960s, when Ernst Mayr and George Simpson openly opposed the growing power of molecular biologists in universities; this was accompanied by an emphatic distinction between two different forms of biology, and between two different types of questions raised by biologists, which was first put forward by Ernst Mayr.[2] Functional biologists, including molecular biologists, asked "how" questions, Mayr argued, whereas evolutionary biologists asked "why" questions. By arguing that evolutionary biology and molecular biology were complementary, Mayr was seeking to defend evolutionary biology against the growing influence of its new rival.

Rather than outlining the transformations of evolutionary biology over the last fifty years, this chapter describes the gradual dovetailing of molecular biology and evolutionary biology, and the difficulties that scientists encountered along the way.

The Heart of the Matter

Mayr took the distinction between functional and evolutionary biology to extremes, considering that the only true biological discipline, the only one that put organisms at the center of its work, was evolutionary biology. Molecular biology, he argued, was a branch of chemistry (and physics) that reduced biological phenomena to chemical and physical explanations. Even the notion of a genetic program, introduced simultaneously in 1961 by François Jacob and Jacques Monod and by Mayr, did not have precisely the same meaning for molecular biologists and evolutionary biologists.[3] Molecular biologists viewed a program as the totality of genetic instructions that control the building and functioning of an organism; for Mayr and the evolutionary biologists it implied that organisms contained, at their center, a legacy of their evolutionary history.

Most molecular biologists were unfamiliar with the Modern Synthesis, and were even less well versed in the complex transformations it had undergone since World War II—mainly because most of them had been trained in other disciplines. They did not appreciate that many evolutionary biologists viewed natural selection as being so powerful that organisms would always find different strategies to adapt to a new environment; as a result, evolutionary biologists were not particularly interested in the details of the precise mechanisms involved in any given adaptation. They considered that adaptation was explained not by the underlying mechanistic transformations but by the nature of the selection pressure that produced the adaptation. For a functional biologist, the precise nature of variations and of the way they modify organisms forms part of the explanation of their evolutionary history.

Ever since Charles Darwin, biologists have repeatedly argued about the relative weights of variation and selection. The Modern Synthesis put the emphasis on natural selection, but molecular biologists, who are primarily interested in describing mechanisms, are naturally focused on the nature of variations.

The apparent opposition between these two views may seem oversimplified—in reality, the nature of variations and the amplitude and direction of selection pressure both contribute to a full explanation of the evolutionary process. Nevertheless, the existence of these contrasting points of view is key to understanding some of the debates that took place during this period. With each new, unexpected, and bizarre mechanism of genetic variation that has been discovered by molecular biologists, these same arguments have emerged afresh, again and again. The list of such oddities is long, including reverse transcription, transposition, horizontal gene transfer, and many more. Every time, the same questions crop up, focusing on the place of the phenomenon in the evolutionary process and raising the enticing possibility that the novel mechanism provides the key to understanding major unexplained issues in evolution.

▌ 1960–1980: A Difficult Meeting of Minds

Ever since the work of Ernst Haeckel in the 1870s, evolutionary biologists have sought to measure the evolutionary distance between organisms and to draw evolutionary trees. In 1965, Emil Zuckerkandl and Linus Pauling proposed the use of protein sequences for this purpose.[4] The idea had first been put forward in 1958 by Francis Crick, in his brilliant lecture "On Protein Synthesis," in which he envisaged that the future study of what he called protein taxonomy would exploit the "vast amounts of evolutionary information" that he predicted would be found in protein sequences from different species.[5] However, Crick had only a handful of sequences to point to and made no attempt to exploit the idea. Zuckerkandl and Pauling's more detailed proposal, a mere seven years later, had a huge impact and led directly to thousands of evolutionary studies that have variously used information contained in protein sequences, immunological cross-reactions, DNA hybridization, and, most recently, nucleotide sequences. This was both a continuation and a development of the relatively primitive comparative work carried out in previous decades by biochemists. Underlying all these studies was a common principle: from a quantitative estimate of the differences between two species it was possible to deduce an evolutionary distance. Comparisons with observations of evolutionary relationships that had been established by studying fossils soon revealed that the degree of divergence between two species at the molecular level was proportional to the time elapsed since they diverged. The evolutionary "molecular clock"— an expression coined by Zuckerkandl and Pauling—was shown to tick at an approximately constant rate.

This approach was criticized by evolutionary biologists, with Ernst Mayr and George Simpson leading the charge.[6] Their initial critique focused on the nature of the characters studied by molecular evolutionists. These molecular characters, they argued, had not been shaped by natural selection and were therefore not important for evolution. The debate ignited when some molecular biologists argued that the regularity of the molecular clock showed that evolution was due mainly to spontaneous variations that were not sieved by natural selection.[7] If natural selection played a major part, they argued, it would give the evolutionary clock an irregular tick.

This debate was linked to another discussion among evolutionary biologists at the time, relating to the importance of heterozygosity—the existence of different allelic forms of the same gene. Those who gave natural selection a major role were inclined to consider that heterozygosity was rare. Richard Lewontin not only argued that heterozygosity could be favored by natural selection, above all he used electrophoresis to show that it was present in natural populations. The debate had huge repercussions for evolutionary theory, as it helped lead to the emergence of the "neutral theory" of Motoo Kimura, in which Kimura argued that most mutations were selectively neutral—for instance, because they occurred in noncoding regions of the genome or had no effect on which amino acid was produced.[8] Whether such mutations were conserved or not was due to stochastic events, not selection, which could not "see" them. Taxonomists were soon drawn into the debate, attracted by the new possibilities opened up by the use of protein sequences; these molecular characters were particularly appropriate for the new principles of cladistic classification developed by Willi Hennig.[9]

Molecular phylogenies occupied an increasingly important place in evolutionary biology. Key problems, such as the occasionally irregular ticking of the molecular clock, and technical issues such as long branch attraction (a mathematical phenomenon that can lead distant taxa to group together when an evolutionary tree is calculated, thereby giving a false result), were slowly recognized, and attempts were made to solve them by increasingly sophisticated models.

An early discovery, made by the American microbiologist Carl Woese through the comparison of 16S ribosomal ribonucleic acid (RNA) sequences, was the existence of two different groups of prokaryotes: bacteria and archaea.[10] This unexpected result demonstrated the power of molecular techniques, contributed to a renewed interest in the microbial world, and prompted a debate that still continues with regard to the relations between the three branches of life—bacteria, archaea, and eukaryotes—and the Last Universal Common Ancestor (LUCA).

Interpreting molecular phylogenies soon became more complicated, in that horizontal (also called lateral) gene transfer can take place between some organisms, mainly through the action of viruses. The phenomenon is quite widespread in prokaryotes and was found in eukaryotes in the early 1980s, but it is rare in animals. It has since been revealed to have been involved in some adaptations, but many other adaptations that were initially proposed to be due to "viruses" turned out to have other explanations. Because horizontal gene transfer is very infrequent in multicellular organisms, it does not undermine the phylogenetic trees that are drawn up; unusual links between species due to this unusual form of genetic innovation can usually be explained (and ignored). The problem is much more serious when it comes to prokaryotes, where horizontal gene transfer seems to have been much more frequent, in particular in the early history of life. This effect blurs the lower branches of the tree of life, prompting an intense debate between those who try desperately to integrate these occurrences into their trees by adopting distinctly un-tree-like diagrams, and those who have given up trying to depict any tree at all because they have doubts about its existence or its significance.[11]

A great exception to this uncertainty is the fact that eukaryotic organisms acquired their mitochondria when one prokaryote ended up living inside another, creating a weird hybrid that eventually gave rise to all multicellular life. This outrageous idea, which was reintroduced by Lynn Margulis in the 1960s after having been initially proposed at the beginning of the twentieth century, is now accepted.[12] Over the billions of years since, some of the genes from the ancestor of the mitochondrion have been integrated into the nuclear genome.

During the 1970s there was a strong reprisal of the debates on the role of natural selection in evolution when Niles Eldredge and Stephen Jay Gould proposed that evolution did not take place gradually but instead proceeded at an irregular pace, with long periods of stasis interrupted by rapid changes—what they called punctuated equilibrium.[13] Gould and Lewontin also produced an article critiquing what they called "the pan-adaptationist program" of evolutionary biologists.[14] They argued that there were important limits to the power of natural selection, produced by developmental constraints within organisms. Some structures were not adaptations but were simply the only way of building the organism. At the same time, the paleontologist Pere Alberch put forward a similar argument.[15] These men were all convinced that the rhythm of evolution was irregular, and used Richard Goldschmidt's distinction between micro- and macromutations to explain the origin of this irregularity.[16]

As Darwin was well aware, the irregularity of the fossil record could have many explanations; new discoveries in the 1970s and 1980s confused matters even further. Gould was not only critical of the suggestion that evolution took place at a constant rate, he also argued that natural selection found its limits in catastrophic stochastic events that interfered with the evolution of organisms, such as meteorite strikes that eliminated a large proportion of all organisms, creating major discontinuities in the fossil record. Evidence for precisely such an event came in 1979, when the father and son team Luis and Walter Alvarez of the University of California proposed that the impact of an asteroid was responsible for the extinction of dinosaurs 66 million years ago.[17] (We now recognize that this event eliminated the nonavian dinosaurs, along with most other large reptiles and many marine organisms, as well as many other groups.)

These attacks on mainstream evolutionary biology came not from molecular biologists but from evolutionary biologists and paleontologists. Strikingly, however, Gould sought support for his ideas in newly discovered molecular mechanisms, emphasizing the possibility that mutations in the genes controlling the rhythm of development might produce significant changes extremely rapidly—a hypothesis that has not been confirmed. Several decades later, with the essential points of both sides now integrated into modern evolutionary thinking, these debates look rather odd.

As mentioned in the introduction to this chapter, molecular biologists were particularly interested in the nature of variations within and between species. There were two aspects to this focus: the functional nature of the variations (for instance, the functions of the genes that changed during a given evolutionary process) and the physical nature of the genetic variation that occurred (whether it was a point mutation, a gene duplication, or some other more drastic modification of the genome). However, discovering exactly what was taking place at a molecular level was hindered by the lack of molecular tools; as a result, there was no major experimental breakthrough at this time. For example, the evolutionary biologist Allan Wilson emphasized the role of regulatory variations and chromosomal rearrangements in producing evolutionary change, but had no strong experimental evidence to back up his views (see Chapter 22).

Molecular visions of genetic change have already been discussed in the chapter on developmental biology. However, the evolutionary implications of these discoveries were equally significant: the Jacob-Monod model of 1961 and the model by Roy Britten and Eric Davidson of 1969 opened the door to the hypothesis that regulatory mutations, whatever their precise nature, might play a major part in evolution. This hypothesis, which accompanied the rapid development of evolutionary developmental biology (Evo-Devo) in the 1980s, was initially developed by Wilson, Antonio Garcia-Bellido, and Gould, Rudolf Raff, and Thomas Kaufman. This emphasis on regulatory changes during evolution was also the reason why some molecular biologists were more open to Gould's criticisms than were the evolutionary biologists. Regulatory mutations suggested that there could be leaps in evolution, a view that seemed to support Gould's theory of punctuated equilibrium.

A second, less visible, molecular contribution to evolutionary biology was François Jacob's emphasis on the role of tinkering in evolution.[18] Crystallins in the vertebrate eye became a key example of this evolutionary *bricolage*. As they were slowly characterized by Joram Piatigorsky's research group, it was discovered that over evolutionary time various preexisting proteins and enzymes had been co-opted to ensure the continuing transparency of the lens, their expression pattern being modified.[19] The "tinkering" model supported the hypothesis that most evolutionary adaptations were associated with a change in gene regulation, not with structural changes or the addition of new genetic material. This emphasis on the importance of gene regulation was amply confirmed by the analysis of genome sequences, in particular by the human genome in the late 1990s, which turned out to contain surprisingly few protein-encoding genes.

Alongside the functional aspects of genetic variation, molecular biologists explored the physical nature of that variation and its causes and came to attribute a major evolutionary role to some of these mechanisms that underlie genetic variation. Gene duplication provides an excellent example. Following duplication, subsequent mutation could either lead to the appearance of a new function for one of the two copies, or to the accumulation of deleterious mutations that eventually led to one of the copies being nonfunctional, the other continuing to carry out the required function in the cell. As early as 1936 this process had been proposed to explain the formation of pseudoallelic gene complexes, including *bithorax* and *antennapedia,* which have major effects on body organization in *Drosophila*.[20] In 1970, Susumu Ohno lent a new theoretical impetus to the hypothesis.[21] But the key breakthrough came in the 1980s with the characterization of multigene families of developmental genes, much like *bithorax* and *antennapedia*. This provided strong support for the evolutionary role of gene duplication. The more global phenomena of chromosomal and genomic duplications were later added to the general category of gene duplication.

The 1970 discovery of reverse transcriptase, which can copy RNA into DNA and insert that DNA into the genome of an organism, added a new mechanism to explain how genomes change over evolutionary time. This mechanism was proposed by Howard Martin Temin in 1971, shortly after his discovery of reverse transcriptase.[22] However, in the years that followed he was unable to obtain clear experimental evidence for the presence of reverse transcriptase activity in normal cells (Chapter 15). The hypothesis surfaced once again at the end of the 1970s, with the discovery of pseudogenes—nonfunctional stretches of DNA that were nevertheless clearly related to known genes. Some eukaryotic pseudogenes were found to be devoid of introns, and probably owed their origin to the mistaken retrotranscription of messenger RNA (mRNA) into DNA, with the resultant DNA then being inserted into the genome as a new, intronless copy of the gene.[23]

The fact that these pseudogenes were generally nonfunctional undermined the suggestion that they had any real evolutionary significance (Chapter 17). Nonetheless, in 1988 John Cairns suggested that the existence of retrotranscription might be adaptive in bacteria.[24] He argued that mRNAs useful for the survival of bacteria were retrotranscribed and the DNA copy inserted into the genome, leading to an increase in the number of these mRNAs. After a long debate and many experiments, it was finally concluded that other, less heterodox mechanisms might explain the observations.[25]

A third process of genome variation that was presumed to play a role in evolution was the possibility of genetic elements moving from one place to another in the genome. The existence of mobile elements, called transposons, was first described in maize by Barbara McClintock in 1949.[26] They were confirmed in bacteria in 1976 by a study of the mechanisms of antibiotic resistance.[27] They also were described in eukaryotes, first in *Drosophila* with the *copia* elements, and later in vertebrates.

McClintock hypothesized that transposition was a mechanism regulating gene activity during development.[28] In 1959, Elie Wollman and François Jacob proposed a similar role for episomes—small genetic elements found within bacteria—which was subsequently abandoned in favor of stable regulation inscribed in the genome (the operon model).[29] Although John Gurdon's cloning experiments on amphibians tended to undermine the suggestion that transposition played a vital role, this hypothesis was not completely abandoned.[30] It was strengthened by the discovery that the genes involved in the production of antibodies have undergone repeated rounds of reorganization.

TRANSPOSONS AND RETROTRANSPOSONS, which move inside the genome through an RNA intermediary, were only one minor factor in the major conundrum created by the discovery of vast stretches of noncoding DNA sequences in the genome.[31] In 1971—long before the first sequencing of any genome—Charles Thomas described what he called "the C-value paradox." This is the fact that the size of a genome is not related to the apparent complexity of the organism that harbors it; for example, we now know that the genome of the onion is about five times the size of that of a human.[32] More puzzling—because measuring complexity is difficult—was the discovery that similar species that were very close to each other on the evolutionary tree could have genomes of very different sizes. At around the same time, DNA hybridization experiments revealed that there were large numbers of repeated sequences in the genome, some of which were later shown to be transposons or retrotransposons.[33]

A major change that occurred in 1977 was the completely unexpected discovery that eukaryotic genes are "in pieces"—they are mosaics of exons and introns (Chapter 17). Both kinds of sequences are transcribed, but only the exons are expressed (hence their name). The intronic sequences are spliced out after transcription and are absent from the mature mRNAs used for protein synthesis. In most genes, introns are far larger than exons.

The initial response to this discovery was to propose that these apparently useless intronic DNA sequences must have some kind of regulatory function. It was suggested that splicing of RNAs was an important regulatory step in the flow of information from DNA to RNA. Britten and Davidson had already proposed that the repeated sequences play a major role in the control of gene expression and in changes in gene expression over evolutionary time.[34]

Another hypothesis, which was regularly proposed, claimed that these useless sequences were a reservoir for evolution, as they could be recruited for new functions. Molecular biologists considered that this hypothesis was plausible, but evolutionary biologists disagreed—Darwinism has no place for future utility, as natural selection is indifferent to the future and merely sieves characters with differing fitness in the present. From this point of view, a huge amount of useless DNA is obviously not adaptive.

In the 1950s and 1960s, when researchers first began to suspect that not all DNA was involved in coding for proteins, they coined the term "junk DNA." This was given a formal definition by Susumu Ohno in 1972: junk DNA can be deleted without affecting the fitness of the organism.[35] The neutral theory of molecular evolution provided an explanation for the existence of this stuff—if the possession of junk DNA is not too onerous for an organism and has no negative effect on fitness, it can escape the purifying action of natural selection. The discovery of inactive pseudogenes at the end of the 1970s was another finding that proved that apparently useless genetic elements can be conserved in the genome.

The role of junk DNA was explored further in two papers published in *Nature* in 1980 by Ford Doolittle and Carmen Sapienza and by Leslie Orgel and Francis Crick.[36] These authors took a rather different view, focusing not on the implications for the organism, but rather on the meaning for the DNA molecule. In the gene-centric vision of George Williams and Bill Hamilton, as popularized by Richard Dawkins, all DNA sequences were now seen as ultimately selfish, propagating for their own sake.[37] What better example of selfish DNA than a stretch that had no function?

Most evolutionary biologists considered that this new model closed the debate over the role of junk DNA. Michael Lynch later showed that there was an inverse relation between the size of the genomes and the size of the populations: genetic drift is more active in small populations, which explained the retention of selfish DNA in the genome despite a certain loss of fitness.[38]

Not all molecular biologists gave up hope of finding a role for junk DNA. In 2012 the ENCODE consortium announced that 80 percent of the genome was functional (that is, it had some biochemical activity, such as being transcribed into RNA or harboring sites for transcription factors); previously most scientists had accepted that 90 percent of the genome was junk DNA.[39] This claim prompted a heated debate—as much due to the ensuing media attention as to the surprisingly high figure—that is still not fully resolved. Whatever turns out to be result of the current dispute, this development showed that not all molecular biologists had been fully convinced by earlier evolutionary arguments.

▌ The Rise of Evo-Devo (1984–2017)

I have described the accumulation of unexpected observations in the period from 1960 to 1980, which often ignited confused debates between molecular biologists and evolutionary biologists. This situation changed at the beginning of the 1980s. With the discovery that developmental genes show strong evolutionary conservation, many biologists focused their attention on these genes and their variations. This was an essential step if evolution was to be fully understood at the molecular level.

The suggestion that studying only one specific type of gene could provide significant insights into evolutionary processes was at odds with the approach of the Modern Synthesis. However, developmental biologists felt that the time was ripe to integrate developmental mechanisms into modern views of evolution. For both good and bad reasons, these mechanisms were not included in the original conception of the Modern Synthesis. These developmental genes, it was now claimed, were a key part of the genetic material with which evolution had tinkered.

The question of whether the driving force of evolution is variation or natural selection reemerged soon after the discovery in the mid-1980s of the main master genes that control development. There was a link with another major scientific observation made just before this: a reevaluation of the fossil animals found in the Burgess Shale in Canada, which revealed the richness and diversity of the animal forms that appeared with great rapidity during the Cambrian explosion, between 540 and 520 million years ago. The richness of this fauna contrasted with the paucity and enigmatic nature of fossils from the earlier Ediacaran period.

It was not a great leap of imagination to suggest that this explosion of animal forms was due to the appearance of developmental genes. However, phylogenetic trees of developmental genes did not support this simplistic scenario, revealing instead that developmental genes appeared long before the Cambrian explosion.[40] The explanation for this explosion is now being sought in environmental changes (in particular the availability of oxygen and trace elements) and ecological factors (such as the disappearance of earlier species and the action of burrowing organisms in the seabed creating new environments) that could have offered organisms new ecological niches in which to expand and evolve.

In 2009 Neil Shubin, Cliff Tabin, and Sean Carroll put forward the notion of deep homology, which implied that morphological innovations arose by mutations in preexisting genetic regulatory circuits.[41] Carroll went further, claiming that these mutations were specifically located in the upstream sequences that control the expression of the genes belonging to these regulatory circuits.[42] This view suggests that morphological evolution has a specific underlying mechanism, different from those processes producing adaptations in bacteria and in other organisms. This clear distinction between two different mechanisms of evolution, generating two different types of evolution, would clearly be at odds with the foundations of the Modern Synthesis, but as ever in science the data determine the theoretical framework, not the other way around. For the moment there is no unambiguous evidence of such effects, and one example put forward by Carroll has turned out to have a simpler explanation.[43]

In a similar way, Eric Davidson has proposed that evolution is molded by variations in networks of developmental genes, in what he has called a gene regulatory network (GRN). Davidson's view is the most ambitious schema to have been prompted by the discovery of developmental genes. He was one of the most active advocates of the sequencing of the sea urchin genome in the early 2000s; he described the precise structure of the highly complex promoter of the *endo-16* developmental gene, and demonstrated that its expression level could be determined from the combination of transcription factors that bound to it.[44] The next step was to describe the gene regulatory network responsible for the formation of the endomesoderm in sea urchins.[45] More recently, the network has been successfully modeled, and this model can fully account for the observed developmental steps in this species.[46]

There is an evolutionary aspect to this work. For Davidson, the GRN can be seen as being composed of different subsystems, some of which are peripheral and others—called kernels—are more central. Kernels are argued to be extremely stable; mutations in their components are rare, but they may have dramatic effects. In a speculative model proposed by Davidson and Douglas Erwin, the nature of the subcircuit modified by a mutation determines the evolutionary consequence of this mutation. A mutation in a kernel subcircuit may lead to a new phylum, whereas a mutation in groups of differentiation genes could lead to the formation of a new species.[47] Davidson and Erwin argue that because the existence of sudden, major morphological change is generally considered to have deleterious effects on fitness, such individuals may survive in small populations, in the absence of major selection pressures.

These new variants correspond exactly to the "hopeful monsters" conjured up by Richard Goldschmidt in the 1930s, which were rejected by mainstream evolutionary biology. Indeed, twenty years before the kernel model was proposed, Erwin had published an article with "hopeful monsters" in the title.[48] This article contained similar ideas, but the mechanism involved—the migration of transposons—was different.

Davidson and Erwin were not alone in putting forward this kind of schema based on rapid morphological change induced by developmental gene mutations—Denis Duboule has proposed a similar role for the *Hox* genes.[49] The fact that different genetic mechanisms—transposons, GRN mutations, *Hox* genes mutations, and others—have all been put forward as ways of powering rapid morphological change shows that there is a common desire among some molecular biologists to give their discipline a central role in answering the most fundamental questions in evolutionary biology.

Whatever the eventual fate of these hypotheses—and for the moment they remain hypotheses—one obvious strength of Evo-Devo is the unprecedented possibility it offers for testing evolutionary scenarios. If a mutation in a component of a gene network is thought to have been involved in a morphological transformation that can be observed in evolution, then it is possible to test the hypothesis by introducing that mutation into a closely related species and studying its consequences. A good example of this can be seen in attempts to understand the changes in beak morphology that Darwin observed in the Galapagos finches in terms of a modification of the calmodulin pathway.[50] More recent genomic studies of Darwin's finches have implicated *ALX1,* a gene that encodes a transcription factor that affects craniofacial development.[51] This suggests that the story may be more complex, with natural selection operating on whole suites of genes and not simply on one or two master genes.

▌ Human Evolution

The study of human evolution provides a striking example of the different perspectives of molecular biologists and evolutionary biologists, and of how they have slowly become dovetailed.

In 1975, Mary-Claire King and Allan Wilson showed that although there are huge morphological, behavioral, and physiological differences between humans and chimpanzees, there are relatively few differences in structural genes; the huge differences were supposed to result from a limited number of mutations in regulatory genes (Chapter 22).[52] Expectations that human genome sequencing would reveal the genetic basis of human nature were rapidly replaced by the idea that understanding what makes us human would emerge from a comparison between the genomes of humans and chimpanzees.[53] Nevertheless, the sequencing of the human genome was not fruitless in this respect; it finally laid to rest the widespread assumption that human psychological and social complexity was reflected in a large number of genes.

Before these genome-wide comparisons appeared, a series of studies appeared exploring variation in candidate genes—genes in which mutations might have had a dramatic role in the emergence of modern humans. The best example of these candidate genes was *FoxP2,* instantly and unfortunately dubbed the "language gene."[54] This gene encodes a regulatory protein controlling the expression of an ensemble of genes in the brain and is mutated in some families that suffer from language disorders. The gene is highly conserved in mammals, except during the evolution of modern humans where two mutations have modified the properties of the regulatory protein it encodes. More recently, then-fashionable microRNAs were proposed as possible candidates explaining our origin and uniqueness.[55] In each case, the story was the same: after an initial wave of enthusiasm, the status of the candidate genes was downgraded from being "main actors in human evolution" to mere participants (among many others) in what was increasingly understood as a highly complex process.

Disappointingly, the comparison of the human and chimpanzee genomes did not provide any immediate insight into human evolution. The small genetic distance between humans and chimpanzees turned out to involve a huge number of genetic variations, and it was (and still is) difficult to determine which of these are significant. When researchers tried grouping the genes into functional categories in order to make some sense of the differences, they made an unexpected discovery: the genes that varied most between humans and chimpanzees were those involved in metabolism and the immune system, neither of which had generally been imagined to be key differences between our species.

The hope—or illusion—of being able to identify a single gene of major effect that is unique to humans has not gone away, but the trend is clearly toward viewing human evolution as a complex genetic process. While regulatory mutations in master control genes that guide brain development may have been involved in the evolution of our unique abilities, it is generally accepted that many other mutations would have been required to give human beings their current characteristics.

In the last decade, the study of human evolution has been transformed by our astonishing ability to genetically characterize long-dead humans and our close relatives from the ancient DNA still present in their fossil bones. In 2010, the complete sequence of the genome of a Neanderthal was published.[56] This tour de force was made possible by the development of high-throughput techniques of DNA sequencing in the mid-2000s.[57] It was also the result of over thirty years of dogged and inspired work by the lead researcher, Svante Pääbo. After a difficult start, Pääbo's group was able to extract and amplify the rare traces of degraded ancient DNA, to take into account the effects of the degradation process and to eliminate contamination from modern sources.[58] Even more astonishingly, these techniques were used to reveal the existence of a previously unknown and unsuspected human relative, still known only by the name of the cave where a tooth and a tiny finger bone were found: Denisova Cave in Siberia.[59] The Denisovan genome, along with that of several Neanderthals, has revealed that there were a series of genetic exchanges between *Homo sapiens* and these two populations of hominins and other, unknown forms, the traces of which can be seen in the genomes of modern humans. The possibility that humans and Neanderthals interbred had been raised in the past but could not be resolved by the study of skeletons, artifacts, and habitats. One particularly remarkable discovery in a period that has become a new golden age of human evolution, thanks to the use of ancient DNA, was the observation that the allelic form of a gene that enables modern Tibetans to live at altitude was acquired by mating with the Denisovans, who also possessed this form.[60] It is possible to interpret genomic or protein data in terms of morphological, physiological, or even sensory function in a way that is simply impossible with fossils. This can reveal hitherto unimagined relations between extinct species, and even enable us to peer back into the sensory world of these long-dead individuals—for example, by investigating the sense of smell in Neanderthals and Denisovans.[61]

These observations have also shown that human evolution looks more like a tangled bush than a neat tree of well-separated lineages, a vision of evolution first proposed by the paleontologist George Gaylord Simpson in the 1940s.[62] It has been made richer and more complex through the power of molecular biology.

Molecular genetic data have also been used to trace the movement of humans across the planet after the emergence of *Homo sapiens* in East Africa. Allan Wilson's work on maternally inherited mitochondrial DNA played an important role in launching this field. Mitochondrial DNA is easy both to isolate and to sequence, and it also shows a high rate of mutations, which makes it possible to spot differences that have occurred over a short period of evolutionary time. Wilson's work led him to identify the existence of a "Mitochondrial Eve," a human ancestor to whom all current human mitochondrial sequences can be traced, who lived over 200,000 years ago.[63] This discovery attracted huge media interest, but its true significance is rarely properly understood. "Eve" was merely the common source of our mitochondria. Many other female ancestors contributed to our nuclear genome, which contains the vast majority of our genes. Soon after, Luigi Cavalli-Sforza expanded early studies of blood groups to provide a general picture of the patterns of migration shown by our species over the planet.[64]

Both the conclusions and the methods of these studies were criticized. The selection of human populations that were considered as genetically different and as representatives of human genetic diversity turned out to be problematic, and—according to the critics—the conclusions regarding the link between genetic and linguistic evolution were drawn too hastily. The Human Genome Diversity Project, launched by Cavalli-Sforza in 1991 to collect genetic samples of different human populations, met strong opposition, in particular from indigenous populations, who felt there was a danger that these studies could lead to genetic discrimination or to the exploitation of their genetic adaptations by external commercial bodies.[65] Over recent years, most of these doubts have evaporated as spectacular studies have been carried out, although some indigenous peoples still oppose the use of their DNA in such research. The whole Icelandic population has been studied, revealing patterns of migration and adaptation; and genomic research in Great Britain has been able to show the patterns of interbreeding between those living in England and the repeated waves of invasion—from the Romans to the Norwegian and Danish Vikings.[66] As genomic sequencing becomes increasingly widespread, we can expect further fascinating revelations about human migration as well as about other evolutionary issues, such as the complex interactions of these populations with pathogens.

From Evo-Devo to a General Place for Evolution in Molecular Biology

Most issues in molecular biology can be addressed without any reference to the evolutionary history of the elements involved. For example, the complex machinery of small interfering RNA production was described without any idea about its evolutionary origin (Chapter 25).

Nevertheless, some molecular biologists are trying to reconstitute the complex history of the macromolecular machines that operate in organisms, such as the ribosome or the bacterial flagellum.[67] Proponents of "Intelligent Design" have frequently pointed to these superb nanomachines, claiming they are too complex to be the result of a Darwinian process driven by random variation and natural selection.[68] Molecular genetics is showing that such views are wrong.

Although molecular data from extant organisms provide only palimpsests, traces of past structures still visible in extant macromolecules, they make it possible to reconstitute the evolutionary history of molecular systems such as the glucocorticoid receptor, which has been intensively studied by Joseph Thornton's research group. A different but complementary approach has been followed for over twenty-five years by Richard Lenski's research team; this involves allowing bacterial populations to evolve in vitro—for instance, to adapt to a nutrient—and studying the consequent molecular changes.[69] This experimental approach has only recently demonstrated its full explanatory power, thanks to the development of high-throughput sequencing technologies that make it possible to sequence the bacterial genome at successive steps during the in vitro evolutionary process. Lenski's project provides a complete description of this evolutionary process, of the spontaneous mutations that have occurred and have been selected for, and of their phenotypic characteristics. This was possible only because samples of the cultures had been stored on a regular basis over the decades, even though at the outset Lenski could not have dreamed of what he would be able to do with them.

These studies are good examples of what Antony Dean and Joseph Thornton have called the "functional synthesis": a modern evolutionary synthesis that is complemented not only by developmental biology, but also, and more generally, by a precise knowledge of the molecular mechanisms involved in evolution.[70]

None of these studies has so far posed a challenge to evolutionary theory, as their results emphasize the role of the twin foundations of neo-Darwinian evolution: mutations (including neutral ones) and natural selection. But they have emphasized the role of rare, contingent variations that have oriented the future development of the structures under study, variations that deserve to be described as innovations.[71] A genetic description of evolutionary innovations has been also proposed from an Evo-Devo perspective, but the complexity of the systems under study makes it difficult to definite exactly what an innovation is.[72]

For the moment, there are still too few studies of the historical development of molecular systems that not only complement molecular studies but can also be considered as necessary for an understanding of the characteristics of these systems. The genetic code is a primary example of the difficulties that are encountered: its characteristics (and in particular the nature and number of the amino acids involved) are the result of the historical or contingent events that generated it. But despite the vast number of studies carried out and the occasionally bold and baroque hypotheses that have been put forward, its origin remains as nebulous as ever. Chaperonins, proteins involved in the folding of other proteins, are another example: their current role in organisms and the characteristics of the proteins that are their targets are the results of the complex evolutionary history of these target proteins, of the mutations that have stabilized or destabilized their structure over deep evolutionary time. Only a knowledge of this history can explain why proteins either rapidly fold without assistance or require the help of chaperonins.[73] Unfortunately, this evolutionary history remains largely a mystery.

Developing a truly functional evolutionary synthesis would be a way of predicting the future—not by designing future organisms but by anticipating how they may evolve, for example, as a consequence of climate change. This is the dream of the researchers trying to build the GRNs of the organisms that they study, as well as of the epidemiologists surveying the mutations in viral genomes that might be at the origin of pandemics. The debate around controversial experiments aimed at creating an avian influenza virus that would be highly infectious for humans must not mask the potential interest and importance of such studies aimed at describing the evolutionary landscape that is accessible to extant organisms.

24

Gene Therapy

FROM THE END OF 2012 ONWARD, the development of CRISPR-Cas9 gene editing has triggered an acceleration in research into gene therapy. The road now seems open to relatively straightforward manipulation both of somatic cells and the germline, although it is too early to know the outcome of this new wave of interventions. Until the development of CRISPR, the very limited use of gene therapy could be argued to be the greatest failure of molecular biology, a field where there was a striking contrast between the huge amount of knowledge accumulated and a relative paucity of practical results.

In his 1958 Nobel Lecture, Ed Tatum stated that biological engineering might lead "to the biosynthesis of the corresponding nucleic acid molecules, and to the introduction of these molecules into the genome of organisms, whether via injection, viral introduction into germ cells, or via a process analogous to transformation. Alternatively, it may be possible to reach the same goal by a process involving directed mutation."[1] With the exception of the synthesis of nucleic acids, these predictions remain virtually completely unrealized in humans, although the first preliminary steps described by Tatum earlier in his lecture—understanding the nature of the relation between genes and proteins, and the mechanisms of gene regulation, both of which were unknown in 1958—were attained in the mid-1960s. Up until now, gene therapy has seen very slow progress, marked at regular intervals by interruptions, obstacles, and major difficulties.

Gene therapy truly began in the late 1970s, when the tools of genetic engineering made it possible to manipulate the genes of eukaryotes. However, the hope of "improving" humans and eradicating genetic diseases had been high throughout the first half of the twentieth century, using methods that are now generally looked on with disapproval. At the time it was believed that these goals could be attained by selecting the "best" individuals and preventing those individuals who were deemed likely to pass on genetic defects from reproducing. Although the eugenicist programs before World War II were sometimes criticized for their brutality and their simplistic conception of gene function, biologists (and many politicians, including those on the left) generally supported the idea of enhancing the quality of human genes, often to avoid the perceived threat that our genetic characteristics were becoming degraded.

As Tatum's prophetic vision shows, molecular biologists did not wait for the development of appropriate molecular tools before exploring the possibilities opened up by the new molecular approach. Among those many possibilities, gene therapy is one of the easiest to describe on paper, if not to put into practice. One of the earliest experiments in molecular biology—the discovery of the chemical nature of genes by Oswald Avery in 1944—showed that it was possible to modify the genetic content of bacteria by the simple addition of DNA into the external medium.[2] Elizabeth and Waclaw Szybalski viewed transformation in bacteria as a model for gene therapy, and were the first to try to integrate DNA into eukaryotic cells.[3] They obtained positive results, but these were not reproducible. Efficient techniques of transfection were developed in prokaryotes in 1973, and later adapted to eukaryotes (Chapter 16).

When molecular biology became increasingly popular and fashionable in the 1960s, some molecular biologists were nonetheless doubtful that gene therapy would be feasible in the near future. Rollin Hotchkiss, a specialist in bacterial transformation who had worked with Avery, was enthusiastic.[4] But other geneticists such as Bernard Davis were more cautious, arguing that with the exception of simple genetic diseases most human traits (and diseases) were under the control of multiple genes; for this reason they would be difficult to modify.[5] In his famous book *Chance and Necessity,* Jacques Monod was even more pessimistic, arguing that the complexity of the human genome prevented it from being modified, possibly ever.[6]

In the mid-1970s, both supporters and opponents of the development of genetic engineering raised the prospect of alleviating or even curing genetic diseases and explained the possibilities and risks of genetically modifying the human genome.

▌ The First Wave of Gene Therapy in the Early 1990s, and Its Problems

In the early 1980s, as the tools of genetic engineering became widespread in laboratories, scientists became optimistic about the prospects for gene therapy. Early attempts rapidly met difficulties, were considered as premature, and stopped. In the years that followed, genetic engineering techniques were used to isolate and characterize the genes involved in the most common genetic diseases that might one day be the focus of gene therapy (Chapter 27). It took ten years to develop vectors that could deliver genes to the affected tissues; three types of vector were the focus of this research—retroviruses, adenoviruses, and liposomes.

At the beginning of the 1990s, the first gene therapy protocols were officially approved and clinical trials began.[7] The simplest of these approaches involved introducing a functional gene that coded for a protein that was nonfunctional in the patient. The genetic diseases chosen for this daring therapy had no effective treatment. The first gene to be transferred coded for adenosine deaminase (ADA); patients with a faulty copy of this gene form toxic derivatives of adenine in the T-lymphocytes in their blood, causing severe combined immunodeficiency disease (SCID). SCID was chosen for several reasons. Because the disease was expressed in the blood, it was relatively easy to isolate the T-cell progenitors from the patients, to introduce a correct version of the ADA gene into these cells using a retroviral vector, and finally to reintroduce the genetically modified cells into the patient. Not only could gene transfer be carried out on isolated cells in a test tube, but it was also possible to increase the population of transfected cells before reintroducing them into the body, making success more likely. In addition, it was expected that the modified cells, which now functioned correctly, would have a selective advantage over the non-modified ones in the patient's bloodstream.

Two other diseases were also the subject of these early clinical trials. One trial introduced a gene coding for a low-density lipoprotein cholesterol receptor found in hepatic cells; patients with faulty copies of this gene have cholesterolemia, early onset atherosclerosis, and heart attacks, symptoms that go under the name of familial hypercholesterolemia. In these trials, part of the liver was surgically removed, the cells were dissociated and put into culture, and they were then transfected by a retroviral vector carrying an intact gene coding for the receptor protein. The transformed cells were reintroduced through the portal vein, and as they passed through the liver they spontaneously returned to their correct location and regained their original function.

Cystic fibrosis, the most common genetic disease in people of European descent, was another disease studied in these trials. The gene underlying this disease had recently been cloned (see Chapter 27), and its molecular function had been characterized. In the trials, complementary DNA corresponding to the correct version of the protein was inserted into a modified adenoviral genome, and the manipulated virus was introduced into the nasal epithelium, which was seen as a model of the tissues principally affected by the disease, the tracheal and lung epithelia.[8] Adenovirus was a suitable vector because it infects the cells of all these tissues.

Gene therapy was not limited to the replacement of nonfunctional genes. Genes were also introduced to mark cells or to eliminate tumors. Gene-marking was initially developed by developmental biologists in the mid-1980s to track the fate of cells in the embryo.[9] But it can also be used for the development and improvement of therapeutic strategies. In leukemia, for instance, treatment consisted of chemotherapy followed by isolation of the hematopoietic stem cells. After intense radiotherapy to eliminate any leukemia cells that had survived the chemotherapy, the hematopoietic stem cells were reintroduced in order to compensate for the elimination of immune cells as a consequence of radiotherapy. Patients often relapsed, and it was thought that this might be due to the presence of leukemia cells among the normal hematopoietic stem cells that were isolated. This hypothesis was confirmed using the cell-marking technique, which allowed researchers to estimate the number of leukemia cells among the isolated cells, and to develop strategies to eliminate them before the cells were reinjected into the patient.

Marking cells also provided a way of understanding the significance of the infiltration of tumors by lymphocytes, and of finding out whether the lymphocytes present in tumors specifically recognized the tumor cells and helped to eliminate them. Lymphocytes present in the tumor cells were isolated, gene-marked, and then reinjected into the patient, after which a significant proportion of them infiltrated the tumor again, thereby confirming that their presence in the tumor was not a random occurrence.

The results of this experiment encouraged research into various strategies designed to combat tumors by stimulating the immune response. One approach involved introducing genes encoding a membrane protein, such as a gene from the histocompatibility gene complex, into the tumor cells to transform them into foreign bodies that would be recognized by the host's immune system. Another method introduced genes encoding interleukin-2 into tumor cells, or into cells in close proximity to the tumor, so as to stimulate the action of lymphocytes against the tumorous cells. Readers who are familiar with recent progress in the immunotherapy of cancer will be struck by the similarity of the objectives pursued then and now, but also by the difference in the methods that are used—monoclonal antibodies have now replaced gene therapy. A third and more direct strategy involved injecting tumor cells with a "death gene," the expression of which would become lethal to the cells that harbor them after introduction of a nontoxic drug, which the gene product would convert into a toxic one. Another approach that was considered involved the introduction of antisense constructions to inhibit the expression of essential genes in tumors or viral genes in infected cells (Chapter 25).[10]

The results of this first wave of clinical trials were mixed. None of the patients involved was cured, but positive effects were observed, even when the expression level of the transgene was far lower than that normally seen. Stimulating a tumor-directed immune response was also shown to be a promising strategy. In the case of two patients, however, the trial caused a dramatic inflammatory reaction that proved fatal.

An obvious source of problems was the vector used to introduce genes into the targeted cells. Retroviruses transform cells with a high degree of efficiency and stability, but the size of the gene that can be inserted into the vector is limited, and at the time they were unable to transform quiescent cells. And there is also a risk that the insertion of the gene into the genome may alter the expression of surrounding genes and activate an oncogene, even if this was not observed in any of the clinical trials. Furthermore, an inflammatory reaction was observed in a patient suffering from a neuroblastoma in which the cells producing the retroviruses had been introduced close to the tumor. The adenoviral vector also caused an inflammatory reaction that had not been observed in previous tests on animals. In addition, genes that are introduced through an adenoviral vector do not insert into the genome: the DNA is gradually eliminated, and the treatment has to be repeated frequently, thereby increasing the risks of an inflammatory response. A third way of delivering gene therapy in which the genes were encapsulated within liposomes (an artificial lipid vesicle) was at the time not very effective.

It was unclear whether it would ever be possible to design a perfect vector—what suited one disease was not necessarily suitable for another. And perhaps the most significant drawback of all these methods lay in the inconsistency of the results obtained from one patient to another and from one trial to another.

After the unexpected deaths of patients, clinical trials were halted. Gene therapy appeared to be too risky. These first trials also underlined that this new technology had not yet been fully mastered. Nevertheless, it did not stop some geneticists from rejecting the cautious attitude of most of their colleagues and raising the possibility of germline modification, a line of research that had been deliberately suspended by most of those involved in gene therapy and was even illegal in some European countries.[11]

In addition to gene therapy, a second strategy was gradually developed by chemists that involved the modification of gene expression through the use of small oligonucleotides that would pair with genes or with messenger RNAs. Its promise was different from gene therapy, but there were also areas in which the two methods might converge. This new approach could not introduce a new function, nor could it compensate for a nonfunctioning gene, but it could inactivate a gene in a tumor or in a pathogenic virus. It is probably no coincidence that many reports on the potential use of these oligonucleotides were published in 1993–1994, at a time when the first trials of gene therapy were yielding disappointing results.[12] This alternative strategy, initially tested only on cells and animals, also had obstacles to overcome. Chemical modification of the oligonucleotides made these molecules stable, but it was unclear how to deliver them to their cellular targets most effectively or to enable them to pass through the cell membrane.

▌ The 2000s: A Second Wave of Enthusiasm and Difficulties

In the year 2000, Alain Fischer and his collaborators in Paris succeeded in curing two children with SCID, caused by mutations in a gene encoding the γ chain of the interleukin-2 receptor.[13] The strategy they adopted was similar to one that had been used in the 1990s versions of gene therapy. They isolated progenitor cells from the patient's bone marrow which were ex vivo transformed by a retroviral vector, into which a DNA fragment coding for the correct version of the γ chain of the receptor had been inserted.

The results were spectacular. Ten months after their gene therapy, the two children were completely well and were able to leave the protective plastic bubbles in which they had each been confined since birth. They both had a normal level of B-lymphocytes, T-cells, and natural killer cells. Their T-cell clones also showed normal levels of diversity, and they reacted normally to vaccinations against polio, diphtheria, and tetanus.

In the months that followed, eight more patients received gene therapy for the same condition, with positive results for seven of them. For the first time, gene therapy was shown to give reproducible results and to be able to cure patients. This cure was symbolized not by any single change in the patients' physiology but instead by their spectacular release from their isolation bubbles. In 2002, Italian research teams obtained similarly positive results for two young patients suffering from ADA deficiency.[14]

The explanation for these successes was complex. The growing T-cell clones were positively selected, but in the previous cases of ADA deficiency this positive selection had probably been blurred by the fact that treatment with injected ADA enzyme had not been stopped when the gene therapy trial began. On the other hand, the isolation of progenitor cells was more efficient, and the level of transformation by the retroviral vectors was higher.

The conviction that gene therapy had entered a new era proved short-lived. In 2003, the French team announced that after two and a half years, two of the nine young patients who had been treated with gene therapy had developed T-cell leukemia, and this figure later rose to four.[15] Although the children recovered after the treatment of their leukemia, this announcement cooled the enthusiasm for this new version of gene therapy.

The leukemia was caused by the retroviral vector inserting close to the *LMO2* gene, which was known to be essential for the development of hematopoietic stem cells. Overexpression of this gene led to the development of leukemia in mice. It was suggested that the random insertion of the vector close to the *LMO2* gene had led to the overexpression of the *LMO2* gene through the action of the promoters present in the retroviral vector. This had encouraged the differentiation of the progenitor cells in which the vector had been introduced at this position, leading to the subsequent development of leukemia.

Advocates of gene therapy were well aware of the risk of retroviral insertional mutagenesis, but it had not been observed in the initial trials. Other risks, such as the reappearance of a replication-competent retrovirus by recombination of the vector with other retroviruses present in the genome, had received more attention.

▌ CRISPR: New Hopes

Where gene therapy involves introducing a gene that is nonfunctional in the host, the ideal solution would be to substitute an active copy for the inactive one at the precise position in the genome where the original is found. This is what CRISPR makes it possible to do.

To understand how and why CRISPR became this long-awaited tool for gene therapy, we need to study the history of this technique in a different way from the traditional focus on the technique's origins in microbiology. The potential of the CRISPR system was revealed by the convergence of two different lines of research: the search for restriction enzymes that recognize a limited number of sites in the genome, and the characterization of mysterious clusters of repeated sequences that were observed in bacterial and archaeal genomes, and which eventually gave the technique its name.

Restriction enzymes had played such a vital role in the development of genetic engineering that the search for new enzymes, with new specificities, was still a priority in the 1990s. Researchers were particularly interested in one category of restriction enzymes: those that recognized long sequences of nucleotides, and which would therefore cleave at only one or a few positions in the genome. These meganucleases, as they were called, were first described in yeast, and it was expected that they would be very useful in the mapping work that would precede genome sequencing, and in particular the ongoing Human Genome Project (Chapter 27).[16]

Evolutionary logic suggested that the probability of finding such meganucleases was low. Bacterial restriction enzymes had been subject to natural selection for their ability to cleave and degrade the DNA of infecting bacteriophages, so there was no selective advantage for the host in having restriction enzymes with a very high specificity. It was far better to be able to seek and destroy any viral DNA, at as many sites as possible.

This impeccable logic prompted Srinivasan Chandrasegaran to adopt a radically different strategy, which in the mid-1990s led to a project involving the construction of new artificial restriction enzymes by combining a nonspecific nuclease with protein motifs that recognized specific DNA sequences. In an early article, Chandrasegaran chose the nuclease *Fok* I, in which the nuclease domain could be easily separated from the DNA recognition domain.[17] He associated this nuclease domain with the homeodomain of the *ultrabithorax* gene, and obtained a restriction enzyme that was specific for the DNA sequence recognized by the ultrabithorax homeodomain.[18]

The final step was to find DNA-binding motifs that could recognize any DNA sequence once they had been appropriately modified. Chandrasegaran selected the three zinc-finger motif that recognizes a nine-nucleotide sequence, the structure of which had recently been determined, and in which the rules of amino acid–nucleotide interaction were gradually deciphered.[19] The combination of two of these motifs made it possible to recognize longer sequences of DNA and to increase the specificity of the enzyme. The first positive results were published in 1996.[20]

Many groups immediately recognized the potential of these artificial restriction enzymes. Five years later, in 2001, their efficiency and specificity were demonstrated by Chandrasegaran while working with Dana Carroll's group.[21] Similar results were obtained by other researchers, and the importance of these new tools became widely recognized.[22]

Homologous recombination involves the replacement of one sequence of DNA by a homologous sequence that is present, for example, on another chromosome, and is frequently observed in yeast. At the end of the 1980s, Oliver Smithies and Mario Capecchi showed that homologous recombination occurs at low rates in mammals and developed a way of isolating such rare recombinants, making it possible to create transgenic animals with targeted gene modifications. The first step in this process was the transformation of embryonic stem cells by a modified copy of a given gene (in what became known as a knockout experiment, this was an inactivated copy). Homologous recombination occurs in a very small fraction of these cells, which can be isolated using genetic tricks. These rare cells are then injected into a blastocyst, which is implanted into a foster mother, leading to mosaicism in the offspring. When mosaicism occurs in germline cells, these animals can generate mice in which one copy of the gene under study has been replaced by a modified copy when they are crossed with wild-type mice. Crossing these heterozygotes leads to the creation of homozygous animals in which both copies of the gene are altered.[23]

This procedure is long and complex; because it involves germline manipulation and the creation of mosaic individuals, it cannot be applied to humans. Together with other researchers, Bernard Dujon, who had been the first to characterize the properties of meganucleases, showed that cleaving DNA at a particular position increased the frequency of homologous recombination at this position—not only in yeast but also in plants and mammals.[24] This discovery opened up a new, easier, and more efficient route to homologous recombination in mammals. And above all, it opened up the possibility of using the technique in humans.

Artificial restriction enzymes now became a tool of choice not for mapping the genome but for modifying it, for cutting it at precise positions, and for replacing a given gene with a variant copy.[25] Researchers raced to dream up medical applications. For example, most strains of type 1 human immunodeficiency virus (HIV-1) require the presence of the cell receptor CCR5 (C-C chemokine receptor type 5) in order to infect immune cells. A small number of people have naturally mutated forms of CCR5 that render them highly resistant to HIV-1 infection. Researchers set out to mimic this situation by extracting T-lymphocytes from patients infected with HIV-1 (only a fraction of their T-lymphocytes are infected), inactivating the receptor by cutting its gene with a zinc-finger nuclease, and injecting these now-resistant lymphocytes back into the patient.[26] First proposed in 2005, this protocol was approved, clinical trials were begun, and over the following years the strategy was shown to be successful, as the number of T-lymphocytes increased in the treated patients.

A new class of artificial restriction enzyme known as transcription activator-like effector nucleases (TALENs) was produced by using the same bacterial nuclease as in the zinc-finger nucleases. But the TALENs technique used a different DNA recognition motif that was present in plant pathogens, which was easier to modify and more specific than the zinc-finger motif.[27]

So efficient and accurate were these new tools that their developers began to use a new term to describe their action on the genome: "editing." In the early 2000s, this was the word that was increasingly adopted to describe the results obtained with zinc-finger nucleases, and at the end of the decade Fyodor Urnov used the metaphor "genome editing" in the title of a widely read review emphasizing the precision of the new method.[28]

This was the context that led the researchers who had long been exploring the structure of CRISPR complexes—determining their role in protecting bacteria and archaea against bacteriophages—to realize the potential of the Cas9 nuclease that was involved in providing bacteria with protection against viral infection. It offered a far better editing system than the artificial zinc-finger nucleases and TALENs painstakingly constructed by molecular biologists.

The realization that CRISPR constitutes a kind of bacterial immune system followed a slow and tortuous route, but without it there would be no CRISPR-based gene editing. The existence of clusters of regularly interspaced short palindromic repeats (hence the acronym CRISPR; the "palindromes" are not very exact) alternating with variable "spacer" sequences was discovered in the 1980s. By the beginning of the 2000s the CRISPR systems were shown to be widespread in microbes. In 2005, three bioinformatic studies of the spacer sequences revealed that some of them were derived from bacteriophage DNA. This led the researchers to propose that CRISPR systems were involved in protecting prokaryotes against viral infection, and this was demonstrated two years later.[29]

This system was immediately likened to RNA interference—RNAi (Chapter 25)—which had been discovered a decade earlier in eukaryotes.[30] In 2011, Emmanuelle Charpentier and her collaborators described the role of small RNAs in the CRISPR system, encouraging the idea that there was an analogy between CRISPR immunity and RNAi.[31] But a series of observations demonstrated that the CRISPR system was targeted directly against phage DNA, which is degraded by the Cas9 nuclease, and not against RNA as in RNAi. In 2012, Jennifer Doudna and Charpentier's groups proposed a new interpretation of the role of those small RNAs, suggesting that they guide the Cas9 nuclease toward its target by pairing with specific sequences in the bacteriophage genome. Once this had been shown, they immediately realized that the system could be programmed to cleave any given sequence. They therefore suggested that Cas9 could be used as an editing system and demonstrated this through a series of pre-liminary experiments.[32]

The CRISPR-Cas9 craze exploded in 2013, promising to bring major medical breakthroughs.[33] The system was rapidly adapted to eukaryotic cells, including human cells, and its site-specificity was demonstrated. The system could be adapted to any new target by simply changing the RNA that guides the nuclease to its site. The efficiency and precision of the new technique opened the door to germline therapy, and the first experiments were begun to this end.

No sooner was the mechanism of action of CRISPR-Cas9 established than the term "editing" was immediately and unanimously adopted, to the point where it was forgotten that other words had ever been used. As many journalists and philosophers of science have pointed out, this was not a neutral choice—editing implies simplicity and, in publishing at least, is carried out to improve a text. After having published the book of life (the sequence of the human genome), biologists were now editing it, by eliminating the errors that it contains. The term "editing" not only conveys the ambition of scientists to rewrite the book of life but also suggests to the public a nonthreatening, merely technical approach to gene manipulation.

Two aspects of the CRISPR story deserve to be underlined. The first is that it is a powerful example of the way in which scientific progress results from a combination of pure, blue-skies research into natural phenomena and applied research with a defined and practical aim. The microbiologists who discovered the CRISPR-Cas9 system had no idea that they were on the verge of creating an apparently perfect tool for editing the human genome. But the reason why they could immediately recognize the potential of the mechanisms of action of Cas9 was because they knew of the work that had been done to build artificial restriction enzymes with a high specificity.

The second interesting aspect of this story is that the recognition of DNA (or RNA) sequences by RNAs or short oligonucleotides was a well-known, naturally occurring mechanism, which was used experimentally by biologists. It was known that small interfering RNAs and microRNAs interacted with messenger RNAs and inhibited gene expression (Chapter 25). It is not immediately clear why molecular biologists did not think of using small RNAs to guide nucleases to specific targets—or indeed why Chandrasegaran chose instead to associate a nuclease with a DNA-binding motif from a protein. The answer probably lies in the history of molecular biology.

The idea that gene expression might be controlled by small RNAs directly interacting with DNA had been put forward a number of times, but always in vain. In François Jacob and Jacques Monod's operon model, the repressor was initially thought to be an RNA, before it was realized two years later that it was in fact an allosteric protein.[34] In 1969, Roy Britten and Eric Davidson proposed that their producer genes—equivalent to the structural genes in the operon model—were activated by small RNAs, a hypothesis that was later abandoned by Davidson in his vision of gene regulatory networks (Chapters 22 and 23). The strategy of using short oligonucleotides to control gene expression in eukaryotes failed to lead to any applications, partly because of the unsolved issue of how to deliver molecules to the nucleus. Of the two strategies that were considered, targeting DNA or RNAs, only the second was successful in vitro, and it was thought that it would perhaps be feasible in vivo in the distant future. These recurrent disappointments and failures may account for the fact that molecular biologists did not come up with the idea of building a nuclease guided by RNA.

▌ Conclusion

This chapter was not intended to outline the potential future developments of gene therapy, but rather to consider the extent to which research in this field has remained within the framework laid out by molecular biologists in the 1960s. In no other field of research have the projects generated by advances in molecular biology fulfilled so little of their potential. One reason for the wave of enthusiasm that greeted CRISPR-Cas9 is that it resolves the problems of insertional mutagenesis, which molecular biologists working on bacteriophage were well aware of in the 1950s. It is remarkable that current proposals with regard to gene therapy are essentially the same as those proposed by Tatum and other molecular biologists at the end of the 1950s. The difference is that, with the eruption of CRISPR-Cas9 onto the scene, ideas that had been put to one side because they were impractical have suddenly became possible.

Germline gene therapy has been the most high profile of the projects to be initiated following the discovery of CRISPR-Cas9 gene editing, but it seems probable that somatic cell therapy, which is more widely acceptable for both scientific reasons and on the basis of societal and ethical considerations, will be the first to benefit from the development of the CRISPR system. Germline gene therapy would have consequences for future generations and would rapidly breach the widely accepted bans on genetic enhancement and modification of the human species.

The international opprobrium that greeted Chinese researcher He Jiankui in November 2018 when he announced (on YouTube!) that he had used CRISPR to modify the CCR5 genes of two perfectly healthy human embryos, which had since been born and were now babies, was based on a number of factors. From a technical point of view, the procedure appeared to have failed, producing babies who were mosaic for the desired characteristic (an alteration that is thought to give some protection against HIV): some cells were affected and others not, and with the introduction of an unintended mutation. Furthermore, there were profound ethical problems related to whether the parent had given informed consent as well as the principle of altering an otherwise healthy embryo. Supporters of the idea of using measured and controlled germline modification to treat genetic diseases were dismayed—this unregulated experiment threatened to set the field back for years.

Just as He Jiankui was seeking to remedy a problem that did not exist, the idea that genetic diseases can be eradicated from the population is an illusion because these diseases constantly arise through new mutations. The birth of babies with familial genetic diseases can already be prevented by early genetic screening of embryos after in vitro fertilization.

In the same way that the development of synthetic biology is the direct consequence of the accumulation of molecular knowledge over the last half-century (Chapter 28), so too the rapid progress in the development of somatic cell therapy, which now seems to be within our grasp, would represent the belated accomplishment of the early dreams of molecular biologists at the end of the 1950s.

25

The Central Place of RNA

ACCORDING TO FRANCIS CRICK'S CENTRAL DOGMA, outlined in 1958, ribonucleic acid (RNA) played the role of an intermediary, a passive messenger, between DNA—genetic memory—and proteins. Over half a century later RNA has taken its place as a molecule of fundamental importance with its own unique activities that are essential for life.[1]

From the Late 1950s to the 1970s

Although RNA's roles as both catalyst and regulator were not fully revealed until the 1990s, all the essential technology and hypotheses that could have led to their discovery were already in place three decades earlier.

By the late 1950s it was realized that RNA could pair with RNA, or with a single chain of DNA.[2] This last phenomenon lay behind the technique of hybridization, and was incorporated into early models of genetic regulation.[3] In François Jacob and Jacques Monod's groundbreaking 1961 article published in the *Journal of Molecular Biology,* the French duo initially suggested that the repressor of the lac operon was an RNA molecule.[4] The repressor was soon discovered to be a protein (see Chapter 14). In the late 1960s, the model by Roy Britten and Eric Davidson proposed that integrator genes produced activator RNAs that, by binding to receptor genes, activated batteries of producer genes.[5] This theoretical model turned out to be wrong about the nature of the activators. A series of other now-forgotten models involving genetic regulation by polynucleotides or ribonucleoproteins were proposed, none of which survived.[6] In other models, it was suggested that the ability of a stretch of a messenger RNA molecule to form a double helix as the molecule bent back on itself led to a structure that would prevent the molecule from being recognized by the ribosome and being translated into a protein. In some bacterial operons, this mechanism regulates gene expression, a process called attenuation.

After the genetic code had been completely decoded in the mid-1960s, Carl Woese, Francis Crick, and Leslie Orgel all produced models of the origins of life that as one of their first steps involved the formation of oligonucleotides capable of self-reproduction.[7]

Despite what an initial impression might suggest, these studies were not the origin of the now widely held hypothesis of an RNA world, which preceded the present world of DNA, RNA, and proteins. In this world, which is thought to have existed over 3.6 billion years ago, RNAs filled the catalytic role now occupied by proteins, and at the same time they were the hereditary material. A version of this hypothesis was first put forward in 1957 by Rollin Hotchkiss at a symposium organized by the New York Academy of Sciences; unlike the RNA world hypothesis, Hotchkiss's concept required that proteins would have existed at this early stage, and his suggestion had little immediate impact.[8] While it was clear to Crick and Orgel that proteins appeared after the first polynucleotides, the exact nature of those polynucleotides (RNA or DNA) did not matter. Although the replication of these polynucleotides was a necessary step toward the origin of life, Orgel suggested that this could have been a spontaneous, noncatalytic chemical process. As with the discovery of the double helix in 1953 (see Chapter 11), the complementarity of the bases did away with the need to invoke catalytic activity to explain replication. Orgel and Crick were not opposed to the idea that the first enzyme to appear might have been an RNA molecule with replicase qualities, but they were not interested in showing that RNA could have a general catalytic activity.

The realization that the *lac* repressor was not an RNA molecule may have subsequently led researchers to downplay suggestions that the control or modification of polynucleotides was effected by other polynucleotides. This dampening effect may have persisted for some time—it took many years before it was realized that RNA guides were involved in CRISPR-Cas9 DNA target recognition. It is striking that none of the earlier projects aimed at creating specific nucleases attempted to exploit such a mechanism (Chapter 24).

In fact, the two first technological developments open to RNA—using its binding ability on the one hand and seeking to find or produce RNAs with a catalytic role on the other—were diametrically opposed. Researchers studying RNA simply were not interested in the three-dimensional structure of RNAs and the functional capacity of the chemical groups carried by nucleotides, while these were of fundamental importance to the catalytic approach.

▌ The Discovery of Antisense RNAs

The ability of antisense oligonucleotides to inhibit the functions of other RNAs was discovered in the late 1970s and early 1980s after the development of a technique that made it possible to link a particular DNA sequence and a specific protein. By hybridizing a preparation of messenger RNA (mRNA) molecules with a fragment of cloned DNA, and by translating the RNA molecules in vitro, before or after hybridization, it was possible to find out which protein had not been synthesized because its RNA had been retained during the hybridization stage.[9] This technique also made it possible to distinguish between the introns and exons in a fragment of DNA—only the exons were present in mRNAs. At the same time, RNA-DNA hybridization was widely used in electron microscopy to demonstrate the existence of splicing, and to identify the order of exons and introns in a given gene (Chapter 17).

In 1978, Paul Zamecnik and Mary Stephenson inhibited the replication of the Rous sarcoma virus and the transformation of fibroblasts by the addition of an oligodeoxyribonucleotide complementary to the repeated 3' and 5' sequences of the virus. In an additional experiment, they showed that this oligodeoxyribonucleotide inhibited the translation of the viral RNA in an in vitro protein synthesis system.[10]

In 1981, it was discovered that the replication of plasmids coding for colicins—bacterial toxins—or for genes responsible for the resistance to antibiotics was controlled by antisense RNAs. Two years later, it was shown that the movement of transposons in the bacterial genome was also regulated by antisense RNAs, which inhibited the translation of transposase, the enzyme that enables transposition. At the same time, Masayori Inouye's group described how the expression of *ompF,* one of the genes coding for the proteins of the outer membrane of *Escherichia coli,* was reduced by the action of a 174-nucleotide antisense RNA produced through the transcription in a reverse orientation of a different gene from the same family *(ompC).* This antisense RNA bound to the 5' end region of the mRNA of the *ompF* gene.[11] This effect turned out to be one of a number of mechanisms that enable bacteria to adapt to changes in osmotic pressure.

In 1984, Jonathan Izant and Harold Weintraub demonstrated that the transfection in mouse L cells of a thymidine kinase gene *(tk)* in reverse orientation reduced the expression of the endogenous *tk* gene: the antisense RNA produced from the transfected DNA bound to the mRNA produced by the transcription of the endogenous gene.[12] The researchers involved in this work were primarily interested in adding new tools to the tool kit of molecular biology, enabling biologists to inactivate genes (and hence the products of genes) to thereby deduce their function. Another approach that was developed in this period involved injecting cells with antibodies to inhibit the function of a particular protein. John Rubenstein and Jean-François Nicolas, working in Jacob's laboratory, obtained similar results demonstrating the potential of antisense constructions.[13]

Biologists now had a new tool, one that would enable them to inhibit the function of genes, and that could be used both for the identification of gene functions and, in principle, for therapeutic applications. However, it is striking that these studies did not directly lead to the discovery of RNA interference or the role of microRNAs. The development of this technology did not generate the widespread idea that a mechanism of this sort might occur naturally in cells.

There were many reasons for this. First, there were few real-life examples of this type of regulation in bacteria. It was a decade before the first description of antisense RNA in eukaryotes (*Lin4* in *Caenorhabditis elegans*) was published.[14] Furthermore, experiments involving the injection of antisense RNA into amphibian eggs to inhibit the expression of genes during early development produced unexpected results, such as the unwinding of the RNA double helix and the modification of its bases. The physiological significance of these effects was unclear.

It seems likely that researchers were focusing their attention on potential future applications, rather than trying to understand the importance of the natural phenomenon. This tendency would have been increased by the fact that the approach that was being used involved the modification of oligodeoxyribonucleotides to increase their stability and make it easier for them to be carried into the cells. Biology was giving way to chemistry.

Although positive results were obtained in vitro, the path to the development of therapeutic applications was a long and rocky one, strewn with pitfalls and setbacks.[15] This doubtless explains why the discovery of RNA interference in the late 1990s was greeted with such enthusiasm—it offered a way of getting around these difficulties. Similar practical problems also limited the development of gene therapy, even though its principles seemed well established (Chapter 24).

Up until this point, in both bacterial gene regulation and eukaryotic transfection experiments, it was generally thought that antisense RNA blocked the initiation of translation by binding to the 5' region of the mRNA, or—less commonly—by distorting the whole structure of the mRNA. Both these models were proved wrong by the discovery of RNA interference.

Discovery of the Catalytic Activity of RNA

Although the possibility that RNAs could have a catalytic activity had been raised in the 1960s, it was two decades later that this was reported, in two virtually simultaneous discoveries that had a major impact. In 1982, Thomas Cech's group showed that a *Tetrahymena* RNA intron was capable of autoexcision—of excising and ligating RNA fragments during splicing without the intervention of a protein.[16] The following year, Sidney Altman's group showed that in ribonuclease P of *E. coli,* formed by an RNA–protein complex, it was the RNA that had the catalytic activity.[17]

These discoveries immediately prompted a renewal of interest in the question of the origin of life because they suggested that catalytic proteins might not have been required at the very beginnings of life—these processes might have been carried out by RNA molecules. Following Cornelis Maria Visser in 1984, Norman Pace and Terry Marsh returned to the idea first proposed by Woese in the 1960s that RNA lay at the origin of life.[18] The following year, Wally Gilbert focused attention on these ideas with his description of the RNA world, while Thomas Cech attempted to define the characteristics of an RNA replicase that could have formed the origin of life.[19]

This surge in speculation about the origins of life may have distracted attention from showing that RNAs were capable of catalyzing a large number of chemical reactions. In the 1980s, the only proof of such catalytic action came from the excision and ligation of RNA in viroids (plant viruses without a protein coat) and viruses. These studies led to the notion of ribozymes, as well as a good understanding of the chemical structures and mechanisms by which ribozymes act.[20] It was only in 1990 with the development of the Selex method of variation-selection in vitro within a large RNA population that the widespread ability of RNAs to act as enzymes was finally demonstrated.[21]

▌ The Complicated History of the Discovery of RNA Interference

Despite many claims, the existence of RNA interference was not first revealed by studies demonstrating the cosuppression of transgenes and endogenous genes in plants. The story is more complex and more interesting. Interference was described in plants as early as 1929, when it was observed that an infection by an attenuated form of a virus could protect against (or interfere with) an infection by a virulent form of the same virus.[22] Although this might have looked like an effect similar to that produced by vaccination in animals, there was a major problem: plants do not have an immune system. It was therefore impossible to explain this effect by the production of antibodies. This did not deter researchers in plant diseases from using the language of immunologists when they referred to "attenuation" or even "immunization." Appropriating the vocabulary of immunology helped mask their ignorance of the mechanisms that could explain such effects.

Over the decades, three hypotheses were proposed to explain this phenomena, none of them either accurate or very convincing. The first was that the development of viruses required a substance produced by the plant—the attenuated form of the virus used up this substance, thereby preventing the development of the virulent form. Interestingly, this was the same hypothesis that Louis Pasteur had used to explain the success of vaccination in animals before the function of the immune system was understood.

According to another hypothesis, plant cells contained a unique virus-producing site, which when occupied by attenuated forms of the virus prevented virulent forms from gaining access to it. This hypothesis was not original either; for instance, it had been put forward by Elie Wollman and François Jacob in the 1950s to explain how bacteria containing a silent form of the bacteriophage were protected against secondary infections by the same type of bacteriophage.

According to the third hypothesis, proteins produced by the attenuated forms of the virus blocked the development of virulent forms. The production of large quantities of capsid proteins, for example, might prevent the deployment of viral RNA from the capsids of virulent viruses during the first stage of infection. This is interesting not because it came any closer to the mechanisms that eventually explained interference, but because it formed the basis for a novel antiviral strategy that was developed in the 1980s (as discussed later).

Despite the absence of a single and convincing explanation, there were large-scale tests of using interference to combat viral diseases in plants, such as on the Isle of Wight off the south coast of England in 1964.[23] In the 1970s and 1980s, despite lingering doubts over its effectiveness and even concerns over its safety, it was put into practice with a variety of plants (tomatoes, lemons, cocoa, and papaya) in many different locations (Brazil, the countries of west Africa, and Taiwan).[24] This method of controlling viral infections in plants was also sometimes "spontaneously" used by producers.[25]

In the mid-1980s, a new generation of researchers trained by specialists in plant biology and familiar with molecular biology set out to find an explanation for interference and to replace more or less empirical practices with evidence-based strategies. In 1986, Roger Beachy's group showed that it was possible to delay the development of tobacco mosaic virus in plants by inoculating them with the coat protein gene for this virus, following a feasibility study a year earlier.[26] Within a few years, similar results were obtained in other plants infected by different viruses.[27] It appeared that scientists had developed a simple method of producing plants that were resistant to viruses.

This approach fell within the wider concept of "pathogen-derived resistance" put forward by John Sanford and Stephen Johnston in the battle against the acquired immunodeficiency syndrome (AIDS) virus and developed in the Qβ bacteriophage that attacks *E. coli*.[28] All these studies made the host express genes coding for proteins produced by the infectious agent, thereby producing protection from infection. Although this approach worked, an explanation remained out of reach. The earlier hypotheses were replaced by molecular scenarios, which seemed more modern but were not necessarily any more convincing. It was possible to imagine, for instance, that viral coat proteins bound to a regulatory site on the viral RNA and prevented it from replicating.

Other more precise strategies were proposed to produce virus-resistant plants by transgenesis.[29] David Baulcombe showed that the expression of an inactive form of the viral replicase rendered transgenic plants virus resistant, as the mutated replicase competed with the normal replicase produced by the infectious virus.[30] In this case, the observed effects could be explained using conventional models. Another suggested strategy involved expressing antisense RNAs to inhibit gene expression—this was first developed for animal cells in 1984 and shown to be effective in plant cells two years later.[31]

A number of observations were incompatible with the proposed explanations, vague though these were. Interference was observed not only in plant viruses but also in viroids, even though these do not code for proteins and are therefore incapable of making the infected cell produce proteins.[32] Worse, in 1992 William Dougherty showed that the transfection of plant cells by a gene coding for viral coat protein could protect plants even in the case of a mutated transfected gene that could not translate into a protein.[33] These two experiments demonstrated that interference in plants could not be explained at the level of proteins.

It was against this background that the first experiments demonstrating the effects of cosuppression in plants were carried out. The aim was to create transgenic plants by transfecting them either with genes from other organisms or with extra copies of endogenous genes. In 1989, A. J. M. and M. A. Matzke showed that the simultaneous transfection of two plasmids carrying different genes for antibiotic resistance could inhibit the expression of the genes carried by one of the plasmids.[34] As this inhibition appeared only when both plasmids were cotransfected, the authors hypothesized that it was the similarity of the two plasmids—their homologous sequences—that caused the inhibition of gene expression, perhaps through the methylation of the promoter of the inhibited gene. The following year, two research groups simultaneously revealed a phenomenon of cosuppression (of both a transgene and an endogenous gene) in petunia plants. Both groups wanted to use transfection to increase the synthesis of anthocyanins, the pigments responsible for petal color. Their results were the opposite of what they expected: the color of the petals in some of the plants was less intense, with white patches (indicating the complete inhibition of the endogenous genes) and even some petals that were completely white. Both teams showed that expression of the transgene was weak while that of the endogenous gene was weaker than normal.[35] It was Carolyn Napoli who coined the term "co-suppression" to describe this phenomenon, hypothesizing that the methylation of DNA was responsible for this effect.

The general characteristics of cosuppression and of its interpretation can be seen in these early publications (even if the first article did not precisely describe cosuppression). The most striking thing was the variability that was observed. The degree of cosuppression varied from one plant to another, and from one petal to another on the same plant, and even between different areas of the same petal. The second characteristic was the rapidly established link between DNA methylation and repression of transcription. This was a consequence of many studies over the previous decade that had demonstrated the relation between DNA methylation and the inhibition of gene expression (Chapter 26). This led to multiple explanations that tended to distract researchers from making the second major discovery relating to cosuppression: the degradation of mRNAs. Soon cosuppression (now renamed gene silencing) was linked to the inhibitory effect of antisense constructions.

In 1988, tomatoes showing delayed ripening were developed through the inhibition of the enzyme involved in the ripening process by transfection with antisense constructions.[36] This prompted considerable ethical debate. For some biologists, these experiments suggested that transgenic plants were being produced for purely commercial reasons, with the aim of cutting producers' losses by extending the fruit's shelf life between harvesting and sale. Two years later, the same team showed that it could obtain a similar result by expressing a sense truncated gene.[37] In hindsight, the similarity of these results can easily be explained by RNA interference. At the time, however, these findings were seen as undermining the originality of cosuppression; when a transgene was inserted into the genome, it was very easy to imagine that it might be transcribed in an antisense direction through the action of promoters close to the insertion site. Cosuppression was therefore believed to be an artifact, produced by the fact that researchers were unable to choose insertion sites in nontargeted transgenesis—a problem that was also encountered in gene therapy (Chapter 24). A final difficulty was caused by the fact that, despite the very many transgenesis experiments on animal cells or whole animals, cosuppression had never been seen.

Gradually, a distinction was drawn between RNA-stimulated transcriptional silencing linked to DNA methylation and the post-transcriptional silencing associated with RNA degradation.[38] From 1995, it was realized that there was something in common between the phenomenon of viral interference described previously and post-transcriptional gene silencing.[39] Both had the same physiological significance, offering protection against the expression of foreign genetic information. It did not, however, reveal the underlying molecular mechanism. In both cases, researchers noted the importance of small RNA and double-stranded RNA (dsRNA), but in and of itself this observation did not give rise to any specific experimental program or to any conceptual breakthrough.

The experiment that finally revealed how interference works was carried out on the nematode *C. elegans*. From the beginning of the 1990s, *C. elegans* researchers routinely injected antisense constructions to silence genes at will—the worm possesses naturally occurring antisense RNA. Curiously, however, in many cases the introduction of sense RNA and the injection of antisense RNA both induced the same effects. In 1998, Andrew Fire and Craig Mello injected sense and antisense RNAs, and a mixture of the two, into *C. elegans*.[40] The dsRNA formed in the mixture had a much greater inhibitory effect than either of the two component parts. The dsRNA was not simply an agent of silencing, it was the trigger for it. This experiment also showed that dsRNA had other functions, in addition to its known role of initiating and mediating the antiviral effect of interferon. RNA double helices could be active in vivo, even if cells were busy places full of enzymes that could unwind such molecules and modify their nucleotides. Fire and Mello's experiment fueled an explosion of studies that rapidly confirmed how widespread this phenomenon was in the living world.

Within a few years, the development of noncellular systems in *Drosophila* made it possible to describe the stages of RNA interference and the proteins and enzymes involved in them.[41] Biochemical studies complemented genetic approaches carried out on the model systems of *C. elegans* and the *Neurospora crassa* fungus. The long molecules of dsRNA were cleaved by the nuclease Dicer to produce small antisense RNAs which, when incorporated into the RNA-induced silencing complex (RISC), guided the cleaving of the mRNA. This could be amplified by an RNA-dependent RNA polymerase. Because it was so easy to use, dsRNA very quickly became an essential tool of basic research even though other methods of gene silencing were available. Therapeutic uses were also soon envisaged.

Once the mechanisms of RNA interference were understood, they revealed the significance of microRNAs throughout the living world. In 2000, a second microRNA was discovered in *C. elegans*. Everything changed when it was shown that these antisense microRNAs had been conserved during evolution, and that they induced the degradation of sense RNA by the same mechanisms as those described for RNA interference.[42]

Studies of microRNAs now took a different direction, looking in genomes for sequences of about twenty conserved nucleotides that were transcribed and complementary to the sequences present in the mRNAs.[43] The experimental approach was thus turned back to front: instead of starting from genes with a known inhibitory effect and showing that this worked through the formation of antisense RNA, the quest was now for complementary gene sequences, with the aim of showing that antisense sequences have a functional role.[44] In a few years, the number of potential antisense RNAs exploded. The subsequent demonstration of the role of these microRNAs was a much lengthier process and has been completed in only a few cases.

From MicroRNAs to Competing Endogenous RNAs

As microRNAs generally have several target mRNAs, and each mRNA carries several different microRNA response elements, it is extremely difficult to assess the role of a particular microRNA on a particular target. This problem is rendered more acute because microRNA binding sites are also found on two other types of RNA, the first of which is produced by the transcription of pseudogenes.

The discovery of pseudogenes in the late 1970s attracted enormous interest. Their existence was subsequently interpreted as the negative face of an important evolutionary process—gene duplication. Pseudogenes were seen as traces of the failure of two existing genes to functionally diverge (Chapter 23). The discovery that the majority of pseudogenes were nonetheless transcribed did not attract attention until the discovery of RNA interference and microRNAs. It then became clear that, because they possessed the same microRNA recognition sites as the genes for which they were nonactive copies, these pseudogene microRNAs recognition sites could distort the regulation of the functional gene.

The same was true of long noncoding RNAs (lncRNA), which were discovered thanks to improvements in the techniques of RNA sequencing. These molecules were often viewed as noise produced by the partly random nature of transcription initiation, but their conservation during evolution suggests that they have a function.[45] The presence in these RNAs of sequences that bind microRNAs may suggest that—like the transcription of pseudogenes—they play a very precise role in gene expression. If this were the case, the level of expression of a specific gene would need to be understood in the context of the expression of all other genes (and transcribed sequences) that carry the same microRNA recognition sites. Only a holistic (systems biology) approach to transcription could enable us to know if this is the case (see Chapter 28).

▌ The Extension of the Catalytic Power of RNAs and Discovery of Riboswitches

After 1990, the Selex method revealed that RNAs were capable of binding large numbers of molecules. These aptamers, as they were called, were soon used in biotechnology as sensors.

From molecular recognition to catalysis is only a short step. Just as antibodies were converted into antibody-enzymes in the 1980s (Chapter 21), so aptamers that could recognize the transition state analogs of an enzymatic reaction were shown to possess enzymatic activity.[46] The catalytic power of RNAs, which had long been limited to the cleavage and ligation of nucleic acid chains and centered around the phosphate group, was rapidly extended to other reactions catalyzed by proteins.[47]

The best confirmation of the catalytic power of RNAs was provided, almost by chance, after the determination of the crystal structure of the ribosome by X-ray diffraction. In 2000, Thomas Steitz's group showed that active sites on the ribosome where peptide bond formation occurred consisted solely of large ribosomal subunit RNA.[48] The same team suggested that RNA alone was involved in the catalytic mechanism, which was rapidly confirmed.[49] Although this was a result that had been predicted by some researchers, it nevertheless had a considerable impact.[50] Could there be more direct proof that before the DNA-protein world we know today there was an RNA world? These results stimulated new research into the origins of life and prompted a great deal of speculation about the RNA world and its transition into the existing living world.[51]

Nevertheless, specialists in protein structure remained convinced that, as well as the more restricted diversity of the functional chemical groups carried by nucleotides as compared to amino acids, the conformational potential of RNA molecules was less than that of proteins. Moreover, the description of the conformation of RNAs was less advanced than that of proteins.[52] All this meant that proteins continued to be considered as playing the main role in molecular recognition and catalysis in cells.

This competition between the two types of macromolecules probably helped push protein specialists to uncover a dynamic that was specific to their favorite molecules. As we have seen, these projects have revolutionized the study of proteins, which is now seen in terms of the study of conformational populations in equilibrium (Chapter 21).

However, an approach that viewed the conformational dynamic of RNAs as intrinsically limited was incompatible with the rapidly growing list of ribozymes. It also seemed increasingly strange that none of the large number of sensors found in cells used the recognition power of aptamers. In 1999, this led Ron Breaker, a specialist in the construction of sensors from RNA, to carry out research into natural RNA sensors.[53] All the bacterial sensors that had been characterized up to that point were proteins, such as the repressors that blocked biosynthetic pathways when their products were present in the environment. Breaker's findings were completely unexpected: it was the mRNAs themselves that bound the ligands. The change in RNA conformation caused by this binding inhibited its translation into proteins. The first example of this was found for vitamin B_{12}, which binds to the mRNA coding for the enzymes needed for its production and blocks its translation.[54] This was soon followed by the discovery of a similar mechanism of regulation by vitamin B1.[55]

Numerous other riboswitches have since been described, both in prokaryotes and in eukaryotes.[56] They may block the initiation of translation and also interfere with transcription termination or splicing. Other studies have confirmed the ability of RNAs to adopt a wide range of configurations. It is clearly tempting to imagine that riboswitches played a major role in the RNA world.

RNA now well and truly occupies the central place referred to at the beginning of this chapter. Unlike its semi-inert cousin DNA, RNA is a jack-of-all trades molecule, endlessly active in the cell in many different forms. The discoveries of the last three decades have substantially strengthened the RNA world hypothesis, but we still do not know which of the current functions of RNA were inherited from the RNA world, and which are the result of the perpetual cycle of evolutionary tinkering.

The emergence of RNA as a major macromolecular actor was a revolution within molecular biology, not outside of molecular biology or against it. This revolution involved endowing RNA molecules with properties that had hitherto been attributed solely to proteins. While this was novel and exciting, at one level it was merely a rebalancing of the macromolecular landscape.

26

Epigenetics

EPIGENETICS PLAYS a major role in modern biological research. But it does more than this: it resurrects forgotten and long-abandoned theories, and it is claimed to shake the foundations of molecular biology. Recently, via the media and popular books, epigenetics has percolated down to a lay audience, raising great and unrealistic expectations. At the same time, debate within the scientific community about the significance of the supposed "epigenetic revolution" has grown increasingly bitter.

Epigenetics cannot be understood, and its ambitions cannot be discussed, unless it is viewed in terms of its historical development. To understand the terms used, and the expectations of its supporters, we need to return to the past, to the confrontation between biochemists and molecular biologists that occurred in the middle of the twentieth century, to the resistance of many embryologists to the appearance of genetics at the beginning of the last century, and even further back to the two contradictory models of embryonic development, epigenesis and preformation, that emerged in the seventeenth century.

The debate about the significance of epigenetics blew up at the beginning of the 2000s. This may seem too recent to have a place in this book, and only some aspects of the debate will be outlined here. But the interactions between molecular biology and the precursors of epigenetics started earlier, when molecular biology was still in its infancy. Examining the beginnings of epigenetics also enables us to better understand the opposition that the models and explanations of molecular biology had to face when the science was in its earliest days.

▌ Epigenetic before Epigenetics

The term *epigenesis* was first used by William Harvey in the seventeenth century to describe the gradual construction of the organism during embryogenesis. As a theory this was not new: it had been proposed in the fourth century BCE by Hippocrates and Aristotle. What was new in the seventeenth century, however, was the emergence of the theory of preformation, which asserted that organisms existed, fully formed but in miniature, either in the male sperm or in the female ovum (according to different naturalists), and merely grew during embryogenesis. A number of factors contributed to the emergence of a theory that at the time was at the cutting edge, but in retrospect appears absurd, and even a regression in comparison with the ideas of the ancient Greek philosophers.[1]

First, the use of microscopes revealed previously invisible organisms and gave the impression that there was no lower limit to life. Second, early embryological studies of insect eggs and of seeds revealed tiny versions of the adult form. Third, despite the discovery of eggs and sperm, there was no recognition that these two components were equivalents—this required the development of cell theory in the nineteenth century. Other contributory factors included the difficulty of proposing a mechanistic or Cartesian model that could account for the construction of organisms and the need of theologians and early scientists to see God's hand in the creation of life and the establishment of natural laws, rather than in constant and miraculous divine interventions in the earthly sphere.

The theory of preformation gradually lost its influence in the eighteenth century, but in the early years of the twentieth century the development of genetics was perceived by embryologists as a return to preformationism. This, for instance, was the position of the embryologist Thomas Hunt Morgan before his conversion to genetics.[2] Genetics showed that the characteristics of an organism, its phenotype, were contained in its genotype. It was not the organism as a whole that was "preformed," but rather its different parts or characters that were somehow contained in the genes.

This explains why, in the early years of the twentieth century, embryologists began to use the adjective "epigenetic" to describe aspects of development that appeared not to be determined by genes. For example, for the Soviet biologist Nikolai Koltsov the determination of the axes in the egg was an epigenetic phenomenon because it was the consequence of the position of the oocyte relative to the surrounding cells in the maternal organism, and not of the action of genes.[3]

This use of "epigenetic" to mean "nongenetic" persisted throughout the twentieth century and continues to the present day. For example, once David Hubel and Torsten Wiesel had demonstrated the plasticity of synapses in the visual cortex of cats in the 1960s and 1970s, the role of genes in the precise wiring of the nervous system and the formation of synapses became a focus of discussion. The stabilization of synapses late in development, which depends on the earlier functioning of the nervous system and the selection of active circuits, was described as an epigenetic phenomenon; the plasticity of the nervous system was seen as the result of this epigenetic formation of synapses.

The same nongenetic meaning of epigenetic can be found in the expression "epigenetic mechanisms of inheritance." The use of this term may be relatively recent, but the famous experiments showing the existence of mechanisms of inheritance independent of genes were carried out on simple biological systems between 1940 and 1960. Max Delbrück showed that if two bacterial metabolic pathways interact, such that the intermediary products of one inhibit the other, the order in which the substrates of the two pathways are added to the environment determines which pathway is active, and, most significantly, this differential activity persists in the bacterial progeny.[4] In the lactose system, the temporary addition (or not) of the inducer will determine for many generations whether the synthesis of lactose-metabolizing enzymes is induced by the addition of low concentrations of lactose (this occurs through the synthesis of lactose permease and its stability in the bacterial membrane, which facilitates the entry of the inducer into the cell). If scientists change the position of some of the rows of cilia present on the external surface of the unicellular organism *Paramecium,* these modifications can be faithfully transmitted to the following generations.[5] Similarly, conformational alterations caused by prions can also be said to be epigenetic because the phenotype is not dependent on genetic variation and is transmissible through cell division (Chapter 20).[6]

▌ Epigenetics According to Conrad Waddington (1942)

The geneticist and embryologist Conrad H. Waddington is universally acknowledged as the father of epigenetics. In 1942, he called for the creation of
a new science, epigenetics, in order to describe the mechanisms by which
genes control development.[7] But although Waddington's definition accurately describes modern developmental biology and developmental genetics,
it does not correspond to the contemporary meaning of epigenetics. For
Waddington to be considered the father of epigenetics, his other work and in
particular his attitude toward molecular biology need to be explored.

Waddington's conception of gene action during development differed
from that of most geneticists. According to his famous drawings of what he
called the epigenetic landscape, differentiation and development are due
to the indirect action of multiple genes, which create the landscape in which
development occurs. In addition, according to Waddington, the environment played an important role in development. Although embryogenesis
was buffered against environmental variations, there were nevertheless critical periods during which it was more susceptible to perturbation. The effects of the environment could be mimicked by the effects of genes—this
was the basis of the famous mechanism of genetic assimilation supported
by Waddington—but the reverse was also true.[8] Variations in the environment could mimic the action of genes and, in the case of what were called
"phenocopies," mimic the effects of specific mutations. (Until molecular biology became a routine tool in studies of eukaryotic organisms, phenocopies were one way of trying to interpret the effects of mutations.)

A focus on the large number of genes involved in development and the
role of the environment are typical features of modern epigenetics. Waddington also opposed the attempt by molecular biologists to use the operon
model of gene regulation, which had been developed by François Jacob and
Jacques Monod in bacteria, to explain the control of gene expression in development. Instead, he supported the very different model by Roy Britten
and Eric Davidson, which emphasized the role of networks of regulatory
genes producing ribonucleic acids (RNAs).[9] For these reasons, and not for
his definition of epigenetics, Waddington may nevertheless be considered
to be the father of epigenetics.

In 1958, David Nanney introduced another meaning of the term
epigenetics. According to Nanney's definition, epigenetics is the sci-

ence that describes the mechanisms by which the expression of genes is controlled.[10] This is one of the modern meanings of "epigenetics"—it is effectively a sexy-sounding synonym for "gene regulation." If Nanney's definition is the right one, the operon model was one of the first significant results of epigenetics, and Waddington was wrong to deny its value!

▌ The Roots of Modern Epigenetics: 1 Histone Modification

I will use the term "modern epigenetics" to encompass all current studies of DNA methylation and histone modifications (including the replacement of major forms of histones by minor ones)—two processes that alter gene expression. Histones are major components of the group of proteins that surround the DNA molecule and form chromatin. Studies of small RNAs and of the structure of the nucleus are often described as being epigenetic studies, but their significance in this respect remains marginal. Taken together, these studies represent most of the research done today under the banner of epigenetics.

The distinction between different forms of chromatin—euchromatin and a compact form, heterochromatin—dates back to the work of geneticists and cytogeneticists in the 1930s. By the 1940s, the position effects resulting from the translocation of genes and the low activity of genes present in the heterochromatin were well-established facts. In 1950, Edgar and Ellen Stedman showed that histones differed from one differentiated cell type to another. The Stedmans—who were dubious about the role of DNA as the genetic material—hypothesized that these histones had a negative effect on gene activity.[11]

In the early 1960s, characterizing and understanding the role of histones became a major topic, notably thanks to new molecular biological tools that made it possible to directly estimate the level of gene expression by measuring transcription in cells, or in cellular extracts. The Stedmans' suggestion that histones had an inhibitory effect was soon confirmed.[12] Even more significantly, it was discovered that histones could take different modified forms—by acetylation or methylation of some of their amino acids—and that these modifications altered their activity; in particular, acetylation blocked their inhibitory action.[13]

These observations led Alfred Mirsky and his collaborators to propose a general model of gene regulation in eukaryotes based on histones (bacteria do not possess histones). According to this model, global gene expression was inhibited by histones, but the selective modification of histones might permit the specific activation of genes.[14]

The emergence of this model needs to be understood in its historical context. In 1961, Jacob and Monod proposed the operon model of gene regulation in bacteria.[15] A few months later they daringly adapted it to account for eukaryotic gene regulation during differentiation and development.[16] As we have seen, this drew a hostile reaction from embryologists (Chapter 22), who argued that the explanation of gene regulation during development required the construction of new models specifically adapted to the phenomenon under study, rather than simply imported from bacteria, which do not develop.

The model proposed by Mirsky and his colleagues was specific to eukaryotes, but it had a serious flaw, as was swiftly pointed out by Eric Davidson, one of his former students: it could not explain the specific pattern of gene activation that was observed during development.[17] The need to overcome this difficulty explains the role attributed to activator RNAs in the Britten–Davidson model.[18]

Though not completely forgotten, the possible role of histones in gene regulation was partly overshadowed by advances in the description of the structural organization of histones in chromatin. In 1974, Ada and Donald Olins suggested that chromatin was organized in spheroid units.[19] Later dubbed nucleosomes, these units were gradually characterized, and after over a decade of work, their structure was precisely determined.[20] A great deal of attention was given to the role of nucleosomes in the compaction of DNA within the nucleus. The study of histone modifications was not wholly abandoned, but early studies failed to show any major effect of these modifications on nucleosome structure. In the late 1970s, the use of nucleases gradually revealed an accurate picture linking the initiation of transcription (and the expression of genes) to the absence of nucleosomes on the promoter. Attention was focused first on the remodeling enzymes driven by adenosine triphosphate (ATP) that were able to displace nucleosomes, before being directed in the late 1980s to histone modifications and their role in stabilizing or destabilizing nucleosomes.

▌ The Roots of Modern Epigenetics:
2 DNA Methylation

The discovery of DNA methylation and of its role in gene regulation belonged to a completely different line of research, which had its origins in the discovery of modified bases in the DNA of some organisms. The first observations were carried out on the DNA of bacteriophages. These modifications attracted a lot of attention, but their raison d'être remained obscure. The first clear role to be attributed to them followed the characterization of restriction / modification in bacterial systems.[21] Bacteria protect themselves against foreign DNA by cleaving it with nucleases (restriction enzymes); to protect their own genetic material from these dangerous enzymes, they make a specific modification (methylation) of one of its bases. As we have seen, the isolation and characterization of restriction enzymes from different bacterial organisms was a major step in the development of genetic engineering.

In the mid-1960s, the characterization of bacterial systems of restriction / modification also generated a renewed interest in base modifications, in particular in eukaryotes, with the mechanism of restriction / modification being used as a model. A protective role for this mechanism, as in prokaryotes, was not ruled out, but the favored hypothesis was that it might play a role in differentiation and development.[22] Eduardo Scarano and colleagues proposed a model in which the modification of DNA during differentiation produced cuts in the sequence, enabling the different cells of the organism to synthesize different proteins.[23] This article foreshadowed current suggestions that DNA methylation might lead to mutations that could play a role in evolution (and not simply in development as in the earlier model)—in other words, that epimutations might be converted into true genetic mutations.

These early attempts to give DNA methylation a functional role in eukaryotes have now been forgotten. Nowadays, the history of DNA methylation is considered to have started in 1975, with the simultaneous suggestion by Robin Holliday and Thomas Pugh and by Arthur Riggs that DNA methylation was an important mechanism for the control of gene expression in eukaryotes.[24] They hypothesized that two different types of methylation enzymes were necessary, one for introducing new methyl groups into DNA, and the other for maintaining and reproducing the pattern of methylation during DNA replication.

Two important things about these articles often go unrecognized. The first is that the methylation hypothesis was not based on any experimental evidence. The second is that, as a result, both articles were highly speculative and most of the hypotheses they contained have been abandoned. (For example, Holliday and Pugh proposed that DNA methylation might be a way to count cell divisions during differentiation and development, and could also form a key component of the genetic program of aging.)

Despite their lack of experimental evidence, these articles were highly regarded, partly because developmental biologists, having discarded the operon model, were now frantically searching for mechanisms that could explain the dramatic changes in gene expression during development. Both 1975 articles mentioned X chromosome inactivation in female mammals, even though there was no evidence to implicate DNA methylation in this process. (Mary Lyon, who discovered X inactivation, had considered many different mechanisms that might explain it.)[25] The issue of specificity, which was raised by the abundance of methylated residues (cytosines) in the eukaryotic genome, was resolved by hypothesizing that sequences recognized by the methylating enzymes and responsible for the control of gene expression were concentrated in certain areas that were close to the genes they regulated.

In the following period, there was growing experimental support for the role of DNA methylation in some examples of gene regulation. It was soon demonstrated that the level of cytosine methylation was higher in heterochromatin than in euchromatin, and that the pattern of methylation was conserved during cell division.[26] Transfection data showed that methylated DNA fragments were expressed at lower levels than non-methylated ones. Islands of demethylated CpG (5' cytidine-phosphate-guanosine 3') were described in the promoters of actively transcribed genes.

The most striking experiments were carried out in the 1980s with 5-azacytidine, an analog of cytidine that cannot be methylated. Adding this compound to cells or organisms led inactive genes to be reactivated.[27] One example of this was the fetal form of the gene coding for hemoglobin, which could be reactivated in adults by treatment with 5-azacytidine. Researchers seriously considered using this technique to reactivate this gene in patients suffering from thalassemias, genetic blood disorders in which hemoglobin genes are in various ways dysfunctional. This strategy had to be abandoned because the drug was not specific in its effects—it reactivated all sorts of genes and was consequently toxic.

Understanding the role of DNA methylation in gene regulation also encountered obstacles. DNA methylation does not occur in either *Drosophila* or nematodes, two of developmental biologists' favorite organisms, showing that this phenomenon is far from universal. When it was found that cytosine methylation induced a change in the conformation of the DNA molecule, from the classic B-form to the newly discovered Z-form (a less elegant right-handed double helix—Chapter 11), this seemed to suggest a mechanism that could explain how methylation worked. This turned out to be wrong, and the suggestion sidetracked this field of research for some years.[28]

The strongest argument in favor of a role for DNA methylation in some cases of gene regulation came later, with the discovery and description of genomic imprinting. In a fraction of mammalian genes, the copies inherited from the father and the mother are not expressed equally, one copy being inactivated while the other is fully active. In any given species, moreover, it is always the same copy, either paternal or maternal, that is active or inactive. The difference in the expression level is associated with the level of methylation in the corresponding genes. This fine-tuning of gene expression was shown to be a requirement for the proper development of the organism.

By the late 1980s, Robin Holliday claimed that his hypothesis that gene expression could be controlled by DNA modification (methylation) had been confirmed.[29] His 1987 article was as speculative as the article of 1975, but the nature of the speculations had changed. His idea of a developmental clock based on DNA methylation had gone, but he kept the focus on aging, which had been his main research topic before his interest in DNA methylation.[30] Holliday suggested that the limit on the number of cell divisions that can occur in cell cultures, known as the Hayflick limit, was determined by DNA methylation, thereby reintroducing the idea of a "methylation clock" in aging, even though he had abandoned the idea of such a clock operating in development.

Holliday changed his model in two significant ways, both of which had important consequences. The first was to relate his DNA methylation observations and hypotheses to Waddington's models, placing DNA methylation in the realm of epigenetics. The second was to insist on the phenomenon of cell inheritance, or the ability of a cell to conserve its pattern of gene expression after cell division. This was explained by the conservation of the pattern of DNA methylation and opened up the possibility that these patterns, or their alterations, might be transferred from one generation to the next, a form of nongenetic inheritance. This began a debate about the role of epigenetics in evolution that continues to the present day.

Epigenetics in the 1990s

In the 1990s, epigenetics emerged as a major field of research. In both methylation / demethylation and histone modifications, progress came from the isolation and characterization of the genes responsible for these modifications. The isolation of mutations in yeast, and the search for homologous genes in mammals, made it possible to describe the enzymes involved in the DNA methylation / demethylation reactions in the early 1990s. Enzymes involved in the modification of histones, such as acetylase-deacetylase reactions, were identified from the mid-1990s.[31]

The first results revealed the extraordinary complexity and diversity of these mechanisms. Earlier hypotheses, such as the distinction between de novo and maintenance methylases in DNA methylation, were confirmed. It also became possible to test the role of these modifications in vivo through the inactivation of the corresponding genes.[32] The results were often more ambiguous than had been expected—the scale of the effects did not correspond to the fundamental role attributed to these modifications. Eventually, sophisticated physicochemical techniques were used to demonstrate the stabilizing or destabilizing effects on the nucleosomal structure of the different types of histone modification. A new technique, combining DNA hybridization and immunoprecipitation and known as chromatin immunoprecipitation (ChIP), provided a precise description of the localization of the different modified forms of histones on the genes, and confirmed the association of certain types of histone modifications with high (or low) levels of gene expression. The role of variant forms of histone was also demonstrated.[33] Multicomponent, ATP-driven, enzymatic machines capable of remodeling chromatin by displacing nucleosomes were described.[34]

The most significant advance at this time was the gradual description of the many interactions between the macromolecules that contribute to transcription regulation and chromatin modification. Among the first and most significant links to be established were those between transcription factors and acetylation / deacetylation enzymes, and between proteins specifically recognizing methylated DNA or certain modified forms of histones, and proteins responsible for chromatin compaction.[35] The diversity of histone modifications, and more generally the complexity of the machinery involved in the control of gene expression, explains the level of interest aroused by epigenetics in the late 1990s and early 2000s.[36]

In the late 1990s, the discovery of RNA interference (see Chapter 25) raised the visibility of epigenetics again. RNA interference explained the paradoxical observation, made on transgenic plants in particular, that an increase in gene copy number could lead to a dramatic decrease in expression level. This was explained at the post-transcriptional level by the inhibition of translation or the degradation of messenger RNAs (mRNAs) by small interfering RNAs (siRNAs), while at the transcriptional level it was associated with chromatin modification. It was soon shown that small regulatory RNAs also played a role in chromatin modification, thereby introducing a new actor into epigenetics, alongside DNA methylation and histone modification.[37]

Enthusiasm and Difficulties

Many researchers found these results particularly exciting, as they held out the promise not only of discovering new phenomena but also of transcending the disappointing aspects of the human genome sequencing program, which had not immediately produced any significant insights. For this reason, epigenetics was considered a component of postgenomics.

Some interesting and unexplained findings were shown to be the product of epigenetic mechanisms. For example, paramutation, where one allele is altered by another allelic form, occurs through histone modification or DNA methylation. Similarly, the complex mechanisms by which protists such as *Tetrahymena* and *Paramecia* generate an active macronucleus from the micronucleus were shown to be epigenetic, while the appearance of different castes in social insects such as honeybees also turned out to depend on epigenetic mechanisms. This last example was not surprising—workers and queens have the same genome, so, by definition, this was a case where differential gene regulation, not sequence variability, was involved in producing phenotypic differences.

For many, the promise of epigenetics flowed from its particular characteristics. Epigenetic modifications regulate gene expression, thereby limiting what some perceived as the threatening, deterministic power of genes. A whole series of experiments showed that epigenetic marks can be induced or altered by the environment: by changes in temperature in plants; by food or toxic chemicals present in the environment in mammals, such as endocrine disruptors; and by the behavior of other organisms, such as the way that rat mothers act toward their pups.[38] These modifications may play a part in the development of diseases and—in particular through the action of hormones—in changing behavior. As we have seen, in some species, in particular in plants, some epigenetic marks can be transmitted to succeeding generations, although they are eventually removed.[39]

The enthusiasm for epigenetics extended far beyond the scientific community, producing a popular but mistaken belief that our behavioral choices can alter our genetic susceptibility to various diseases, changing our destiny and even that of our children. The mechanisms underlying obviously sensible public health advice were blurred by unwarranted extrapolations from poorly established facts.

THERE IS NO SINGLE FIELD of epigenetic research, nor has there ever been.[40] Our understanding of the role and mechanisms of DNA methylation and histone modifications grew gradually, but until the late 1990s these two mechanisms were generally studied by completely different groups. Researchers working on DNA methylation did not refer to those working on histones, and vice versa. Today, the effects of the two mechanisms cannot be separated: DNA methylation and some types of histone modification (as well as small regulatory RNAs) may contribute to chromatin condensation. But much remains to be discovered about the direct interactions that might occur between these different regulatory mechanisms. For example, we still do not know how small regulatory RNAs might contribute to the transmission of epigenetic marks across generations in some organisms.

One reason for this lack of unity in the field is that from the very beginning, DNA methylation was studied by molecular biologists, while histone modifications were investigated by biochemists and developmental biologists. The rise of epigenetics has not led to the emergence of a new discipline. Instead it involved a revival of biochemical studies in a field that had previously been dominated by molecular biology, and therefore also constituted a continuation of the conflicts between the two disciplines that had emerged in the 1960s.

In the early 2000s, many epigenetic researchers began to refer to a so-called histone code.[41] When different chemical modifications are introduced at different positions of the polypeptide chains and in the different types of histones that form a nucleosome, this can affect the structure of the nucleosome, and hence the way the gene is expressed. This histone code describes the effects on transcription and on the interactions with other proteins that are produced by different combinations of these modifications. The term "histone code" obviously refers to the well-established genetic code that involves the translation of genetic information into amino acid sequences, and implies that there is an equivalent level of coding, or information, in the structure of histone modifications.

The histone code is sometimes referred to as the "second genomic code."[42] But two decades after its existence was first suggested, the characteristics and nature of this supposed code remain elusive. By coining this term, with its bold implications, epigeneticists trained in biochemistry were attempting to demonstrate that their discipline was somehow equal to molecular biology. But behind the boldness there was a major scientific error: these phenomena were completely different from the rules underlying the translation of genetic information, and it was not necessary to invoke the existence of a code in order to explain them. In addition, whereas the genetic code is the same in all organisms (with a few exceptions; see Chapter 17), epigenetic modifications are highly variable, both in their nature and their functions, in the different branches of the tree of life. This nonuniversality of the epigenetic code, which is rarely mentioned, creates a key difference, and a hierarchy, between the genetic code and the supposed epigenetic codes.

The difficulties encountered by epigenetics found their origin in its ambiguous terminology. "Epigenetic" can simply mean "nongenetic," or it can describe the mechanisms of gene regulation, such as DNA methylation or histone modification. In the popular imagination, however, it is synonymous with nongenetic transmission of hereditary information. In reality, when a character is described as being epigenetically transmitted because its transmission did not obey Mendel's laws, this does not necessarily mean that epigenetic marks were involved; other factors, such as hormonal effects, may also be involved.

Epigenetics is not alone in such ambiguity—there is a similar vagueness about "inheritance," which may describe what happens either during cell division or in the transfer of hereditary information to progeny. Interestingly, DNA methylation is directly inherited during cell division in those species that show this effect, whereas it is still unclear if the same things occurs for histone modifications. In contrast, the direct transmission of methylation marks across the generations in mammals—something that popular accounts of epigenetic inheritance in humans would suggest is commonplace—is challenged by the fact that mammals possess mechanisms that erase epigenetic marks during the earliest stages of embryogenesis.[43]

Some researchers, sometimes egged on by journalists and popular writers, have argued that genetics needs to be replaced by epigenetics, and that the Darwinian view of evolution needs to be abandoned in favor of a neo-Lamarckian one, based on the intergenerational transmission of epigenetic marks acquired during life. Epigenetic mechanisms have an important part to play in explaining development as well as some diseases, but epigenetics is not genetics. Indeed, the word literally means "above" or "on top of" genetics. Epigenetics cannot exist without genetics, and a purely epigenetic world would be impossible because genes and epigenetic modifications have different roles. Genes determine the nature of cellular products; epigenetic marks help regulate their synthesis.

Furthermore, epigenetic modifications are often the consequence of the genetic events that preceded them, so they could legitimately be described as genetic. Scientists now accept that, in species that show the effect, DNA methylation stabilizes inactive states of gene expression—states that were initiated in the absence of methylation. Because histone modifications are so complex, their significance remains unclear. It is probable that in most cases these modifications can also be initiated by genetic events, such as the binding of transcription factors. This view of the mechanisms of gene regulation is rather less dramatically novel than the more extreme accounts of epigenetics that can sometimes be found in the press.

The enduring legacy of epigenetics will probably be that it will have contributed to the nuancing of the old gene-centered view of organismal function, replacing it with a framework in which genes are neither at the top of the system nor at its heart. Rather, genes are fundamental components whose activity is not determined but instead is regulated by both the environment and by other genes. Despite the more exaggerated claims of some of its proponents, in particular among the nonscientific community, epigenetics does not sound the death knell of genetics. It will not overthrow Darwinism and the Modern Synthesis, and it will not fundamentally alter molecular biology.

27

Sequencing the Human Genome

▍ The Origins of the Human Genome Project

The story of the Human Genome Project (HGP) has been told by both participants and observers.[1] The project traces its roots to a number of key events. First, in 1985 Robert Sinsheimer, chancellor of the University of California at Santa Cruz, organized a conference to discuss reallocating part of the funds that had been raised for developing a new telescope to a major biology project of unprecedented scale: the sequencing of the human genome. The same year, Charles DeLisi, director of the Office of Health and Environmental Research at the U.S. Department of Energy (DoE), proposed that sequencing the human genome could be a new objective for the department's many laboratories. (These DoE laboratories had hitherto been studying the effects of radiation on human cells, but this research had been rendered obsolete by the gradual reduction of Cold War tensions and by the ban on atmospheric nuclear tests.) The following year, 1986, Renato Dulbecco published an article in *Science* in which he argued that it was time to abandon the dominant piecemeal approach to studying cancer in favor of a systematic search for the genes involved in tumor formation.[2] At the same time, Wally Gilbert led discussions at Cold Spring Harbor Laboratory to evaluate the feasibility and cost of such a project. Finally, also in 1986, Applied Biosystems announced the development of a sequencing machine in which radioactivity was replaced by fluorophores, and which made it possible to automate Sanger sequencing.[3]

Strictly speaking, even these near-simultaneous suggestions were not entirely novel—a project like this never comes out of the blue. In 1980 David Botstein, Ray White, Mark Skolnick, and Ron Davis published an article in which they proposed the building of a precise genetic linkage map of the human genome, using restriction fragments length polymorphisms (RFLPs).[4] This proposal—which never came to fruition—can be seen as an early step toward the HGP, involving a marriage of traditional genetic mapping techniques and new molecular tools.

The foundation stone of human genome sequencing had been laid much earlier, in 1953, when Jim Watson and Francis Crick described the structure of the double helix. Once it had been hypothesized (in their second *Nature* article of that year) and then demonstrated that the sequence of nucleotides in DNA contained the information to synthesize proteins, sequencing the human genome became a possibility and a goal. What was needed was a catalyst, and this was provided by those convergent events of 1985 and 1986.

One immediate consequence was the birth of a new discipline, genomics, a combination of genetics and molecular biology, supported by computational biology.[5] The comparison of genomic sequences and an understanding what they could reveal about the evolution of genomes were immediately perceived as an important new development.

In 1988 the responsibility for the HGP was transferred from DoE to the National Institutes of Health and the newly created National Human Genome Research Institute (NHGRI). Jim Watson was initially appointed to head the program (he was succeeded by Francis Collins in 1992). In parallel with this U.S.-based initiative, in 1988 the international Human Genome Organization (HUGO) was created. National initiatives were soon launched in the United Kingdom, Japan, and France.

▌ Debates about the Value of the Project

The idea of sequencing the human genome attracted criticism from its very beginning. Everyone accepted that genetic and physical mapping would be useful, especially when it came to characterizing the genes involved in human diseases. In a physical map of the genome, genetic characteristics are linked to specific DNA sequences; this makes it possible to localize DNA fragments in the genome. But sequencing the full genome was widely considered to be pointless, partly because around 95 percent of it was thought to be junk DNA (see Chapter 23). Other strategies were suggested. For example, Sydney Brenner argued that the highly compact genome of the pufferfish (fugu) could be a model for the study of vertebrate genomes. People also argued about what sequencing methods should be used.

The cost of the project and the negative impact this would have on the funding of other research programs were also a focus of argument. Furthermore, to many observers the project appeared to be driven by political imperatives rather than by the spirit of scientific enquiry. Most biologists viewed the conversion of biology into Big Science as a threat to their discipline. And to add to all this, it was hard to see how sequencing the genome might be appropriate training for graduate students.

In 1990 and 1991, the project underwent a crucial redesign. Sequencing—which was proving lengthy and complex—was postponed, and the primary objectives were now the production of both an overall genetic map of the human genome (with a resolution of 2 centimorgans [cM], or about 1 million base pairs), and a corresponding physical map to make it possible to localize the fragments of the genome. Efforts would be directed toward bringing down the cost of sequencing and developing the computer programs necessary to assemble the millions of sequences that would be produced.

In parallel, the genomes of other organisms were to be studied, either to test the methods to be used for the human genome (for example, sequencing *Escherichia coli* and *Bacillus subtilis*), or to help interpret the data that would eventually be generated by human genome sequencing (for example, studying the mouse genome). The mapping and sequencing of genomes such as archaea microbes or the plant *Arabidopsis thaliana* was also begun.[6] A significant part of the HGP was also to be devoted to addressing the legal, social, and ethical issues that its development would inevitably raise.

While this strategy was still being discussed within the HGP, a 10 cM genetic map of the human genome was published by the Paris-based Centre d'Etude du Polymorphisme Humain, founded by Jean Dausset and directed by Daniel Cohen.[7] This organization was a legacy of work that Dausset had performed on the human leukocyte antigen (HLA) human histocompatibility complex and was funded by a French charity devoted to the study of muscular dystrophies (Association Française contre les Myopathies), which hoped that new genetic and molecular tools might help combat these debilitating conditions.

Pretty much from the beginning, scientists recognized the need for "a common language for physically mapping the human genome"—in other words, for all groups to use the same approach.[8] The recently developed polymerase chain reaction (PCR) was felt to be best adapted for the project, and microsatellite sequences (short repeated and variable stretches of DNA) were seen as good candidates for creating the map. In 1991, Craig Venter suggested that instead the project should concentrate on complementary DNA sequences: DNA sequences established from ribonucleic acid (RNA) that had been transcribed from genomic DNA by the cell, which were therefore presumably of some biological significance.[9] He proposed using these sequences to physically map the genome.

Eventually, agreement was reached, and the different approaches were grouped around the notion of sequence tagged sites (STS). Radiation hybrids—hamster cells containing long fragments of human chromosomes produced by radiation—were very useful for the mapping process. The cloning vectors used to constitute the libraries of genomic fragments also changed gradually; cosmids derived from bacteriophages were initially used, followed by "artificial chromosomes" of human genetic sequences, which were created using bacteria (BAC) or yeast (YAC), although the latter were initially found to be unstable.

As a result of all this work, genetic and physical maps of the human genome were produced, which meant that the human genome sequencing effort could be coordinated and divided between laboratories.[10] The percentage of errors in the sequence that would be acceptable was the subject of much discussion. A consensus was reached that for a sequence to be released, it should be 99.99 percent accurate. When the draft sequence of the human genome was finally published in 2001, the level of accuracy was ten times higher.

Despite the modifications of the initial project, the criticisms did not immediately go away.[11] Many biologists complained that the funding of general research had decreased after the launch of the HGP. Many still considered sequencing the whole genome, along with its billions of bases of junk DNA, to be a waste of time and money. Some of the national genome projects developed outside the United States focused on genes rather than on whole genomes. As Philippe Kourilsky of the Pasteur Institute put it in a reference to *pain au chocolat,* the traditional teatime snack for French children, their aim was to "eat the chocolate but not the bread"—that is, to sequence genes rather than huge tracts of junk DNA.[12] The hype surrounding the promised results of the HGP also attracted harsh criticism: suggesting that scientists would discover the holy grail that would reveal the essence of human life and provide a blueprint for the understanding of human biology was viewed as an exaggeration and a deliberate manipulation of public opinion.[13]

The Race to Find "Disease Genes"

The late 1980s and early 1990s saw a race between a number of laboratories to isolate and characterize the genes and mutations involved in some of the most common and most visible human genetic disorders. These projects were independent of the HGP both organizationally and in terms of their aims, but they shared a common set of methods.

One of the first of these "disease genes" to be characterized was the gene that codes for the protein dystrophin—mutations in this gene are responsible for Duchenne muscular dystrophy. In some cases, this disease is due to a deletion of the gene or to a chromosomal rearrangement; both conditions make it easier to physically locate the gene. Complementary DNAs (cDNAs) corresponding to a part of the gene were cloned and sequenced.[14] The gene's exon / intron structure was described, and the protein was first characterized and then localized to a cellular structure called sarcolemma.[15]

The most celebrated discovery was that of the gene involved in cystic fibrosis (CF), the most common genetic disease in people of European descent. The complex symptoms of the disease made it impossible to predict the function of the mutated gene. At the end of 1985, three articles were published simultaneously in *Nature* and one in *Science* reporting the localization of the *CF* gene on chromosome 7, in close proximity to a DNA sequence that could be used to walk along the chromosome, sequencing along the way, to find the gene in question (see Chapter 22 for the first uses of this method).[16] The race was widely considered to be nasty, with rivalries and premature (and wrong) announcements.[17] Finally, a collaboration to carry out this walk was set up by Lap-Chee Tsui, Francis Collins, and John Riordan, and three articles were published simultaneously in the September 8, 1989, issue of *Science*.[18]

There was similarly tough competition to locate the *BRCA1* gene, which when mutated is responsible for a predisposition to early-onset breast and ovarian cancer. In the end, this race was won not by traditional academics but by researchers working for the Myriad Genetics company, thus enabling it to monopolize the genetic tests for this gene for over fifteen years.[19] The characterization of the gene involved in Huntington's disease was less bad-tempered and had lower financial stakes; it was published in 1993 by the Huntington's Disease Collaborative Research Group, an academic consortium.[20]

Genes involved in other genetic diseases were also isolated and characterized during this period. The list was not limited to monogenic diseases; complex multigenic diseases were also studied.[21] As a result of all this work, it became clear that the painstaking technique of positional cloning had to be replaced by the development of precise genetic and physical maps, and by a complete human genome sequence.

Characterization of the genes involved in diseases opened the way to prenatal diagnosis for these most common mutations. But it did not immediately explain the symptoms of the diseases, nor did it offer a means of fighting them. This was particularly true for cystic fibrosis, in which the precise nature of the defect in the most common mutation—the improper folding of an ion channel and problematic delivery of the protein to the cell membrane—was understood only after years of work.

Gene therapy did not require any understanding of how the gene worked in disease or in health. You simply had to isolate the healthy version of the gene and find a safe way of delivering it to the relevant tissues. Nevertheless, gene therapy encountered many obstacles that delayed or (in most cases) prevented the development of this new therapeutic approach (Chapter 24).

One "discovery" that gave rise to a huge debate was the claim that a gene had been identified that was responsible for homosexuality in humans.[22] This had been preceded by the publication of evidence that supposedly revealed a "gay brain," characterized by a difference in the size of hypothalamus.[23] Both the anatomical evidence and the genetic data were criticized.[24] Interestingly, at the time many gay rights activists embraced these mistaken claims because it reinforced their feeling that they were "born this way," which, they believed, was key to being accepted in society.

The search for genes controlling behaviors started long before high-resolution genetic maps in humans had been obtained. Seymour Benzer's project to characterize the genes involved in *Drosophila* behavior, begun in the 1960s, eventually led to the isolation and characterization of genes involved in circadian rhythms and learning.[25] The molecular study of the circadian genes, carried out principally at Brandeis University and Rocke-feller University, was eventually rewarded by the 2017 Nobel Prize in Phys-iology or Medicine. Studies of mice had localized genes involved in emo-tivity and drug and alcohol dependency.[26] There were problems with many of these reports because of the striking differences in the results from dif-ferent laboratories and when the experiments were repeated after a delay of several years.[27]

But despite these recurrent criticisms and the difficulties in establishing solid links between genes and behavior, the search for behavioral genes con-tinued. Observations of twins, or of behavioral differences due to different alleles involved in brain functions (for example, genes that code for recep-tors of neurotransmitters), seemed to support the ambitions of the behav-ioral geneticists. Too often, researchers leaped from identifying a gene that was somehow involved in a given disease to claiming that it played a major role in the functions affected by the disease, without sufficient evidence. For example, when mutations were found that were thought to be responsible for dyslexia or speech disabilities, the genes involved were immediately claimed to be responsible for the development and evolution of reading or speech (see the discussion of the *FoxP2* gene in Chapter 23).[28]

In historical terms, this period can be seen as one of regression. Molec-ular biology had gradually recognized the gap that often exists between the direct product of a given gene and the function—or more often functions—of that gene in the organism. In the late 1980s and early 1990s, the isolation of genes involved in diseases or in particular behaviors led to a resurgence of the naive early twentieth-century view of the gene, in which gene products were identified with specific characteristics of the organism. This view, in which there was a "gene for" virtually every behavior or phenotype, was in fact a late twentieth-century version of preformationism.

Advances in Genome Sequencing, and Controversies around the Right Strategies

It is worth looking in detail at two early genome sequencing projects. The first was organized by André Goffeau, who studied the genes involved in drug resistance in yeast; his ambition of sequencing the genome of the yeast *Saccharomyces cerevisiae* was supported by the biotechnology division of the European Commission.[29] The idea was to distribute the sequencing task between the various laboratories working on this organism (there were seventy-nine in all). Yeast was chosen for this pioneering work because it is a eukaryote, and its genome was expected to reveal molecular mechanisms specific to this major group of organisms, which includes humans. The yeast genome is also relatively small and compact, promising to minimize the time spent sequencing junk DNA. The fact that yeast has eighteen chromosomes made it easy to divide the work between the different laboratories. The different yeast chromosomes were sequenced successively: chromosome III in 1992, chromosome XI in 1994, and the full genome in 1996.[30] This project could also be easily sold in terms of its significance for the development of European biotechnologies.

The project soon revealed unexpected results. Over 40 percent of genes turned out to be hitherto unknown, while the fit between the genetic and physical maps was not perfect. These findings underlined the importance of drawing up precise genetic and physical maps, and of sequencing full genomes in order to discover the many genes that—even in an organism as well-studied as yeast—had not yet been identified, despite decades of painstaking work by classical geneticists. Following the completion of the sequencing program, new methods were developed to explore the functions of these newly discovered genes.[31] In the ensuing period, yeast remained a model for this work, which was called genome annotation, and for the development of what became known as postgenomic technologies (Chapter 28).

The *Caenorhabditis elegans* project followed a completely different logic and organization. This flowed from the very reason why the worm had been chosen as a model by Sydney Brenner and others in the first place—to obtain a truly complete description of the animal. The complete physical map of the genome of *C. elegans* had been drawn up early in the project, making it possible to share the work between different laboratories, and to initiate the sequencing program. The size of the genome (100 megabases [Mb]) was fifty times larger than that of the yeast genome, but thirty times less than the human genome—a good intermediate point for checking the protocols before initiating human genome sequencing.

In 1992, the Sanger Institute was set up by John Sulston at Hinxton, near Cambridge, in the United Kingdom. It was financed by the Wellcome Trust and the Medical Research Council. In parallel, Robert Waterston established a genome sequencing center at Washington University in St. Louis, Missouri. Sulston and Waterston had worked together—the

C. elegans community was small, with a tradition of tight collaboration. The work in these genome centers was performed by highly skilled technicians, each of whom was responsible for the sequencing of a small part of the genome. In 1994, the sequence of a 2.2 Mb fragment of a chromosome was published.[32] In 1998 the full sequence of the genome of *C. elegans* was completed.[33] Meanwhile, the two sequencing centers in Cambridge and St. Louis had turned to the sequencing of the human genome.

The first organisms to be fully sequenced were neither yeast nor *C. elegans,* but the bacteria *Haemophilus influenzae* and *Mycoplasma genitalium,* which were sequenced in 1995 by Craig Venter, who headed a privately funded group, The Institute of Genomic Research (TIGR).[34]

The sequences had been established by what was called the shotgun approach: random breaks were generated in the genome, and all the resulting fragments were systematically sequenced. Overlaps between the fragments made it possible to assemble the sequences into the whole genomic sequence; any gaps could be filled by additional, targeted experiments. The shotgun strategy was not new—it had been invented in 1981.[35] What was new, however, was the ambition to use it to sequence full genomes. This strategy did not require preliminary genetic and physical mapping, implying that previous efforts in this direction had been a waste of time and money. The emphasis was instead on newer and faster DNA sequencers, on arrays of powerful computers, and on the creation of sophisticated software that was designed to assemble the fragments.[36]

It was not clear whether the shotgun approach would work on huge genomes containing vast stretches of repeated sequences, and opinions differed on this vital question.[37] As it turned out, both approaches produced results. The sequence of *Bacillus subtilis,* obtained through a mapping strategy, was released two years after the first bacterial genomic sequences.[38] *Escherichia coli,* which because of its role in the development of molecular biology had been expected to be the first organism to have its genome sequenced, also provided early results, but its full genome was not sequenced until 1997.[39] The *Drosophila* genome, meanwhile, was the result of a mixed strategy in which shotgun sequencing played a major part.[40]

Venter was convinced that the shotgun strategy could be used to sequence the human genome, and he got his opportunity when the major pharmaceutical company, Perkin-Elmer, decided to set up a new company,

Celera Genomics, with Venter at its head and involving financial and personnel links with TIGR.

A huge debate erupted around the feasibility of shotgun sequencing the human genome, and the issue of free access to the data generated by the project.[41] Venter was accused of surreptitiously using information from the HGP and of not being sufficiently collaborative.[42] The price to be paid for being funded by a biotechnology company was that information had to be held back in order to develop diagnostic and therapeutic tools and obtain patents on them. This debate clearly gave the publicly funded HGP a powerful impetus to quicken its pace, with the result that in June 2000 it announced it had attained its aim. After complex political negotiations, the draft sequence of the human genome was published simultaneously in February 2001 by the public consortium and by Celera Genomics—in *Nature* for the HGP and in *Science* for Celera.[43]

By 2003, fifty years after the discovery of the structure of DNA, the program was viewed as complete. Interestingly, and contrary to expectations when the program was launched, this achievement had not required any technological breakthroughs in DNA sequencing.

▌ Genomics and Postgenomics

Once the human genome had been sequenced, it did not instantly reveal its secrets—not just because it was necessary to identify genes among long sequences, but also because the functions of the sequenced genes were not immediately obvious.[44]

The most striking finding was that the human genome contains surprisingly few protein-encoding genes. In the mid-1990s, the expected value was between 50,000 and 100,000; by 2000, when the program was nearly finished but the results not yet published, the predicted value had drastically decreased, to between 30,000 and 40,000.[45] Over the next few years, it was further reduced to little more than 20,000, and some current estimates are even lower. This came as a shock (*Drosophila* has around 13,000 genes and a mouse around 21,000), and it has been seen as a lesson in humility for our species.

Scientifically it was more significant: it was a confirmation of the tinkering action of evolution, which invents by recombining preexisting pieces, not by creating new ones from scratch. It also showed that genes homologous to those involved in human diseases might usefully be sought in organisms such as yeast or *Drosophila*, making it possible to exploit the richness and simplicity of the genetic tools available in these models to cast some light on the functions of these genes and the effects of mutations. The surprisingly low number of genes emphasizes the importance of the cellular processes that generate different products from a single DNA sequence, such as splicing, and protein processing and modification. Complexity lies not only in the number of components, but also in the way these components associate to generate structures and functions. Later, this became an argument in favor of what became known as systems biology (see Chapter 28).

One major challenge raised by the sequencing of the first genomes was the large number of genes with unknown functions—this was compared to the discovery of a new continent of life. This emphasis on the scope of our ignorance was also a way of distracting attention from the fact that the sequencing of the genome had not revealed the secret of life, and of arguing for a continuation and expansion of the programs. In the language of the time, it was the moment to shift from structural to functional genomics.

The rise of postgenomic technologies such as transcriptomics was one result of this shift in focus and is discussed in Chapter 28. But genomics itself could also help to discover these unknown functions. One of the main tools that was used was comparative genomics: a gene function that was identified in one species could often be assumed to be valid for a homologous gene in a different species. It was not only the conservation of genes that showed the significance of some genes, but also the conservation of the sequences surrounding the gene that revealed information about their function and evolution.[46] Further insight was gained from the observation that two proteins with unknown functions had homologs in another organism fused into a single chain, suggesting their involvement in a common pathway in both species.[47] Evidence that different proteins showed correlated evolution was also taken to indicate that they might function together in a pathway or in a structural complex.[48] Projects involving comparisons between closely related species in model systems, such as yeast and worms, or between humans and chimpanzees, were of particular importance.[49] The comparison of genomes immediately raises issues about the evolutionary history of the lineages in the period since they diverged from their common ancestor. These comparisons confirmed the importance of gene duplication, and also of the duplication of large stretches of chromosomes or even of the whole genome. Although genome sequencing was seen as a form of functional biology, once the first wave of sequencing had been successfully completed it rapidly opened up a way for evolutionary biology to come center stage.

As a result, the choice of organisms to be sequenced gradually shifted. In the late 1990s the objective had been to map the sequences of one or a few representatives of the major branches of the tree of life, each of which could be considered as models—Archaea, plants *(Arabidopsis thaliana)*, and so on.[50] In the subsequent period, the choice became both broader and more considered, focusing on organisms that lay at crucial positions in the tree of life, and which might therefore provide information about major evolutionary steps. For example, the genome of a tunicate was sequenced to gain a better understanding of the split between chordates and invertebrates.[51] This shift was concomitant with a transformation of the notion of model organisms in both animals and plants. Previously, the bacterium *E. coli* and bacteriophages had been at the core of the developments of molecular biology.[52] Now genome sequencing and postgenomics led to an increase in the number of models by allowing ideas, examples, and insights to be rapidly transferred from one system to another.

THIS PERIOD was followed by a second and opposite development. Instead of starting from the premise that a given species has a unique genome sequence, the objective now became to explore the diversity of genome sequences within a single species. The favored object of these studies was the human genome. From 2006 the number of such projects exploded, as new technological developments in DNA sequencing offered the hope that it would soon be possible to sequence any human genome for a thousand dollars (we are nearly there).[53]

An early objective, explored by the Human Diversity Genome Project, was to characterize the genetic diversity of different human populations (briefly described in Chapter 23). But the goal that became the focus of most efforts involved relating differences in sequences to differences in susceptibility to disease, or—with the creation of what was called pharmacogenomics—in response to drugs.[54] A first method, launched in 2005 and called HapMap, was rapidly replaced by genomewide association studies (GWAS), in which researchers systematically looked for associations between single-nucleotide polymorphisms (SNPs) and diseases. The first major result from this approach related to retinal macular degeneration and was published in 2005.[55] GWAS made it possible to characterize the genes involved, directly or indirectly, in any disease. This became the preferred way of developing new tools for what became known as precision medicine.

In one field of research—cancer—characterizing the differences between genomes concerned the cells of the same organism. Initially, cancer genomics research set out to sequence the genomes of tumor cells in order to characterize all their mutations and thus select the therapeutic approaches and drugs best adapted to each type of tumor.[56] But it was soon realized—or more accurately, it was soon rediscovered—that a tumor is not genetically homogeneous.[57] Rather, it is formed of different competing cell clones each bearing different mutations. The complete genetic description of a tumor would therefore require the characterization of the many cellular clones in which mutations had occurred, and which are competing for growth within the tumor.

There were two other important ways in which DNA sequencing technology was deployed to study the diversity of life. The first of these was the search for the origins of infectious diseases—in particular, new infectious diseases. Using recently acquired samples as well as preserved samples from the past, it became possible to use sequencing to reconstitute the origins and characteristics of the infectious agents responsible for historic pandemics, such as the Spanish influenza of 1917–1918, or for more recent pandemics such as acquired immunodeficiency syndrome (AIDS).[58] Once again, genome sequencing helped provide a richer description of evolutionary history, in this case of viruses.

The most interesting and significant application of the new DNA sequencing technologies, and especially of shotgun sequencing, began in 2004 with what Craig Venter called the "environmental genome sequencing" of samples taken from the Sargasso Sea. The project's breakthrough came from the immediate sequencing of the genomes present in the samples without any culture, thus avoiding the accidental overlooking of organisms that do not grow in the media traditionally used for bacterial culture. The result was a striking demonstration of the hitherto unsuspected diversity of organisms present in the ocean, most of them unknown.[59] This approach has recently revealed a whole new branch of eukaryotic life, the Lokiarchaeota, which may shed light on the origin of eukaryotes.[60] A similar diversity and abundance of forms was also described for bacteriophages.

What emerged from these observations was evidence of rich communities of microorganisms interacting together, accessible only through what was called metagenomics, which was to lend fresh impetus to microbiology.[61] One of these communities was to become the focus of a great deal of attention: the microbiota present in the human gut. The role this plays in immunity and in a wide range of human diseases, and its potential influence on human mood and behaviors, are still the focus of a great deal of excitement.

IN 1991, Wally Gilbert wrote a one-page article in *Nature* in which he defended the genome sequencing programs but also tried to understand the "malaise" felt by so many biologists: the feeling that molecular biology was becoming boring, filled with repetitive DNA sequencing tasks, and reliant on the use of kits or recipes described in detail in Tom Maniatis's book *Molecular Cloning,* which could be found on every laboratory bench.[62] Gilbert interpreted this as the result of a "paradigmatic shift in biology." This shift was not limited to biology but was the result of a more general phenomenon—the rapid increase in the amount of information stored in databases, combined with the possibility of immediate access to these data.

Gilbert was not pessimistic, however. Biology had not been transformed into Big Science, in which individuals and their insights would play little part. Once biologists had access to the information stored in databases, he argued, they would continue to work much as they had before: elaborating hypotheses and conceiving experiments to test them within small research groups. As Peter Medawar would have put it, biology (and science in general) remained "the art of the soluble," the art of using the best techniques to solve the right scientific issues of the time.[63] The use of DNA sequences was a technological bottleneck, not a conceptual one.

Gilbert was right, although he underestimated the risk that the forces who develop powerful technologies or design databases might impose their own agendas.[64] Nor did he see that in order to be fully interpreted, sequences would require comparisons between as many different organisms as possible. With the explosion of genomic sequencing, molecular biology resumed a naturalistic and collecting tradition that it had never completely abandoned.[65] As the early years of the twenty-first century have shown, comparative genomics and the exploration of vast data sets reaching across the tree of life and even back deep into time will once more put evolution at the heart of biological explanations.

28

Systems Biology and Synthetic Biology

SYSTEMS BIOLOGY EMERGED IN THE LATE 1990S, with synthetic biology appearing immediately afterward at the beginning of the century. The development of systems biology was linked to the Human Genome Project (HGP) and other sequencing programs. These genome projects produced vast amounts of information, which meant that to some researchers it appeared no longer appropriate to attempt to characterize the functions of an organism gene by gene, protein by protein. Biology was ripe for a more global, holistic approach to life.

The origins of synthetic biology are often traced to an article published in 1999, which emphasized its close relations with systems biology.[1] Synthetic biology was dependent on one major result of systems biology—the revelation that macromolecular components of the cell are organized in functional modules.

Both disciplines were based on new technologies, and both were emblematic of the postgenomic era. Both depended for their development on the contributions of new types of scientists trained in physics, in computational biology, or in bioinformatics, who could view organisms with the eyes of engineers. Both disciplines also embraced modeling and through it a revival of theoretical biology. For this reason, this chapter also examines a third branch of research, a component of both systems biology and synthetic biology which has its own agenda: a return to a theoretical and physical approach to biological phenomena. One of the main focuses of this third branch has been the study of "noise"—apparently random variations in the activity of cells and their components.

The aim here, as in other chapters of this fourth part, is not to offer a detailed history of the wealth of developments that took place, but rather to situate them in relation to the molecular vision that was introduced in the 1960s. At first glance, these new disciplines appeared to threaten the dominance of molecular biological explanations. Systems biology sought to explain biological phenomena at a higher level of organization than the molecular, while on the other hand it was suggested that certain physical phenomena could explain biological characteristics independently of molecular descriptions.

▮ Systems Biology

A systemic view of biological phenomena existed long before the development of systems biology. Some philosophers of science have suggested that the roots of this approach can be traced to Georges Cuvier and even to Immanuel Kant. A more recent reference is Ludwig von Bertalanffy, who is unanimously acknowledged as the leading figure behind the emergence of a system theory in the mid-twentieth century.[2] But this historical affiliation is a retrospective reconstruction; von Bertalanffy developed his system theory in the 1930s, in a context that was philosophically, scientifically, and politically very different from that of the 1990s, and long before the emergence of molecular biology. The origins of modern systems biology were more directly linked to the development of genome sequencing programs and their results, and even more so to the expectations they raised. It was believed that tens of thousands of new genes, with totally unknown functions, would be discovered—a conviction that derived from the results of the yeast sequencing program in 1996. Although yeast had been studied extensively by biologists for decades and was believed to have a relatively small genome, the sequence revealed that over a third of yeast genes had completely unknown functions.

New research strategies had to be developed that were completely different from the piecemeal approach of the past because all the genetic information of the organism was now known. The possibility of finding answers in hitherto unknown bits of genetic material no longer existed.

The elucidation of unknown gene functions required new technologies, and yeast was the model in which many of these new technologies were developed. The first and most striking of these was transcriptomics.[3] The initial technique was based on the use of microarrays (DNA chips), which were first developed by Affymetrix. These are small plates to which thousands of short oligonucleotides have been attached, each corresponding to part of the sequence of a particular gene. The experiment involves extracting messenger ribonucleic acid (mRNA) molecules from cells or organisms and labeling them with a fluorescent probe. These molecules are then added to the chip, where they hybridize with the oligonucleotides. After a series of washes, the level of hybridization for each oligonucleotide can be determined by scanning the plate and measuring the fluorescence at each spot. Ideally, the level of hybridization is proportional to the amount of mRNA in the sample—that is, to the expression level of the corresponding genes. Although it occasionally led to overinterpretation, this technology vastly accelerated the rate at which information about gene expression could be obtained, easily outstripping the results that had been gleaned from painstaking Northern blot experiments. Nowadays it has been replaced by direct RNA sequencing.

These experiments could also reveal the existence of groups of genes that were coactivated or coinhibited in particular physiological conditions or during a particular phase of development.[4] Applying the principle of "guilt by association," expression profiling made it possible to develop hypotheses about gene function that could be tested experimentally.[5] Transcriptomics could also help in understanding the differences between different populations of cells. Applied to tumor cells, this strategy made it possible to distinguish different types of cancers that traditional cytological observation had confused, thereby opening the way to more precise diagnoses and more appropriate treatments.[6]

Other technologies designed to provide a functional global picture of cells and organisms were also developed in the form of proteomics and the description of interactomes. Proteomics aimed at doing for proteins what transcriptomics had done for RNAs, but despite making constant progress, this technology has yet to attain the power and simplicity of transcriptomics. Interactomes were claimed to be a global description of the interactions occurring between proteins in cells or in organisms.[7] Each of these interactions was revealed in yeast by the "double-hybrid" technique and then had to be confirmed in the relevant cell or organism.[8] Guilt by association was even truer for proteins connected in interactomes than for genes coexpressed in transcriptomics. Then results of these different techniques could be combined. Between 2000 and 2003 there was a period of a remarkably rapid progress; in 2003, departments of systems biology (under this or similar names) were created in a number of U.S. universities.

Patrick Brown and David Botstein had anticipated these developments when in 1999 they published an influential article describing the possibilities raised by transcriptomics and outlining a series of bold goals.[9] The aim of the new approach was nothing less than the discovery of a new logic of life that had so far eluded biologists. The process of discovery would be data-driven rather than guided by preexisting hypotheses, and it would emerge "spontaneously" from the data, they claimed. The new logic they expected to be revealed would be utterly different from what was previously known. This search for a new logic of life was paralleled by a renewed interest in the nature of life and its definition. For many researchers, the answers provided by molecular biology were no longer valid.

Before considering whether a new logic of life has indeed emerged from these experiments, it is important to focus on two aspects of the explosion of systems biology. The first is that the claims made for this approach had a dual purpose. Following the sequencing of the human genome, a degree of disillusionment had set in; despite the promises of leading researchers and politicians, there had been no immediate applications. Systems biology could help to overcome this apparent impasse in discovery, holding out the promise of new and profound insights. It could also secure the continuation of the massive funding that had been devoted to the HGP.

The second aspect was that the proclaimed systems biology revolution involved a highly disparate cast. This included scientists who had worked on the HGP and other sequencing programs, those who looked back nostalgically to the golden age of theoretical biology that had extended from the early twentieth century to the 1970s and who admired system theory, and the many biologists who were unhappy with the long-standing emphasis on genes and "the book of life." Other key participants included bioinformaticians, computational biologists, and physicists for whom the new technologies seemed to offer ways of applying their expertise to the study of biology. It would have been impossible for such an eclectic group of researchers to truly share either outlook or expectations.

Other factors also help explain why the late 1990s and early 2000s were so important in defining the new landscape in which some biological research now takes place. The rise of the internet led to a renewed interest in network theory and graph theory. In 1999 Albert-László Barabási discovered that some networks, such as the World Wide Web, are not random but are instead characterized by a "scale-free" structure, in which, to put it simply, a few nodes are highly connected while the majority are only poorly connected.[10] This characteristic was rapidly shown to be shared by some biological networks (as discussed later). There was also a new interest in complexity and the search for the laws of complexity, as highlighted by a special issue of *Science* on complex systems, audaciously introduced as taking science "beyond reductionism."[11]

The first result of systems biology was a description of the modular organization of molecular biological systems, which provided strong support for the work of synthetic biologists.[12] Many pieces of evidence supported the existence of this modular organization. Modules were identified in cell signaling pathways and gene regulatory networks, in the control of developmental steps, and in the structure of proteins. Previously described pathways and macromolecular complexes were reinterpreted as modules. The claimed new logic of life was situated at a supramolecular level, which had previously been neglected by molecular biologists. "Modules" were not the only novel description that was put forward at this time. In 2002, Uri Alon and his colleagues described the motifs found in gene regulatory networks and demonstrated that some motifs had been favored by evolution.[13]

The suggestion that biological networks such as cell signaling pathways and gene regulatory networks were scale-free received a great deal of attention. Although these descriptions in terms of modules and of networks were not contradictory—a network can be subdivided into different subcircuits (or modules)—they were often made independently of each other. The aim of these descriptions was to reveal the principles of supramolecular organization that could guide the work of modelers, and to explain some characteristics of organisms that were still not understood. The first characteristic to be explored was *robustness*—the capacity of organisms to resist perturbations, whether from the environment or from mutations. The source of robustness was variously located in the redundancy of either genes or modules, in modular organization itself, or in the structure of the networks.[14]

There were two difficulties with this, beginning with the variety of meanings attributed to the concept of robustness. Resistance to variations in the environment and to mutations are two different phenomena; furthermore, it is hard to see what the link might be between robustness and a given form of supramolecular organization. It is not even clear which way the arrow of causality points—does the existence of modules favor robustness, or vice versa? Similarly, are hubs, the highly connected nodes in networks, the core of the network or its Achilles' heel?[15] Some studies tried to correlate aging with the alteration of hubs in biological networks.[16] Even if this correlation had been verified, it is not clear if it would have provided any greater insight than the obvious observation that dysfunction in essential organs such as the heart and the liver has more serious consequences than in less important organs such as the gall bladder.

Biological systems are not only robust, they also evolve. The questions raised about what was clumsily called evolvability were similar to those asked about robustness: is there a supramolecular organization that favors evolvability? It is not easy to reconcile these two properties of biological systems. The widespread existence of regulation and control intuitively increases robustness, but it might also reduce evolvability.[17]

There was no agreement at the time about the general characteristics of the kind of supramolecular organization that systems biologists were looking for, nor of its significance. The evidence for the generality of scale-free networks in biological systems was soon questioned.[18] Even if it were true, it was not clear what its significance might be. It could be linked with the function of the network, or it could be a consequence of the way the network was built.

Descriptions of systems in terms of modules faced similar difficulties. There are many different types of modules, from protein domains to subcircuits in gene regulatory networks. Modules are not independent of each other or of the network, even if they have a certain autonomy. From this point of view, it appears that modularity is more epistemic than ontological, a way for biologists to provide a simple description of organisms, but not for evolution to create the complex functions of organisms. This raises the possibility that these supposedly new descriptions might be nothing more than a new language to describe familiar phenomena. What exactly did the description of p53 as a hub in the cell regulatory network add to the previous observations that p53 was an essential player in the control of oncogenesis?[19]

It is quite possible that the structure of biological networks or of modules has been selected and optimized to fulfill certain tasks. For example, Uri Alon showed how certain types of motifs present in biological networks were better adapted to generating a rapid response. In some cases, it was even possible that a unique way of organizing a network or module could give rise to a particular function.[20] But this clearly did not mean that a new logic of life had been discovered, or that molecular descriptions had become useless.

The proof that systems biology has not replaced molecular explanations can be seen in the way molecular processes, and in particular gene regulatory and cell signaling networks, have been modeled. If some kind of supramolecular organization was clearly significant, this would inevitably have led to the rules of modeling being simplified to take into account this new, high-level insight. Instead, most current models are still based on interactions between individual macromolecules. The only significant simplification that has occurred is the possibility in some (but not all) gene regulatory networks of using Boolean categories (active / inactive) instead of determining the precise degree of activity of a given pathway.

From the outset, systems biology has pursued two very different goals. It set out to describe a supramolecular order that molecular biologists had not detected, without taking account of existing molecular descriptions. According to von Bertalanffy, and more recently to Kunihiko Kaneko, these organizational principles would not be unique to biological systems but would be shared by all complex systems.[21] Less ambitiously, it could be seen merely as the next step, once the molecular description of organisms was virtually complete. In other words, having separately analyzed the components of life, now it is time to see how those pieces work together. Or, to use the terms of René Descartes, after analysis comes the time for synthesis.

Since the appearance of systems biology and its first results at the turn of the century, only this second aspect has made any progress. It has revealed interactions that had not been identified by an analysis of the individual components, and in some cases has made it possible to deduce previously unknown functions. In addition, it has revealed complex dynamics in biological systems that help explain emergent phenomena such as cell death or cell differentiation.[22] Systems biology offers a bird's-eye and dynamic view of molecular events, but it has not, as yet, changed the logic of life.

Synthetic Biology

References to synthetic biology before 2000 were few and far between, but they dramatically increased in number with the new century.[23] A paper published in 1999 foreshadowed the results to come; it was written in the language of modules and suggested that synthetic approaches might help reveal modular function in the laboratory, and also, through this knowledge, build new functional modules.[24] This was soon followed by a flurry of spectacular results showing that it was possible to introduce a circadian rhythm into *Escherichia coli* (this was called the repressilator), to construct a genetic toggle switch generating bi-stability in *E. coli,* and even to engineer *E. coli* into the equivalent of a photographic film—unfortunately at a time when photographic films were being gradually abandoned in favor of digital media.[25]

Following these early, astounding results, the student-focused International Genetically Engineered Machine (iGEM) competition was set up, in which groups of students from around the world competed to build the best synthetic biology project. At the same time, the Massachusetts Institute of Technology (MIT) set up the Biobrick Parts Registry, which contains fragments of DNA that can be used and reused by researchers in synthetic biology. The iGEM competition was an immediate success, rapidly raising the visibility of synthetic biology on a planetary scale. Grasping the opportunity of publicizing their own field, the small group of researchers who were working on the origins of life also joined the new discipline.

Soon afterward, Jay Keasling and his colleagues succeeded in synthesizing the antimalarial drug precursor artemisinic acid in yeast—this substance previously had to be extracted from a plant.[26] Many projects aimed at producing biofuels were launched, as was Craig Venter's headline-grabbing replacement of a bacterial chromosome by a chemically synthesized modified form of the same chromosome.[27] Since then, similar experiments have been carried out on eukaryotic chromosomes.

The social, ethical, and economic implications of synthetic biology, however important they may prove to be, are not the subject of this book.

Nevertheless, the fact that the production of artemisinic acid has been abandoned for economic reasons is not a good sign. What concerns us here is whether the science in question is genuinely novel, and whether it has led to a breakthrough in how we think about organisms.

As with systems biology, the heterogeneous nature of the field means there can be no simple answer. Artemisinic acid was synthesized thanks to the complexity of the metabolic pathway introduced into yeast cells and to the time and energy invested in the project, not to the particular methods that were employed. Craig Venter's results are more striking for the way they were presented—the alleged creation of a new form of life— than for the principles that guided the work. Other projects, such as the repressilator, were more significant in terms of their focus on modeling in the conception and realization of experiments.

The reality is that, despite some claims, synthetic biology was not born in 2000. Srinivasan Chandrasegaran began his work on creating specific artificial restriction enzymes, which can with hindsight be placed among the finest projects in synthetic biology, in the mid-1980s (see Chapter 24).

Driven by their claims to radical novelty, many synthetic biologists began to contrast the apparent perfection of the devices they built to the messy inefficiency of organic cellular nanomachines, highlighting the difference between the tinkering action of evolution and the engineering approach behind their designed constructions. However, not all their designs proved to be so perfect. For example, it was difficult to build a stable repressilator, and synthetic biologists found themselves obliged to introduce one or more steps of directed evolution to optimize the genetic circuits that they had built.[28] The contrast between a tinkerer and designers, between the action of evolution and the engineering of synthetic biologists, has gradually faded.[29]

In addition to its spectacular applications, there are two further aspects of synthetic biology that are not given sufficient recognition, but which nevertheless show the importance of this discipline in the production of biological knowledge.

Synthetic biology opens up the possibility of exploring the limits of life. Many features of organisms are considered by chemists and biologists as frozen accidents of evolution, the result of contingent events, not unique or optimal solutions to the functional challenges of life. Some examples of such apparent accidents include the number of naturally occurring amino acids in proteins, the nature of the genetic code (the rules of correspondence between trinucleotides and amino acids), the limited number of structural families of known proteins, and even the chemical nature of genes. All of these fundamental features of life could have been different and better. We can presume that the potential space occupied by the objects that we describe as being alive is far larger than the one that has been explored by life on Earth. Xenobiology, the creation of such novel forms of life, may enable us to explore this new biological space, guiding the work of astrobiologists in their search for extraterrestrial life.

Richard Feynman's pronouncement, "What I cannot create, I do not understand," has been adopted as a banner by synthetic biologists. It contains an element of truth, which gives synthetic biology its full place in the construction of a scientific knowledge of organisms. Synthetic biology is a key way of testing our descriptions of what it is to be alive, in particular the type of descriptions that emerged from molecular biology. Synthetic biology can be seen as the peak of molecular biology: simultaneously a test of the value of this form of knowledge and a set of applications of that knowledge. The failures of synthetic biologists may be as instructive as their successes, just as the failure of a systems biology model to account for the observed functioning of a cell or an organism can help reveal some unknown component contributing to the phenomenon being studied—a remarkable example of this can be found in the completion of the gene network regulating the formation of segments in *Drosophila* by George von Dassow and colleagues.[30]

Given the significance of synthetic biology, it is legitimate to ask why it did not develop decades earlier, when the main principles of molecular organization were described in the 1960s, and also to wonder why it emerged around the year 2000. Each of these questions has its own answer. In the 1960s, molecular biology provided a general framework, but most molecular functions remained unknown. Genetic engineering technology was developed in the 1970s, but it still took over twenty years to put some flesh on the molecular descriptions of organisms. The tools required for the appearance of synthetic biology were simply not available.

The year 2000 did not represent an end point in the accumulation of molecular data. Quite the opposite: the recently sequenced genomes revealed many genes with unknown functions, while the discovery of RNA interference and of the importance of microRNAs showed that the molecular description of cellular activity was far from complete. But the events that occurred around the new millennium were a milestone, a catalyst for the rise of synthetic biology, just as it was for systems biology. With the imminent completion of the Human Genome Project, there was a general feeling that a period that had begun in 1953 with the description of the double helix structure of DNA, or in 1944 with the work of Oswald Avery, was drawing to a close. It was time to develop radically new strategies (systems biology), and to exploit the huge amount of knowledge that had been obtained in the previous decades (synthetic biology).

Physics and Molecular Biology

As systems and synthetic biology developed, so physics gradually assumed a larger role in biology. This was most clearly visible in the new and growing part played by models and modeling. Instead of being used to summarize results, as had been traditional in molecular biology, modeling now preceded and guided the work of experimenters (Chapter 29). Modeling is now seen as the only way to predict and explain the behavior of systems that are too complex to be understood intuitively.

"Physics" is too broad a term to fully describe the new techniques and approaches that have been incorporated into molecular biology. There are many different types of physicists, and they use different methods and provide different insights. From its very beginnings, molecular biology benefited from this richness, with different types of physics and physicists contributing to its emergence (Chapters 7 and 9). Quantum physicists, notably Niels Bohr, Erwin Schrödinger, and Max Delbrück, brought their ambitious objectives to the new science, while other physicists constructed machines such as electrophoretic apparatuses and ultracentrifuges to enable their colleagues to characterize macromolecules. It would be impossible to describe all the ways that physicists were involved in the development of systems biology and synthetic biology, so I will focus on some of their early contributions.

The first attempts to model genetic circuits were made at the beginning of the 1960s. The Belgian biologist René Thomas was a pioneer in the field, followed by many others such as Stuart Kauffman and Brian Goodwin. The highly influential *Journal of Theoretical Biology* published many of the key articles in the field. Some systems were the focus of particular attention, such as the way bacteriophages flip between lysogeny and a lytic cycle, the operon model and its derivatives, and the phenomenon of transdetermination observed in *Drosophila*—a rare switch between different fates shown by imaginal disk cells. These early efforts were greeted by a complete lack of interest among most molecular biologists, but the key problem was that the models lacked data. This came later.

Genetic engineering enabled observations of key cellular processes to be made, but the initial results were complex. It took many years before researchers could gain even a partial understanding of the mechanisms of eukaryotic gene regulation (see Chapter 20). Meanwhile, modelers had abandoned molecular biology for other fields, such as sociology and economics, where their expertise was more readily appreciated.

The 1980s and 1990s saw considerable progress in cell imaging. Existing obstacles were overcome, mainly through the improved statistical processing of images. Immunofluorescence was replaced by the use of chimeras between cell proteins and green fluorescent protein (GFP), making it possible to localize the cellular position of a given protein or RNA molecule, and its movement within the cell.

This produced a more dynamic vision of molecular phenomena and made it possible to describe a phenomenon whose existence had been predicted but not explored—the stochastic nature of many cellular processes. This was described as "molecular noise," and it affected many fundamental mechanisms, including the one that is probably the most significant for molecular biologists: gene transcription. Stochastic phenomena had been repeatedly observed from the earliest work on bacteriophages, β-galactosidase expression, and the flagellar motor, but the implications were not drawn out.[31] The early years of the twenty-first century saw rapid progress by measuring the levels of a protein in genetically identical cells and studying how they vary over time, or by directly observing single protein synthesis events.[32]

Researchers from different disciplines flocked to study noise, including physicists, who were accustomed to the study of stochastic phenomena in simpler, nonbiological systems, and many biologists, who viewed this phenomenon as a challenge to the deterministic—and simplistic—models of molecular biology. The origin of molecular noise, which can be compared to Brownian motion, was explained by the small number of macromolecules present in a cell and the slow rate of occurrence of some reactions, such as transcription initiation. But beyond these general observations, we still await a precise explanation of the origin and nature of these stochastic variations.

The issue that focused the attention of most observers was not so much the mechanisms at the origin of molecular noise but its physiological significance. It exists, and cells and organisms have to cope with it, argued many biologists. But "coping" can mean many things. Cells might fight against noise, limiting its impact on essential regulatory mechanisms. Or organisms could instead exploit the existence of noise, using it to create diversity among cells or for other reasons.[33] These opposing interpretations led to the study of molecular noise occupying a growing place in systems biology and synthetic biology. In synthetic biology noise was a challenge and had to be controlled. In systems biology, noise required a function to be attributed to it.

It was discovered that evolution had shaped some cellular circuits to limit the impact of noise. In other cases, noise seems to enable a population of cells to adapt to varying conditions. The most interesting observations have been those relating the precise nature of stochastic variations to the type of dynamic behavior they generate. For instance, in eukaryotic cells transcription initiation occurs in bursts, instead of the isolated events that might be expected; this might generate transient changes in the cellular functional state or induce irreversible processes such as cell death or cell differentiation.[34]

▌ Conclusion

Systems biology has expanded into developmental biology and evolutionary biology (see Chapters 22 and 23). However, much of the initial excitement around systems biology and synthetic biology has dissipated, although it has not completely disappeared. Both disciplines have found a place in the landscape of biological research, but they have not replaced molecular descriptions, nor have they led researchers to abandon reductionism. Perhaps the most interesting studies are those that link precise molecular descriptions with the emergence of complex dynamic phenomena.

Researchers continue to propose models and hypotheses, and research is not uniquely data driven. But the rapid accumulation of data is endless and seemingly overwhelming. This has led some researchers to suggest that it will be possible to bypass a long process of explanation and instead directly predict the behavior of biological systems, and above all to act on them from the simple observation of "markers." Correlations would become sufficient. This current of thought is particularly powerful in medicine, where action often has to precede knowledge.[35] Medicine, however, would remain molecular, not by the nature of its explanations—that would disappear—but by the nature of the phenomena that are observed.

29

Images, Representations, and Metaphors in Molecular Biology

OPEN A BOOK on molecular biology or glance at an online article, and you will be struck by the quantity and quality of the graphical representations of the topic being discussed—the pages will teem with illustrations of macromolecules, intracellular pathways, and networks. This is not true of most other disciplines, and this wealth of images contributes to the high visibility—literal and metaphorical—of molecular biology. It also provides an important clue to understanding why so many students and young researchers are attracted to the subject.

The study of representation—which refers both to imagery and to the underlying concepts and metaphors—is a growing area in the history and philosophy of science. The nature of the representations used in molecular biology—be they representative or symbolic—and their role in the way research develops and is presented have been the focus of a number of investigations.[1] But the way in which these images emerged, their roots and the historical context that favored their development and gave them their unique characteristics, have remained largely unexplored. This historical approach is particularly appropriate for this last part of the book, which aims to look at the transformations of molecular biology since its rise and the establishment of a dominant position that it acquired in the 1960s. In this case, the imageries that were successively used at each period of the development of molecular biology were a legacy of previous forms of representation, and the result of their recent transformations and alterations.

▌ Roots

The roots of the imagery used by molecular biologists can be traced to chemistry. The representations used in nineteenth-century physiological chemistry or early twentieth-century biochemistry were essentially the same as those used in chemistry, and provided two key components of the representations now used in molecular biology: arrows to represent a series of steps (reactions) and shapes to represent chemical structures.

In the 1930s, Linus Pauling had a major impact on the way that chemical phenomena were represented, in particular through his book *The Nature of the Chemical Bond*.[2] He emphasized the importance of the shape of molecules for understanding intermolecular interactions, and he replaced the traditional "lock and key" model with a vision of two complementary surfaces interacting through weak bonds. Pauling shaped how molecules and macromolecules were represented in chemistry and biochemistry, with biochemists being particularly influenced by his approach (Chapter 1).

Genetics provided molecular biology with one of the most powerful forms of representation and metaphor: the map (Chapter 1). The first genetic maps, drawn up a century ago by Sturtevant in Morgan's laboratory on the basis of his studies of *Drosophila*, were abstract things that represented recombination frequency in terms of physical space. By the 1930s, these maps were replaced by real, representative maps of gene positions on the giant banded chromosomes found in the salivary glands of *Drosophila*, which could be studied under the microscope. This mapping of recombination frequencies onto real tissue went hand in hand with the growing conviction that the gene was effectively a physical particle (and not, say, the amount of a particular substance). When Morgan gave his Nobel Prize lecture in 1934, he illustrated it with chromosome maps, soon after the first such map had been drawn up.[3] More recently, mapping was also a central issue in the Human Genome Project (Chapter 27). Mapping has always been a way to organize, but in the early years of genetics it was more: researchers were expecting there to be a relation between the map position of genes and their functions.

Shortly after the precise chromosomal localization of genes was reported in the 1930s, a second transformation of genetic representation took place, following Boris Ephrussi and George Beadle's work on the link between genes (and mutations) and embryological development, which involved a precise analysis of the genetic determination of eye color in *Drosophila* (Chapter 2). Ephrussi and Beadle used drawings to show the links between a set of developmental steps (which were increasingly conceived of in chemical terms) and the way that genes affected them.[4] In subsequent decades, our understanding of these relationships was deepened until, at the beginning of the 1960s, a genetic map of an operon more or less corresponded to the sequential chemical reactions catalyzed by its products.

A third influence on how molecular biology used imagery came from the linked fields of cybernetics and information theory. By the 1950s, arrows were no longer used exclusively for chemical reactions or gene actions but also to indicate the transfer of information. It is particularly revealing to study the gradual introduction of these informational arrows into the figures used by molecular biologists. One striking example concerns the relations between DNA and ribonucleic acid (RNA). This transformation, initially conceived as the direct chemical conversion of DNA into RNA, was gradually reconceptualized as the transfer of information—that is, of a particular order of nucleotides in DNA to the same order in RNA. This new meaning of arrows in biology was not completely novel. There was a long tradition of using arrows to represent actions—for example, that of a hormone on a tissue, or the direction of neuronal transmission. It was also traditional to represent the steps in an experimental protocol by arrows. From action to information was only a small step.

Images and Representations in the First Wave of Molecular Biology

The double helix structure of DNA is emblematic of the representations used in early molecular biology because of the impact it had on the development of this science (Chapter 11). Some representations of the double helix were "purely" chemical; that is, apart from the fact that DNA was a macromolecule, the image was similar to the representations of molecules drawn by chemists around the same time. The same conventions were used—for example, the color-coding of the different types of atom.

But another type of representation was also used—one that can be found in Jim Watson and Francis Crick's first publication, in 1953.[5] This was drawn by Odile Crick and is strikingly different from the classic representations of chemistry: deoxyribose and phosphates were replaced by a ribbon, and the pairs of bases by rods. This image foreshadowed the use of ribbons to represent proteins, which came into vogue twenty-five years later, although the existence of any direct link is unclear. It also began a tradition in which molecular biologists present multiple representations of the same macromolecule that differ in their degree of detail. In general, simpler images are used to establish a connection between chemical structure and biological function; in the case of Crick and Watson's first DNA paper, the relation was between the regularity of the double helix as shown by the perfect complementarity of its two strands and the replication of the genetic material. These representations of the double helix remained unchanged over subsequent decades, partly because until the first three-dimensional structures of transfer RNA (tRNA) were published in the mid-1970s the structure of DNA was the only polynucleotidic structure to have been determined. The double helix structure of DNA had no immediate influence on the iconography of similar macromolecules, simply because there was no other polynucleotide that needed to be represented.

The first iconic representations of proteins were globular images, based on the descriptions provided by physical chemists and driven by the need to represent structural changes that can occur in proteins. At the end of the 1930s, John Yudkin proposed that the structure of the active site of an enzyme was produced by its interaction with different potential substrates.[6] This hypothesis generated a type of representation that was reproduced many times; it became a prototype of representations of the changes in protein shapes that proliferated after the emergence of the allosteric theory in the 1960s (Chapters 14 and 21).[7] These drawings were effectively models rather than precise structures—they were not intended to represent the true changes in protein structure, which at the time were unknown and were experimentally inaccessible. Nevertheless, the information in the drawings, such as the distance separating the different sites on the protein, or the nature of the junctions between the different subunits, was considered as an important way of understanding the mechanisms of protein transconformations. It was widely expected that the details of these allosteric changes would be revealed in the future by X-ray diffraction studies.

It was not easy to see how these precise three-dimensional structures should be represented. The first model of myoglobin made by John Kendrew in 1957—the aptly named sausage model—gave only the global form of the protein.[8] Some years later, Max Perutz built a precise model of hemoglobin in which bonds and atoms were represented, but despite its accuracy, this model was hard to understand. Neither the space-filling models nor the skeleton molecular models looked satisfactory. It was also hard to make an informative two-dimensional reproduction (on a printed page, for example), and few laboratories had the technical skills and resources necessary to construct such a model. Brilliant as these models were, their direct influence was largely limited to those who were privileged enough to actually inspect them physically. The poor legibility and accessibility of this style of representation, and of others that were proposed at the end of the 1960s and in the 1970s, was a serious obstacle to the transmission of understanding.

In the 1960s, another type of model emerged in molecular biology, representing the mechanisms of gene regulation. The first of these was the operon model and its numerous developments.[9] In 1969, Roy Britten and Eric Davidson proposed another model of gene regulation, which they considered to be more appropriate for multicellular organisms.[10] Although both models were inspired by cybernetics, the Britten–Davidson model had an engineering touch that the operon model did not. The way the genetic circuits were represented in the Britten–Davidson model was subsequently incorporated into systems biology and synthetic biology. In the case of the operon model, the initial abstract representation was slowly supplemented with structural information: the structural characteristics of the repressor or of the ribosome were gradually added to the drawings as soon as they were known.

This enrichment of an abstract model that initially merely showed the relations between the different components was not novel. In 1953, André Lwoff drew the relations between a temperate bacteriophage and a bacterium, adding to his drawing information about bacteriophage morphology that had recently been acquired through electron microscopy.[11] Representations of the sodium channel found in nerve cell membranes also changed gradually, and this transformative trend became increasingly obvious for a wide range of molecules.[12]

There was a long tradition of representing cells, ever since the acceptance of cell theory in the mid-1800s. This tradition was renewed by the development of electron microscopy at the end of the 1940s and in the 1950s, which eventually transformed our vision of the cell. But this was not immediate: the earliest electron microscopy images were not easily understood. As a result, scientists gradually adopted the habit of publishing, side by side, the electron micrograph and a corresponding drawing that outlined the organelles that were barely detectable on the original image. The next step was to combine the two, with labels, arrows, and drawings overlaid onto the micrograph, mixing the seen and the imagined, producing an apparently authoritative interpretation of the image that brooked no alternative.[13]

▌ New Molecularization, New Representation

The second wave of molecularization that occurred at the beginning of the 1980s with the development of genetic engineering led to an acceleration of research on microorganisms and the beginning of work on multicellular organisms. More scientists were trained, more research groups were set up, and more research grants were awarded. The resultant rapid accumulation of data required new forms of representation.

This new wave of molecularization was accompanied by an increased interest in cell structure. Through the development of new techniques such as immunofluorescence, researchers discovered a complex network of protein and vesicle trafficking and the existence of intracellular pathways that enabled signals arriving at the cell membrane to cascade down to the nucleus, thereby modifying gene expression. This all led to a renaissance of cell biology (Chapter 20), and to a radical transformation of representations of the cell and its components.

The first elements of cellular representation to change were membranes. In the 1940s, observations made with the electron microscope roughly confirmed the existing model by James Danielli and Hugh Davson, according to which the membrane was a lipid bilayer sandwiched by proteins.[14] But the data did not completely support the model. The most serious issue was that according to the model there was a rigid barrier between the interior and the exterior of the cell, whereas research had now revealed a rich and complex exchange between cell and environment. Furthermore, the Danielli–Davson model did not explain the roles that macromolecules (proteins) might have.

In 1972, S. Jonathan Singer and Garth Nicolson proposed a new membrane model that dramatically changed the picture, placing freely moving proteins inside the membrane (Chapter 15).[15] New, beautifully designed images were immediately produced to support and explain this model. Painstaking research using the freeze-etching technique of electron microscopy indirectly revealed the presence of proteins within the lipid bilayer, providing further evidence for the Singer–Nicolson model. The beauty of the drawings that accompanied these publications made up for the absence of direct observations!

The representation of proteins also changed dramatically at the beginning of the 1980s (Chapter 21). Unusually, this change can be attributed to the influence of a single person: the protein crystallographer Jane Richardson. The first three-dimensional structures of proteins were disappointing—they did not reveal simple rules of protein folding, and it was impossible to compare different structures in any meaningful way. In the 1970s, researchers gradually realized that protein folding and structure could nevertheless partly be explained by focusing on secondary structures (α-helices and β-strands), in particular by describing how they were positioned in the polypeptide chain and how they were combined (this was termed "topology").

These new findings required a description that emphasized secondary structures, and their topological organization. In 1976, Michael Levitt and Cyrus Chothia used a ribbon to represent protein structure for the first time, but the drawings were frankly poor, and Levitt and Chothia did not emphasize the significance of their novel representation—indeed they abandoned it in later publications, reverting to more abstract imagery.[16]

Jane Richardson discovered that different proteins contained four linked and sequential protein strands that run up and down, in an antiparallel structure known as the Greek key motif ("Greek key" refers to the zig-zag lines seen in classical Greek decorative borders). This analogy probably encouraged Richardson to develop a new way of representing proteins. She grasped the potential of the ribbon metaphor (it is not clear whether either she or Levitt and Chothia were influenced by Odile Crick's drawing of twenty-five years previously), and she improved it, introduced colors, and in 1981 gave it a central place in an influential review that appeared in *Advances in Protein Chemistry*.[17] Scientists rapidly adopted this new way of representing proteins, and the rules laid down by Richardson were applied to all newly characterized proteins. Twenty years later, Richardson could rightly claim that "a whole generation of scientists see protein structures through my eyes."[18]

This ribbon imagery was subsequently adapted for protein modeling on computers, and it remains unchanged more than thirty years later—it has become an obligatory way of representing proteins. The dominance of the ribbon model is not the product of inertia, of scientists being too lazy to challenge the traditional way of doing things, but instead reveals the power of Richardson's original insight, which was itself the result of a long period of reflection. Richardson was originally trained in philosophy before becoming a technician in her husband's laboratory, and she was convinced of the importance of representations in science. To encourage scientists to adopt the imagery she proposed, she spent a lot of time increasing the aesthetic impact of her figures by using pastel colors, taking advantage of innovations in the publishing industry that made color printing more widely available—it had previously been both expensive and restricted to a crude palette.

Richardson was well aware that beauty was a particularly important criterion used by many scientists when faced with a new model or a new form of imagery. The influence of her vision was reinforced by the emphasis her drawings placed on secondary structures and their topology. It was a good choice that subsequently turned out to be particularly well adapted to the description of the conformational changes that occur in proteins, and to the design of new proteins.

At around the same time, a major new type of representation emerged: imagery associated with cell-signaling pathways and networks. This development flowed from the description of protein kinase cascades in the 1970s. During this period a series of important discoveries underlined the significance of these pathways and required new forms of representation to describe them: cyclic adenosine monophosphate (cAMP) was identified as a second messenger, other second messengers were discovered alongside an abundant and diverse set of protein kinases (the rapidly growing new family of tyrosine kinases was identified at the end of the 1970s and beginning of the 1980s), and many of these pathways turned out to be involved in the formation of tumors (Chapter 18).

The new imagery was employed not only to describe these discoveries but also to publicize them—literally, in some cases. Some biotech companies that were hoping to make money out of these results produced posters that compared cell-signaling pathways to a subway map, with the components represented as stations, and the pathways as the lines. The implication was that understanding the complexity of the cell was as simple as going from A to B. These representations were also important in the education and training of students because they made it possible to summarize, in a single image, information that was otherwise dispersed in several publications. They differed from previous complex representations of biological processes, such as the tangle of arrows and names to be found in images of the Krebs cycle, because molecular biologists often combined symbolic shapes to denote molecules with arrows to indicate the relevant process. Many images in molecular biology give such a hint about the potential role of structure in function.

The growing commercial textbook market played a crucial role in the design and spread of this new imagery. The highly successful *Introduction to Protein Structure* by Carl Branden and John Tooze, first published in 1991, popularized the new protein representations.[19] Probably the most important textbook in the field, *Molecular Biology of the Cell*, which first appeared in 1983, not only reproduced existing models and representations but also actively contributed to the creation of new imagery.[20] This book was both the sign of the emergence of a new molecular biology focused on cell structure and a shop window for new forms of imagery that were consciously designed to represent this new vision.

ALTHOUGH MANY DIVERSE REPRESENTATIONS emerged in this decade of innovation, they all shared a number of features. The first is that they were beautiful, something that was a conscious aim of those who commissioned and designed them. This was partly made possible by changes in printing technology, which made it easier to use colored figures in books and articles. But this does not explain why molecular biology exploited these new possibilities more than other sciences, such as physics.

The images used were mixed—sometimes they were literal and sometimes, when they represented processes not structures, symbolic. Representations corresponding to different levels of complexity were often presented in parallel, emphasizing the multilevel explanations that were an essential feature of this new form of molecular biology.

Although these illustrations implied that they were a direct translation of what could be seen through electron microscopy or other observational techniques, they were not intended to be realistic. It goes without saying that representing the secondary structures of proteins by simple helices and arrows is a mere convention: they are not actual stable and perfect helices, nor, of course, are they accompanied by anything like an arrow. Illustrations of cell-signaling pathways do not accurately present the dimensions of the different components—a relatively massive organelle may be shown at the same size as a single protein molecule. The components of a pathway are represented as though they exist in a vacuum, with all the other molecules and macromolecules present in the cell rendered invisible if they are not directly involved in the given pathway, while the stoichiometry of the pathway components is also not presented.[21]

Above all, such images are a way of summarizing our accumulated knowledge of these pathways. The form that they take is simultaneously abstract and generalized—there is no attempt to present the different parts of the pathway that are active in one cell type but not in another. The undoubted value of these representations is offset by their limitations, and by the invisibility of what they exclude, for good reasons and sometimes for bad.

Despite the many changes, these representations from the second age of molecular biology did not break with existing trends that had been visible in the first age of the discipline. They were characterized by a mixture of shapes and processes and simultaneously included different levels of representation; the meaning of the ever-present arrows was distinctly ambiguous.

▌ The New Revolution: Representation and Imagery in Postgenomics

The most striking characteristic of the new representations and images produced by postgenomics and the new disciplines of systems biology and synthetic biology is their highly abstract character. These images are not primarily icons; instead, they are tools and steps toward mathematical and computational modeling.

The first change that occurred was not visual but metaphorical, as "pathway" was gradually replaced by "network." Although these pathways became increasingly complex after the discovery of numerous feedback loops within them and of cross-talk between different pathways, this was not the reason for the change in terminology. Metabolic pathways were also converted into networks, even though our understanding of them had not been dramatically transformed by any new degree of complexity. The word *network* had become so fashionable and omnipresent that it more or less had to be incorporated into biology, as had been the case with information in the 1950s (Chapter 7).

Genomic sequencing itself did not lead to any major new styles of images or representation. However, the new technological developments associated with postgenomics (from the end of the 1990s to the beginning of the 2000s) did provoke rapid changes (Chapter 28). Postgenomic data, generated through computers, led to many new types of representation.[22] The systematic use of the two-hybrid system made it possible to reveal elements of the interactomes of a cell or of an organism—that is, to describe a network representing the interactions between the proteins present in the cell or organism. Unlike, say, cell-signaling pathways, interactomes can be considered as illustrations of the output of a new technology more than of a finished state of knowledge. Only the links in the diagram are significant, not the positions of the various proteins, which are represented as mere nodes.

Another feature of this style of imagery is that it may not necessarily bear any resemblance to reality. The existence of the interactions in vivo was not initially confirmed before the interactome was published, even where there were good reasons why such interactions might not exist. (For example, the proteins are never in contact because they are found in different cell compartments, or the protein domains used for the two-hybrid experiments are hidden inside the native structure of the proteins.) Furthermore, nothing in the image indicates whether these interactions are stable or not, or whether they are weak or strong. In contrast with biological tradition, this form of image is not a cornerstone indicating the state of current knowledge, but a hypothesis, a heuristic tool pointing to what we still do not understand.

The images used in transcriptomics have much in common with those employed in the case of interactomes. Starting with an estimate of the different kinds of messenger RNA (mRNA) molecules present in a cell or an organism in different environmental, functional, or developmental conditions, it is possible to generate networks showing the correlations between the expression of different genes. These illustrations represent a further degree of abstraction: the links no longer refer to physical interactions but are instead correlations detected in expression patterns that may or may not correspond to actual physical interactions between gene products.

Statistical correlations may be a useful heuristic tool for detecting pathways and networks, but it is not clear that this is the way such diagrams are being employed. When data collection has an immediate applied objective, as in the case of medical research, correlations may be sufficient justification for new therapeutic approaches, the value of which will come from proof of their efficiency, not from a mechanistic explanation of how the therapy works (Chapter 28). These abstract networks are also well adapted to reveal degrees of resemblance between different types of cell and the existence of different categories of cellular processes. For example, they have been used successfully to identify different categories of tumors (Chapter 28). Such distinctions may be important for prognosis and for treatment choice, irrespective of their explanatory power. New types of imagery have been especially created to facilitate the recognition of similarities and differences (such as Circos).[23]

One feature of these new forms of representation is that they have retained similarities with more traditional and literal imagery. A network drawn up on the basis of transcriptomic data looks pretty much like a network that was created in the pregenomic era. Both kinds of network contain "hubs" and "modules," but the words have changed their meanings. A hub is no longer a protein interacting with many partners, but instead is a gene the expression of which covaries with those of many other genes, while a module is simply a part of a network and not a functional unit with an input and an output.

Representations of proteins, and of macromolecules in general, have also evolved, and now emphasize the existence of an ensemble of different conformations and of their interconversion dynamics (Chapter 21). However, representing dynamics and populations of molecules in a two-dimensional image is not easy. Because all these phenomena are grounded in thermodynamics, the landscape imagery traditionally used in this discipline has been adopted by structural molecular biology and transformed to fit its new purposes.

The best example of this can be found in protein folding. The traditional imagery, with arrows separating the different steps in the folding pathway, was no longer valid because each step represents a population of macromolecules with different conformations. The image therefore has to be able to show the many different paths toward the native state of the protein. In 1993 a new style of representation, the protein folding landscape, was introduced.[24] It looks like a deformed funnel, where the narrow bottom represents the native state and the broad end corresponds to the many unfolded states. The deformations of the funnel show that folding is not a regular process. The landscape can also contain crevices in which protein folding intermediates may become trapped.

Similar types of images have also been used to represent the ensemble of native protein conformations and different functional cellular states, together with their interconversion during differentiation or tumor formation. Interestingly, these supposedly new representations are reminiscent of the epigenetic landscape proposed by Conrad Waddington in the 1950s, which in turn was an adaptation of a metaphor of evolutionary processes (the evolutionary landscape) that was introduced in the 1930s by Sewall Wright.

▌ Metaphors in Molecular Biology

Underlying all these different forms of visual representation are the verbal metaphors used by scientists to explain, understand, and present their new outlook. The pages of this book are full of metaphors, but you have probably not noticed some of them because they are so rooted in our way of thinking about molecular biology that much of their original power has been long lost—signals, messengers, information, code, machines, locks and keys, landscapes, motors, factories, power plants, networks, programs, regulation, and even the oldest of them all: "cell."

Philosophers have long been interested in what exactly metaphors are and how they do what they do, in particular in science where they appear to be a necessary, unavoidable part of how scientists work. Writers such as Evelyn Fox Keller and, more recently, Andrew Reynolds have explored the thought-provoking idea that metaphors had a decisive role in the definition of research programs in molecular and cell biology.[25] Scientists have also begun to examine the role of metaphor not only in molecular biology but also in neuroscience and evolution.[26]

For the historian, the most interesting question is how the choice of a particular metaphor may have affected the course of scientific discovery—how the decision to carry out, or interpret, a particular experiment was channeled and framed by the underlying metaphor. My hope is that the reader will have found many implicit or explicit answers in the different chapters of *The Black Box of Biology*.

These examples sometimes reveal that the metaphor was in fact unclear, showing that differences between scientists over how to understand a particular phenomenon were hidden, or smoothed over, by the adoption of a common metaphor. The development of molecular biology provides two, interlinked, examples that continue to resonate down to the present day: the adoption by Watson and Crick in 1953 of the two key metaphors "code" and "information."[27]

The idea that what Watson and Crick called the genetic code was an actual code, composed of logical and consistent correspondences between genotype and phenotype, beguiled mathematicians and physicists in particular, beginning with George Gamow. The result was that for much of the 1950s, a great deal of effort and ingenuity was devoted to trying to "crack the code" theoretically, coming up with a logical and consistent set of correspondences. As was eventually discovered, the genetic code is anything but logical and consistent—or rather, if it does contain such rationality, that is still hidden from us. By taking the metaphor of "code" literally, scientists took a wrong turning and largely wasted their time. The fact that, as Crick later recognized, the "code" metaphor is technically wrong—the phrase should have been "genetic cipher"—reinforces the quixotic aspects of this period.[28]

Equally, the use of "information," which is so important to our understanding of how organisms work at every level, is largely metaphorical. Although the term might seem to have been directly co-opted from the mathematical approaches developed by Claude Shannon and by Warren Weaver during World War II, it soon became apparent that its use was primarily metaphorical, at least in molecular biology. Crick inadvertently revealed this in his 1957 lecture "On Protein Synthesis" when he defined genetic information not in terms of abstract formulas but much more simply as the specification of the amino acid sequence of a protein.[29] Since that point, molecular biologists have generally used "information" in a metaphorical sense, but that has not stopped enthusiastic—and naive—mathematicians from trying, say, to calculate the information content of genomes. As with the "code," all the theoretical attempts to use mathematical concepts of information to advance our understanding of gene function and organization have had little practical consequence.

The transformation of scientific knowledge can be compared to the displacement of a phagocytic cell that projects part of its cytoplasm forward but is ready to deviate from this course if an obstacle is met. Similarly, research programs have to be reoriented when they are less productive, and the metaphors that supported them have to be replaced by new ones. When scientists realize that metaphors both provide insight and limit how we think, they often get excited, wondering how we could adopt new metaphors that would open the road to new discoveries.

Perhaps the new range of graphical representations that have recently appeared is the first sign that molecular biology is undergoing a slow but profound transformation. If these new abstract types of representation were to become firmly established and extended as the traditional iconic representations disappeared, and if there were to be further dramatic changes in the metaphors used by molecular biologists, this would indeed suggest that the oft-announced death of molecular biology was coming about. We are not at that point. In this respect, it is worth recalling that, from the very beginnings of molecular biology right up to the most enthusiastic advocates of postgenomics, bold thinkers have proclaimed that this science will transcend itself, revealing new laws of physics and new ways of understanding life itself. Who knows what tomorrow will bring? But for the moment, we are still waiting.

General Conclusion

WRITING THE HISTORY of any subject is obviously much simpler when the field in question is dead. Whether this is the case for molecular biology has been the central focus of this book.

The term *molecular biology* is now rarely used in the names of university departments or degrees, or in the titles of scientific societies and journals. Its popularity in the 1960s and 1970s proved short-lived, whereas terms such as genetics and biochemistry have survived and thrived. As discussed in Chapters 15 and 20, molecular biology was never a stand-alone science, but it rapidly infiltrated other biological disciplines. Most of these new research fields did not adopt the name of molecular biology, but they did adopt its methods and its models.

Most philosophers of biology would argue that the rise of molecular biology coincided with a period in which strong reductionism came to dominate scientific thinking, but today most biologists have adopted more global, holistic approaches to their subject (Chapters 9 and 28). From both social and philosophical points of view, the time of molecular biology seems to be over.

Previous studies of the recent period have not sufficiently explored biologists' work and their explanations of biological phenomena. They have paid too much attention to fashions and arguments that were more strategic than programmatic. They have looked for revolutionary science and have neglected normal science. They have emphasized the strong, transient forms of reductionism that appeared in the 1960s and 1970s, and have failed to understand that the dominant form of reductionism in recent decades has been what Nils Roll-Hansen has called "biological reductionism," in which biological phenomena do not disappear, but are explained by molecular mechanisms.[1]

These are the reasons why Part IV explored different fields of biology, including some allegedly revolutionary recent developments, in order to identify the place occupied in current biological research by the explanatory framework of molecular biology.

Some of these recent conceptual "revolutions" nonetheless remained fully molecular. For example, even though regulatory ribonucleic acids (RNAs) were not predicted by classical molecular biology they nevertheless were easily inserted into the molecular framework. The same is true for epigenetic modifications, despite claims from researchers that this phenomenon will lead to a new science.

The most striking feature of the recent period has been the gradual dovetailing of molecular explanations with evolutionary and physical frameworks. These do not replace the molecular approach but complement it. They have clearly transformed some areas of research, such as structural biology and embryology. It may surprise the more enthusiastic advocates of these new technologies, but this study has not revealed any field of research in which molecular descriptions and explanations have become obsolete. Even in the most extreme cases, they have merely had to make space for other types of explanations.

If anything is currently dying, it is not molecular biology, but rather the dream that there may be a new and higher logic of life that would replace the molecular logic described by François Jacob.[2] This does not mean that modern biology is not very different from that of the 1980s and 1990s. The increasing use of data banks and the exponential growth in computational power have dramatically transformed the work of many biologists. Equally, this does not mean that we may not be on the eve of a real revolution in the life sciences. All we can say is that we still cannot describe what this new paradigm, if it emerges, might look like. Like it or not, molecular biology is still here, and its history remains of fundamental importance for all biologists.

Appendix
DEFINITION OF TERMS

Notes

Index

Appendix
DEFINITION OF TERMS

Proteins

The most important components of organisms are *proteins*. Proteins are *macromolecules:* large molecules made up of several thousand atoms. Proteins have a mass of between ten thousand and several hundred thousand *daltons* (a unit of mass equal to one-twelfth of the mass of a carbon atom). Proteins are formed of one or several *polypeptides*. Polypeptides are chains formed by stable chemical *bonds—covalent bonds—*between components that are called *amino acids.* Polypeptides are made up of twenty different amino acids, including *cysteine, proline,* and *phenylalanine.* Polypeptides are thus *polymers* formed by the *nonmonotonous* repetition of amino acids with similar but nonidentical structures. Each polypeptide is characterized by the number and nature of the amino acids that it contains, but above all by their order in the chain—the *sequence.*

Because of the *primary structure* (the amino acid sequence) of the polypeptides forming them, proteins should be long, straight threads; this is not the case, however, because no sooner have they been synthesized than the polypeptide chain *spontaneously* folds, providing the protein with a more compact structure, which is often spherical (the *globular structure*).

Proteins have essential functions in organisms.

1. As *enzymes,* they *catalyze* (accelerate) chemical reactions that take place within cells (the totality of these reactions forms the *metabolism*). Enzymes are extremely effective catalysts: they may accelerate a

reaction more than 10 billion times. There are more than 10,000 enzymes in humans, each one specific to a particular chemical reaction. The catalytic power and the specificity of enzymes come from the fact that they form a precise complex (a *stereospecific* complex) with the molecule they transform (the *substrate*). The enzyme molecule surrounds the substrate molecule and establishes with it a series of weak chemical bonds: *van der Waals bonds, hydrogen bonds,* and *ionic bonds.*

2. Proteins play a structural role: they contribute to the architecture of the cell and thus, indirectly, to the form of the organism.

3. Proteins can bind to genes to control their activity. They can also act as *receptors* (of hormones, growth factors, or antigens), *channels* (in the cell membrane), or *transporters* (in the blood, for instance), or they can participate in establishing connections between different proteins.

▌ Genes and DNA

An organism is thus characterized by the nature of the proteins that form it, that is, the amino acid sequence of its proteins. This information is transmitted from generation to generation by *genes.*

Humans have around 20,000 genes that code for proteins. These genes make up the *genome* and form microscopic rodlike structures called *chromosomes.* In virtually all organisms that reproduce sexually, each gene is present in two copies: one from the father, the other from the mother. An organism that contains two copies of each gene is *diploid.* Organisms that contain only one copy of each gene are *haploid.*

Each gene contains the information necessary for making one (or, through splicing, many—as discussed later) polypeptide chain(s) or a ribonucleic acid (RNA). These RNAs may have a structural role, a catalytic role as the RNA present in the large ribosomal subunit that catalyzes the formation of the peptide bond between the amino acids forming a protein, or a regulatory role by destabilizing messenger RNAs, or inhibiting their translation.

Genes are made of *deoxyribonucleic acid (DNA).* DNA is formed of two molecules, each one a long chain (a *polynucleotide*) that is coiled around the other in a double helix. Each polynucleotide is a polymer formed by the non-monotonous repetition of four elementary motifs called *nucleotides* (deoxyribonucleotides).

A nucleotide contains three parts: a *phosphate group,* a *sugar* (deoxyribose), and a *base.* DNA contains four bases: *adenine, thymine, guanine,* and *cytosine.* Bases are small cyclic molecules with a relatively simple structure, each with a mass of between 100 and 200 daltons. In a polynucleotide chain the nucleotides are linked together by covalent bonds between the sugar (of one nucleotide) and the phosphate (of the next nucleotide). Each end of the polynucleotide chain is thus different, giving the chain an orientation.

In a DNA molecule, the two polynucleotide chains are in opposite orientations. Each base on one chain is paired by two or three hydrogen bonds to a base on the other chain: an adenine is always opposite a thymine, and a guanine is always opposite a cytosine. The two strands of DNA are thus *complementary:* the base sequence on one strand makes it possible to deduce the base sequence of the other.

From DNA to Proteins

The sequence of nucleotides (bases) in DNA (that is, in the gene) determines the order of amino acids in the corresponding protein: the sequence of bases *encodes* the sequence of amino acids. The passage from DNA to RNA is called *transcription.* RNA is a single-chain polynucleotide that differs only slightly from the chains of DNA (the sugar is a *ribose* instead of a deoxyribose, and thymine is replaced by *uracil).*

In prokaryotes, DNA is copied into an RNA molecule called *messenger RNA* (mRNA). The RNA molecule is made up of a single chain. It is complementary to the DNA strand on which it was synthesized. In eukaryotes, DNA is copied into a long chain of RNA, from which different fragments are eliminated by a complex process called *splicing.* DNA sequences that are present in the mRNA are called *exons,* whereas those that are eliminated are called *introns.* Splicing can be regulated, and one unique RNA can give rise to different mRNAs, and therefore to different polypeptides.

The mRNA molecule is translated into a polypeptide chain. Each base *triplet (codon)* has a particular meaning, corresponding to a different amino acid. These correspondences are called the *genetic code.* Translation takes place on a particle formed of RNA and proteins, called a *ribosome.* (The first name given to this particle was *microsome.* Microsomes, isolated by ultracentrifugation, were a mixture of ribosomes and fragments of endoplasmic reticulum membranes. In vivo, the association of ribosomes with endoplasmic reticulum membranes is essential for the production of membrane and secreted proteins.)

The ribosome slides along the messenger RNA. Each time the ribosome encounters a new base triplet, a small RNA called *transfer RNA* (tRNA) is fixed onto the messenger RNA. The tRNA bears the amino acid that corresponds to this base triplet, a relationship that flows from the genetic code. The different amino acids are then linked up. The information carried by genes is thus decoded on the ribosome. After their synthesis, different polypeptides eventually interact by weak bonds to form a multichain protein.

From Proteins to DNA

Some proteins bind to DNA. These proteins (for instance, histones in eukaryotes) can play a structural role and allow the long DNA molecule to fold and form chromosomes. Some of these proteins have a more specific role: by binding on the DNA molecule, they can control the initiation of transcription of DNA into RNA—that is, the *expression* of genes. These proteins are called *transcription factors*. They can act as activators or as repressors. The DNA–protein complex is called *chromatin* in eukaryotes. Transcription factors can also interact with enzymes that chemically modify histones, alter the structure of chromatin, and thus indirectly regulate gene expression. Modifications of chromatin are studied by researchers working in *epigenetics*.

Prokaryotic and Eukaryotic Cells

All cells are surrounded by a membrane that restricts their exchanges with the external medium. There are two kinds of cells: *eukaryotic* cells, in which there are a number of intracellular structures, each isolated by a membrane and called *organelles*, and *prokaryotic* cells, which do not contain either intracellular membranes or organelles. Bacteria and *archaea* are prokaryotic cells. Eukaryotic cells form single-cell organisms (such as yeast) or are integrated into multicellular organisms (plants, animals, and some fungi).

A eukaryotic cell has a *nucleus*, which contains the chromosomes, and the *cytoplasm*. Gene transcription (DNA → RNA) takes place in the nucleus, whereas the translation of messenger RNA into protein takes place in the cytoplasm, where the ribosomes are found.

A eukaryotic cell also contains other organelles: *mitochondria*, where most of the molecules of *adenosine triphosphate* (ATP, which serves as an energy source for metabolic reactions) are produced; the *endoplasmic reticulum;* and the *Golgi* apparatus—intracellular vesicles where proteins are transported before being inserted into the plasma membrane or secreted. Plant cells also have *chloroplasts*, where light energy is converted into chemical energy.

Viruses are not autonomous organisms. They are formed of a genetic material (RNA or DNA) that is protected by membranes or protein structures. They hijack the cell's machinery in order to express their genetic information.

Notes

Introduction

1. This explains why, for example, this book does not deal to any extent with research on photosynthesis. Although this topic experienced a similar "reduction" to the molecular level at around the same period as the research described here, strictly speaking it was not part of molecular biology. For an opposite point of view, see Doris T. Zallen, "Redrawing the Boundaries of Molecular Biology: The Case of Photosynthesis," *Journal of the History of Biology* 26 (1993): 65–87. Although molecular biology gives genes a central role in life and in the development of organisms, it cannot be reduced to a study of the structure and function of genes, that is, to molecular genetics. See Richard M. Burian, "Technique, Task Definition, and the Transition from Genetics to Molecular Genetics: Aspects of the Work on Protein Synthesis in the Laboratories of J. Monod and P. Zamecnik," *Journal of the History of Biology* 26 (1993): 387–407.

2. The name and status of molecular biology are discussed in Robert Olby, "The Molecular Revolution in Biology," in *Companion to the History of Modern Science,* ed. R. C. Olby, G. N. Cantor, I. R. R. Christie, and M. J. S. Hodge, 503–520 (London: Routledge, 1990).

3. "Paradigm" is used here in a simple sense, without entering into the various debates that have ensued since the term's "invention" by Thomas S. Kuhn. The molecular paradigm is the new vision of life produced by molecular biology. See Thomas S. Kuhn, *The Structure of Scientific Revolutions* (Chicago: University of Chicago Press, 1970).

4. These studies will be cited wherever necessary. An early list of autobiographical documents can be found in Nicholas Russell, "Towards a History of

Biology in the Twentieth Century: Directed Autobiographies as Historical Sources," *British Journal for the History of Science* 21 (1988): 77–89. See also Pnina G. Abir-Am, "Noblesse Oblige: Lives of Molecular Biologists," *Isis* 82 (1991): 326–343; Jan Sapp, "Essay Review: Portraying Molecular Biology," *Journal of the History of Biology* 25 (1992): 149–155.

5. Robert Olby, *The Path to the Double Helix,* 2nd ed. (London: Macmillan, 1994).

6. Horace Freeland Judson, *The Eighth Day of Creation: The Makers of the Revolution in Biology* (New York: Simon & Schuster, 1979; 2nd ed., Cold Spring Harbor, NY: Cold Spring Harbor Laboratory Press, 1996).

7. Lily E. Kay, *The Molecular Vision of Life: Caltech, the Rockefeller Foundation and the Rise of the New Biology* (Oxford: Oxford University Press, 1993).

8. Matthew Cobb, *Life's Greatest Secret: The Race to Crack the Genetic Code* (London: Profile Books, 2015). See also the useful collection by Jan Witkowski of short historical articles first published in *Trends in Biochemical Sciences:* Jan Witkowski, ed., *The Inside Story: DNA to RNA to Protein* (Cold Spring Harbor, NY: Cold Spring Harbor Laboratory Press, 2005).

9. James D. Watson and John Tooze, *A Documentary History of Gene Cloning* (San Francisco: W.H. Freeman, 1981); Stephen S. Hall, *Invisible Frontiers: The Race to Synthesize a Human Gene* (New York: Atlantic Monthly Press, 1987); Sheldon Krimsky, *Genetic Alchemy: The Social History of the Recombinant DNA Controversy* (Cambridge, MA: MIT Press, 1982); Nicolas Rasmussen, *Gene Jockeys: Life Science and the Rise of Biotech Enterprise* (Baltimore: Johns Hopkins University Press, 2014); Doogab Yi, *The Recombinant University: Genetic Engineering and the Emergence of Stanford Biotechnology* (Chicago: University of Chicago Press, 2015). Excluding the history of genetic engineering from that of molecular biology would be a convenient solution, but it would end up making molecular biology a "theoretical" science and would obscure the fact that the molecular vision of life that developed from the 1940s to the 1960s was, by its very nature, operational and applied.

10. Dominique Pestre, "En guise d'introduction: quelques commentaires sur les 'témoignages oraux,'" *Cahiers pour l'histoire du CNRS* 2 (1989): 9–12. For a more positive appreciation of the consequences of face-to-face contact with key actors, see Nathaniel Comfort, "When Your Sources Talk Back: A Multimodal Approach to Scientific Biography," *Journal of the History of Biology* 44 (2011): 651–669.

11. Frederic L. Holmes, *Meselson, Stahl, and the Replication of DNA: A History of "the Most Beautiful Experiment in Biology"* (New Haven, CT: Yale University Press, 2001).

12. In this book, the balance has at times been tilted in favor of previously neglected work, at the expense of more well-known events.

13. Donald Fleming and Bernard Bailyn, eds., *The Intellectual Migration: Europe and America, 1930–1960* (Cambridge, MA: Belknap Press, 1969); in particular, see Donald Fleming, "Émigré Physicists and the Biological Revolution,"

pp. 152–189. Also see David Nachmansohn, *German-Jewish Pioneers in Science, 1900–1933* (New York: Springer-Verlag, 1979); Paul K. Hoch, "Migration and the Generation of New Scientific Ideas," *Minerva* 25 (1987): 209–237.

14. Fernand Braudel, *La Méditerranée et le monde méditerranéen à l'époque de Philippe II* (Paris: Armand Colin, 1949).

15. Frederic L. Holmes, "The *Longue Durée* in the History of Science," *History and Philosophy of the Life Sciences* 25 (2003): 463–470.

16. Michel Morange, "Science et effet de mode," in *L'État des sciences et des techniques,* ed. Nicolas Witkowski, 453–454 (Paris: La Découverte, 1991).

17. Scientific publications are claimed to be belated products of scientific activity, in which "strategies" are deliberately obscured. But experience suggests the following arguments in favor of a detailed study of such documents. Only such a study can unravel whole sections of the history of a science that may have been forgotten. A study of publications leads to a quantitative appreciation of the relative importance of various models or experimental approaches at a given time; publications are much less likely to be "censored" than many historians of science seem to think, especially during periods of rapid advances in knowledge—a mine of new information may await the alert reader. Finally, publications are "richer" than the work they describe: editing them is an essential part of creative scientific activity. See Frederic L. Holmes, "Scientific Writing and Scientific Discovery," *Isis* 78 (1987): 220–235. In particular, writing for a public larger than fellow specialists can sometimes lead to "semantic shifts," which can be the starting point for scientific revolutions. See Christiane Sinding, "Literary Genres and the Construction of Knowledge in Biology: Semantic Shifts and Scientific Change," *Social Studies of Science* 26 (1996): 43–70.

18. David Bloor, *Knowledge and Social Imagery* (London: Routledge and Kegan Paul, 1976).

19. This a priori position often causes historians of science to value studies that were secondary, simply because they did not lead to the development of experimental systems that could be studied reproducibly. A number of examples of such studies of "bad choices" are provided in Chapters 2 and 8.

20. The first constraint is the choice of the organism to be studied. This initial selection influences subsequent research, often in unexpected ways. See Richard M. Burian, "How the Choice of Experimental Organism Matters: Epistemological Reflections on an Aspect of Biological Practice," *Journal of the History of Biology* 26 (1993): 351–367.

1. The Roots of the New Science

1. Garland E. Allen, *Life Science in the Twentieth Century* (New York: Wiley, 1975); Marcel Florkin, *A History of Biochemistry* (Amsterdam: Elsevier, 1972); Joseph S. Fruton, "The Emergence of Biochemistry," *Science* 192 (1976):

327–334; P. R. Srinivasan, Joseph S. Fruton, and John T. Edsall, *The Origins of Modern Biochemistry: A Retrospect on Proteins* (New York: New York Academy of Science, 1979); Robert E. Kohler, *From Medical Chemistry to Biochemistry* (Cambridge: Cambridge University Press, 1982).

2. Robert E. Kohler, "The History of Biochemistry: A Survey," *Journal of the History of Biology* 8 (1975): 275–318. On the importance of this discovery, its roots and consequences, see Robert E. Kohler, "The Background to Eduard Buchner's Discovery of Cell-Free Fermentation," *Journal of the History of Biology* 4 (1971): 35–61, and "The Reception of Eduard Buchner's Discovery of Cell-Free Fermentation," *Journal of the History of Biology* 5 (1972): 327–353; Joseph S. Fruton, *Proteins, Enzymes, Genes: The Interplay of Chemistry and Biology* (New Haven, CT: Yale University Press, 1999).

3. Keith J. Laidler, *The World of Physical Chemistry* (Oxford: Oxford University Press, 1993).

4. Robert Olby, *The Path to the Double Helix* (London: Macmillan, 1974), chap. 1; Robert Olby, "Structural and Dynamical Explanations in the World of Neglected Dimensions," in *A History of Embryology,* ed. T. J. Horder, J. A. Witkowski, and C. C. Wylie, 275–308 (Cambridge: Cambridge University Press, 1986); Neil Morgan, "Reassessing the Biochemistry of the 1920s: From Colloids to Macromolecules," *Trends in Biochemical Sciences* 11 (1986): 187–189; Ute Deichmann, "'Molecular' versus 'Colloidal': Controversies in Biology and Chemistry, 1900–1940," *Bulletin for the History of Chemistry* 32 (2007): 105–118.

5. J. D. Bernal, "Structure of Proteins," *Nature* 143 (1939): 663–667.

6. Karl Landsteiner, *The Specificity of Serological Reactions* (Springfield, IL: Charles C. Thomas, 1936).

7. Lily E. Kay, "Molecular Biology and Pauling's Immunochemistry," *History and Philosophy of the Life Sciences* 11 (1989): 211–219.

8. Linus Pauling, *The Nature of the Chemical Bond,* 3rd ed. (Ithaca, NY: Cornell University Press, 1960; first edition published in 1939); Linus Pauling, "Modern Structural Chemistry," *Science* 123 (1956): 255–258; Linus Pauling, "Fifty Years of Progress in Structural Chemistry and Molecular Biology," *Daedalus* 99 (1970): 988–1014; Anthony Serafini, *Linus Pauling: A Man and His Science* (New York: Simon & Schuster, 1989); Alexander Rich and Norman Davidson, *Structural Chemistry and Molecular Biology* (San Francisco: W. H. Freeman, 1968); Ahmed Zewail, *The Chemical Bond: Structure and Dynamics* (Cambridge, MA: Academic Press, 1992); Thomas Hager, *Force of Nature: The Life of Linus Pauling* (New York: Simon & Schuster, 1995).

9. Linus Pauling, "Nature of Forces between Molecules of Biological Interest," *Nature* 161 (1948): 707–709. Among the weak bonds, Pauling emphasized the role of hydrogen bonds, attributing to them a higher energy value than is given today. He paid little attention to hydrophobic interactions or to what are now called changes in entropy. These errors were probably useful in the develop-

ment of the concept of the weak bond because they made things simpler. See Howard Schachman, "Summary Remarks: A Retrospect on Proteins," in *The Origins of Modern Biochemistry,* ed. P. R. Srinivasan, Joseph S. Fruton, and John T. Edsall, 363–373 (New York: New York Academy of Sciences, 1979).

10. Alfred E. Mirsky and Linus Pauling, "On the Structure of Native, Denatured and Coagulated Proteins," *Proceedings of the National Academy of Sciences of the United States of America* 22 (1936): 439–447.

11. Alfred H. Sturtevant, *A History of Genetics* (New York: Harper & Row, 1965); Elof A. Carlson, *The Gene: A Critical History* (Philadelphia: Saunders, 1966); Garland E. Allen, *Life Science in the Twentieth Century* (New York: Wiley, 1975); Ernst Mayr, *The Growth of Biological Thought: Diversity, Evolution, and Inheritance* (Cambridge, MA: Harvard University Press, 1982); Peter J. Bowler, *The Mendelian Revolution: The Emergence of Hereditarian Concepts in Modern Science and Society* (Baltimore: Johns Hopkins University Press, 1989); Lindley Darden, *Theory Change in Science: Strategies from Mendelian Genetics* (New York: Oxford University Press, 1991); Jean-Louis Fischer and William H. Schneider, *Histoire de la génétique: pratique, techniques et théories* (Paris: ARPEM et Sciences en situation, 1990); Robert Olby, *The Origins of Mendelism* (Chicago: University of Chicago Press, 1966; 2nd ed., 1985); Robert E. Kohler, *Lords of the Fly:* Drosophila *Genetics and the Experimental Life* (Chicago: University of Chicago Press, 1994); Elof A. Carlson, *Mendel's Legacy: The Origin of Classical Genetics* (Cold Spring Harbor, NY: Cold Spring Harbor Laboratory Press, 2004); Raphael Falk, *Genetic Analysis: A History of Genetic Thinking* (Cambridge: Cambridge University Press, 2009); Hans-Jörg Rheinberger and Staffan Müller-Wille, *The Gene: From Genetics to Post-Genomics* (Chicago: University of Chicago Press, 2017). Even the idea that Mendel's results were forgotten and then rediscovered is disputed by contemporary historians. First, Mendel's results were not unknown; second, the context in which they were rediscovered was very different from that in which they were first stated. See Robert Olby, "Mendel No Mendelian?" *History of Science* 17 (1979): 53–72; and Augustine Brannigan, "The Reification of Mendel," *Social Studies of Science* 9 (1979): 423–454.

12. Thomas H. Morgan, Alfred H. Sturtevant, Hermann J. Muller, and Calvin B. Bridges, *The Mechanism of Mendelian Heredity* (New York: Henry Holt, 1915).

13. Barbara A. Kimmelman, "Agronomie et théorie de Mendel: La dynamique institutionnelle et la génétique aux Etats-Unis (1900–1925)," in Fischer and Schneider, *Histoire de la génétique,* 17–41; Robert Olby, "Rôle de l'agriculture et de l'horticulture britanniques dans le fondement de la génétique expérimentale," in Fischer and Schneider, *Histoire de la génétique,* 65–81.

14. Bowler, *Mendelian Revolution;* Daniel J. Kevles, "Genetics in the United States and Great Britain, 1890–1930: A Review with Speculations," *Isis* 71 (1980): 441–455; Jonathan Harwood, "National Styles in Science: Genetics in Germany and the United States between the World Wars," *Isis* 78 (1987): 390–414;

Jonathan Harwood, *Styles of Scientific Thought: The German Genetics Community, 1900–1933* (Chicago: University of Chicago Press, 1993); Richard M. Burian, Jean Gayon, and Doris Zallen, "The Singular Fate of Genetics in the History of French Biology," *Journal of the History of Biology* 21 (1988): 357–402.

15. Harwood, "National Styles in Science"; Harwood, *Styles of Scientific Thought.*

16. Raphael Falk, "The Struggle of Genetics for Independence," *Journal of the History of Biology* 28 (1995): 219–246.

17. Hermann J. Muller, "Artificial Transmutation of the Gene," *Science* 66 (1927): 84–87; Elof Axel Carlson, "An Unacknowledged Founding of Molecular Biology: Hermann J. Muller's Contributions to Gene Theory, 1910–1936," *Journal of the History of Biology* 4 (1971): 149–170; Elof Axel Carlson, *Genes, Radiation, and Society: The Life and Work of H. J. Muller* (Ithaca, NY: Cornell University Press, 1982).

18. Olby, *Path to the Double Helix,* chap. 7. DNA had been discovered in 1869 by Johann Friedrich Miescher. See Franklin H. Portugal and Jack S. Cohen, *A Century of DNA: A History of the Structure and Function of the Genetic Substance* (Cambridge, MA: MIT Press, 1977).

19. Hermann J. Muller, "The Gene," Pilgrim Trust Lecture, *Proceedings of the Royal Society of London: Series B, Biological Sciences* 134 (1947): 1–37, quotation from p. 1.

20. This is shown by the few detailed models of genes that were proposed in the 1930s and 1940s. See, for example, Dorothy M. Wrinch, "Chromosome Behaviour in Terms of Proteins Pattern," *Nature* 134 (1934): 978–979.

21. N. K. Koltsov, "Les molécules héréditaires," *Actualités scientifiques et industrielles,* no. 776 (Paris: Hermann, 1939).

22. Hermann J. Muller, "Resume and Perspectives of the Symposium on Genes and Chromosomes," *Cold Spring Harbor Symposia on Quantitative Biology* 9 (1941): 290–308; Olby, *Path to the Double Helix,* chap. 7.

23. Jordan's model was part of his larger project of "quantum biology." Richard H. Beyler, "Targeting the Organism: The Scientific and Cultural Context of Pascual Jordan's Quantum Biology, 1932–1947," *Isis* 87 (1996): 248–273.

24. Linus Pauling and Max Delbrück, "The Nature of the Intermolecular Forces Operative in Biological Processes," *Science* 92 (1940): 77–79.

25. Allen, *Life Science;* Ernst Mayr and William B. Provine, eds., *The Evolutionary Synthesis: Perspectives in the Unification of Biology* (Cambridge, MA: Harvard University Press, 1980); Mayr, *Growth of Biological Thought;* Fischer and Schneider, *Histoire de la génétique;* Jean Gayon, *Darwinism's Struggle for Survival: Heredity and the Hypothesis of Natural Selection* (Cambridge: Cambridge University Press, 1998); V. B. Smocovitis, "Unifying Biology: The Evolutionary Synthesis and Evolutionary Biology," *Journal of the History of Biology* 25 (1992): 1–65.

26. Theodosius Dobzhansky, *Genetics and the Origin of Species* (New York: Columbia University Press, 1937; George G. Simpson, *Tempo and Mode in*

Evolution (New York: Columbia University Press, 1944). Dobzhansky played a key role in testing the models of Sewall Wright, J. B. S. Haldane, and R. A. Fisher with experimental data, thus helping to reconcile field scientists and theoretical biologists. See Robert E. Kohler, "Drosophila and Evolutionary Genetics: The Moral Economy of Scientific Practice," *History of Science* 24 (1991): 335–375.

2. The One Gene–One Enzyme Hypothesis

1. George W. Beadle and Edward L. Tatum, "Genetic Control of Biochemical Reactions in Neurospora," *Proceedings of the National Academy of Sciences of the United States of America* 27 (1941): 499–506.
2. Harriet Zuckerman and Joshua Lederberg, "Postmature Scientific Discovery?" *Nature* 324 (1986): 629–631.
3. Robert Olby, *The Path to the Double Helix* (London: Macmillan, 1974); Krishna R. Dronamraju, "Profiles in Genetics: George Wells Beadle and the Origin of the Gene-Enzyme Concept," *Journal of Heredity* 82 (1991): 443–446; Arnold W. Ravin, "The Gene as Catalyst; the Gene as Organism," *Studies in History of Biology* 1 (1977): 1–45.
4. Alexander G. Beam, *Archibald Garrod and the Individuality of Man* (Oxford: Clarendon Press, 1993).
5. Cited in Olby, *Path to the Double Helix,* 130.
6. Archibald Garrod, *The Inborn Errors of Metabolism,* rev. ed. (London: Oxford University Press, 1909; original edition published by London: H. Harris, 1963).
7. Ephrussi and Beadle's work, which used the embryological method of trans- plantation, was one of the first attempts to reconcile genetics and embryology. Beadle and Ephrussi continued to work on this system for many years. Robert E. Kohler, "Systems of Production: *Drosophila, Neurospora* and Biochemical Genetics," *Historical Studies in the Physical and Biological Sciences* 22 (1991): 87–130. For a study of Ephrussi's career, see Richard M. Burian, Jean Gayon, and Doris T. Zallen, "Boris Ephrussi and the Synthesis of Genetics and Embryology," in *Developmental Biology,* vol. 7, *A Conceptual History of Modern Embryology,* ed. Scott F. Gilbert, 207–227 (New York: Plenum Press, 1991).
8. Garland E. Allen, *Thomas Hunt Morgan: The Man and His Science* (Princeton, NJ: Princeton University Press, 1978); E. A. Carlson, *The Gene: A Critical History* (Philadelphia: Saunders, 1966); Peter J. Bowler, *The Mendelian Revolution: The Emergence of Hereditarian Concepts in Modern Science and Society* (Baltimore: Johns Hopkins University Press, 1989).
9. Richard B. Goldschmidt, *Physiological Genetics* (New York: McGraw-Hill, 1938); Garland E. Allen, "Opposition to the Mendelian-Chromosome Theory: The Physiological and Developmental Genetics of Richard Goldschmidt," *Journal of the History of Biology* 7 (1974): 49–92. Goldschmidt's criticisms were aimed at the idea of the particulate gene; Michael R. Dietrich, "Striking the

Hornet's Nest: Richard Goldschmidt's Rejection of the Particulate Gene," in *Rebels, Mavericks, and Heretics in Biology,* ed. Oren Harman and Michael R. Dietrich, 119–136 (New Haven, CT: Yale University Press, 2008).

10. Scott F. Gilbert, "Induction and the Origins of Developmental Genetics," in *Developmental Biology,* vol. 7, *A Conceptual History of Modern Embryology,* ed. Scott F. Gilbert, 181–206 (New York: Plenum Press, 1991); Scott F. Gilbert, "Cellular Politics: Ernest Everett Just, Richard B. Goldschmidt, and the Attempt to Reconcile Embryology and Genetics," in *The American Development of Biology,* ed. Ronald Rainger, Keith R. Benson, and Jane Maienschein, 311–346 (Philadelphia: University of Pennsylvania Press, 1988).

11. Scott F. Gilbert, "The Embryological Origins of the Gene Theory," *Journal of the History of Biology* 11 (1978): 307–351.

12. Jane Maienschein, "What Determines Sex? A Study of Converging Approaches, 1880–1916," *Isis* 75 (1984): 457–480; Muriel Lederman, "Research Note: Genes on Chromosomes—The Conversion of Thomas Hunt Morgan, *Journal of the History of Biology* 22 (1989): 163–176.

13. Jan Sapp, "The Struggle for Authority in the Field of Heredity, 1900–1932: New Perspectives on the Rise of Genetics," *Journal of the History of Biology* 16 (1983): 311–342.

14. Allen, *Thomas Hunt Morgan,* chap. 9.

15. Lily E. Kay, "Selling Pure Science in Wartime: The Biochemical Genetics of G. W. Beadle," *Journal of the History of Biology* 22 (1989): 73–101; Kohler, "Systems of Production"; George W. Beadle, "Recollections," *Annual Review of Biochemistry* 43 (1974): 1–3; Paul Berg and Maxine Singer, *George Beadle: An Uncommon Farmer* (Cold Spring Harbor, NY: Cold Spring Harbor Laboratory Press, 2003).

16. These conclusions were anticipated by the work of Jean Gayon and Hans-Jörg Rheinberger. Gayon showed that Ephrussi explained his data by a model that was very different from that later proposed by Beadle and Tatum, whereas Rheinberger has demonstrated how Kühn interpreted his observations on *Ephestia* in terms of hormone action. See Jean Gayon, "Génétique de la pigmentation de l'oeil de la drosophile: la contribution spécifique de Boris Ephrussi," in *Les sciences biologiques et médicales en France: 1920–1950,* éd. Claude Debru, Jean Gayon, and Jean-François Picard, 187–206 (Paris: CNRS Editions, 1994); and Hans-Jörg Rheinberger, "*Ephestia:* The Experimental Design of Alfred Kühn's Physiological Developmental Genetics," *Journal of the History of Biology* 33 (2000): 535–576; Michel Morange, "Resurrection of a Transient Forgotten Model of Gene Action," *Journal of Biosciences* 40 (2015): 473–476.

17. Kay, "Selling Pure Science"; Lily E. Kay, "Microorganisms and Macromanagement: Beadle's Return to Caltech," in *The Molecular Vision of Life: Caltech, the Rockefeller Foundation, and the Rise of the New Biology,* 194–216 (Oxford: Oxford University Press, 1993). The microbiological determination of amino

acids was already a "classic" method. See, for instance, the review by Esmond E. Snell, "The Microbiological Assay of Amino Acids," *Advances in Protein Chemistry* 2 (1945): 85–118.

18. George W. Beadle, "Genetic Control of Biochemical Reactions," *Harvey Lecture* 40 (1945): 56–211, p. 193.

19. Ernest P. Fischer and Carol Lipson, *Thinking about Science: Max Delbrück and the Origin of Molecular Biology* (New York: W. W. Norton, 1988), 169–173.

20. Olby, *Path to the Double Helix*, chap. 9; Ravin, "Gene as Catalyst," 1–45.

21. Jan Sapp, *Where the Truth Lies: Franz Moewus and the Origins of Molecular Biology* (Cambridge: Cambridge University Press, 1990); Jan Sapp, "What Counts as Evidence, or Who Was Franz Moewus and Why Was Everybody Saying Such Terrible Things about Him?" *History and Philosophy of the Life Sciences* 9 (1987): 277–308. In his provocatively titled book *Where the Truth Lies,* Sapp suggested that Beadle and Tatum's results were predated by the work of Franz Moewus, who was thus the true founder of molecular biology. Sapp argued that the importance of Moewus's research was not recognized because accusations of fraud discredited his ideas and work. Despite the interesting material it contains and its thought-provoking analyses, Sapp's book is unsatisfying. Strikingly, he did not really address the question of whether or not Moewus manipulated his results. For Sapp, even if the accusation were true, it would simply mean that Moewus was no different from other scientists, all of whom "correct" their experimental data. This a priori judgment prevented Sapp from dealing with a number of questions relating to the validity of Moewus's results, such as whether the various hormones he claimed to have characterized have been shown to exist, and if so, whether they have the structure and function that Moewus attributed to them. For Sapp, "discoveries do not exist independently of their inventors," and "when Moewus was excluded from this scientific domain, his discoveries vanished with him" (Sapp, "What Counts as Evidence," 307).

3. The Chemical Nature of the Gene

1. Oswald T. Avery, Colin MacLeod, and Maclyn McCarty, "Studies on the Chemical Nature of the Substance Inducing Transformation of Pneumococcal Types," *Journal of Experimental Medicine* 79 (1944): 137–158.

2. Alfred D. Hershey and Martha Chase, "Independent Functions of Viral Proteins and of Nucleic Acids in the Growth of the Bacteriophage," *Journal of General Physiology* 36 (1952): 39–56.

3. Gunther S. Stent, "Prematurity and Uniqueness in Scientific Discovery," *Scientific American* 228 (1972): 84–93.

4. H. V. Wyatt, "When Does Information Become Knowledge?" *Nature* 235 (1972): 86–89.

5. René J. Dubos, *The Professor, the Institute, and DNA* (New York: Rockefeller University Press, 1976); Olga Amsterdamska, "From Pneumonia to DNA: The Research Career of Oswald T. Avery," *Historical Studies in the Physical and Biological Sciences* 24 (1993): 1–40; David M. Morens, Jeffery K. Taubenberger, and Anthony S. Fauci, "Predominant Role of Bacterial Pneumonia as a Cause of Death in Pandemic Influenza: Implications for Pandemic Influenza Preparedness," *Journal of Infectious Diseases* 198 (2008): 962–970.

6. Pierre-Olivier Méthot has placed the transforming experiment of Griffith in the context of his work on epidemiology. Pierre-Olivier Méthot, "Bacterial Transformation and the Origins of Epidemics in the Interwar Period: The Epidemiological Significance of Fred Griffith's 'Transforming Experiment'," *Journal of the History of Biology* 49 (2016): 311–358. See also Michael Fry, *Landmark Experiments in Molecular Biology* (London: Academic Press, 2016).

7. Maclyn McCarty, *The Transforming Principle* (New York: W.W. Norton, 1985).

8. Avery, MacLeod, and McCarty, "Studies on the Chemical Nature of the Substance."

9. Dubos, *The Professor.*

10. Wyatt, "When Does Information," 86–89; Matthew Cobb, "Oswald Avery, DNA, and the Transformation of Biology," *Current Biology* 24 (2014): R55–R60.

11. Horace F. Judson, *The Eighth Day of Creation: The Makers of the Revolution in Biology* (New York: Simon & Schuster, 1979), 60.

12. Robert Olby, *The Path to the Double Helix* (London: Macmillan, 1974), chap. 13.

13. François Jacob, *La Logique du vivant* (Paris: Gallimard, 1970); *The Logic of Life: A History of Heredity,* trans. Betty E. Spillmann (New York: Pantheon Press, 1974), 262.

14. Olby, *Path to the Double Helix,* chap. 9.

15. Max Delbrück, "A Theory of Autocatalytic Synthesis of Polypeptide and Its Application to the Problem of Chromosome Reproduction," *Cold Spring Harbor Symposia on Quantitative Biology* 9 (1941): 122–124.

16. Dubos, *The Professor.* Some historians have noted that even in his 1944 paper, Avery described the transforming principle not as a gene but as a specific mutagenic agent. See Bernardino Fantini, "Genes and DNA," *History and Philosophy of the Life Sciences* 10 (1988): 145–151. The distinction between gene and specific mutagenic agent, so important for modern genetics, was not at all clear for Avery and his contemporaries, for whom genes and mutagens had the same "catalytic" action. See H. V. Wyatt, "Knowledge and Prematurity: The Journey from Transformation to DNA," *Perspectives in Biology and Medicine* 18 (1975): 149–156.

17. Quoted by Ilana Löwy, "Variances in Meaning in Discovery Accounts: The Case of Contemporary Biology," *Historical Studies in the Physical and Biological Sciences* 21 (1990): 87–121, quote from p. 112.

18. Michel Morange, "La révolution silencieuse de la biologie moléculaire," *Débat* 10 (1982): 62–75.

19. Colin MacLeod felt that the subject was so problematic that it would be difficult to publish any study of transformation. For three years he turned to a topic that was more "rewarding" in terms of publication. See McCarty, *Transforming Principle,* 96–100.

20. For an elegant "counterfactual proof," see Matthew Cobb, "A Speculative History of DNA: What If Oswald Avery Had Died in 1934?" *PLoS: Biology* 14 (2016): e2001197.

21. Erwin Chargaff, *Heraclitean Fire: Sketches from a Life before Nature* (New York: Rockefeller University Press, 1978), 86–100. Chargaff's work is also described in Olby, *Path to the Double Helix,* 211–219.

22. Rollin D. Hotchkiss, "The Identification of Nucleic Acids as Genetic Determinants," in P. R. Srinivasan, Joseph S. Fruton, and John T. Edsall, *The Origins of Modern Biochemistry* (New York: New York Academy of Sciences, 1979), 321–342.

4. The Phage Group

1. The publication in 1966 of a Festschrift in honor of Max Delbrück provided a belated baptism for the group, which was soon followed in 1969 by the attribution of the Nobel Prize to its three founders. See John Cairns, Gunther S. Stent, and James D. Watson, eds., *Phage and the Origins of Molecular Biology* (Cold Spring Harbor: Cold Spring Harbor Laboratory Press, 1966; expanded ed., 1992); Nicholas C. Mullins, "The Development of a Scientific Specialty: The Phage Group and the Origins of Molecular Biology," *Minerva* 10 (1972): 51–82; D. Fleming, "Émigré Physicists and the Biological Revolution," *Perspectives in American History* 2 (1968): 176–213, reprinted in Donald Fleming and Bernard Bailyn, eds., *The Intellectual Migration: Europe and America, 1930–1960* (Cambridge, MA: Belknap Press, 1969).

2. Lily E. Kay, "Conceptual Models and Analytical Tools: The Biology of Physicist Max Delbrück," *Journal of the History of Biology* 18 (1985): 207–246; Ernst P. Fischer and Carol Lipson, *Thinking about Science: Max Delbrück and the Origins of Molecular Biology* (New York: W.W. Norton, 1988); Thomas D. Brock, *The Emergence of Bacterial Genetics* (Cold Spring Harbor: Cold Spring Harbor Laboratory Press, 1990), chap. 6.

3. Niels Bohr, "Light and Life," *Nature* 131 (1933): 421–423, 457–459.

4. Nikolaï W. Timofeeff-Ressovsky, Karl G. Zimmer, and Max Delbrück, "Über die Natur der Genmutation und der Genstruktur," *Nachrichten von der Gesellschaft der Wissenschaften zu Göttingen, Mathematisch-Physikalische Klasse* 6 (1935): 190–245. The text was translated and reedited with comments in Phillip R. Sloan and Brandon Fogel, eds., *Creating Physical Biology: The Three-Man Paper and Early Molecular Biology* (Chicago: University of

Chicago Press, 2011). The life of Nikolaï Timofeeff-Ressovsky illustrates the difficulties faced by all those who could not or would not choose the right camp in the Second World War. Having left Russia in 1926 to work with the German psychiatrist and neurophysiologist Oskar Vogt, Timofeeff-Ressovsky stayed on to work in Germany during the war. At the end of the war, he was sent to a labor camp in the Soviet Union. Frédéric Joliot-Curie's intervention and Timofeef-Ressovsky's scientific abilities helped to free him, after which he was put in charge of a radiation biology laboratory in Sverdlovsk, in the Urals. He was later charged with having participated in research on prisoners during the war and of having tested the effect of radioactive compounds on them. This alleged collaboration with the Nazis—if it happened—did not enable him to rescue his son, an anti-Nazi resister, from the Mauthausen concentration camp, where he died. Timofeeff-Ressovsky's scientific contribution was very important. In addition to his collaboration with Delbrück, he made several fundamental discoveries in population genetics. He helped draw attention to Chetverikov's Russian school of genetics, and hence played a major role in reconciling genetics and evolutionary biology. His influence seems to have been particularly strong on Theodosius Dobzhansky. See Diane B. Paul and Costas B. Krimbas, "Nikolaï W. Timofeeff-Ressovsky," *Scientific American* 266 (1992): 64–70; Max F. Perutz, "Erwin Schrodinger's 'What Is Life?' and Molecular Biology," in *Schrödinger: Centenary Celebration of a Polymath,* ed. C. W. Kilmister, 234–251 (Cambridge: Cambridge University Press, 1987); Bentley Glass, "Timofeeff-Ressovsky, Nikolaï Wladimirowich," in *Dictionary of Scientific Biography,* vol. 18, suppl. II, ed. F. L. Holmes (New York: Charles Scribner's Sons, 1990).

5. Keith L. Manchester, "Exploring the Gene with X-Rays," *Trends in Genetics* 12 (1996): 515–518.

6. Erwin Schrödinger, *What Is Life?* (Cambridge: Cambridge University Press, 1944).

7. Salvador E. Luria, *A Slot Machine, a Broken Test Tube: An Autobiography* (New York: Harper and Row, 1984).

8. Max Delbrück and Nikolaï W. Timofeeff-Ressovsky, "Cosmic Rays and the Origin of Species," *Nature* 137 (1936): 358–359.

9. Elof A. Carlson, *The Gene: A Critical History* (Philadelphia: Saunders, 1966).

10. Lily E. Kay, "Quanta of Life: Atomic Physics and the Reincarnation of Phage," *History and Philosophy of the Life Sciences* 14 (1992): 3–21.

11. William C. Summers, "How Bacteriophage Came to Be Used by the Phage Group," *Journal of the History of Biology* 26 (1993): 255–267.

12. The bacteriophage is one of the best examples of the importance of the "right tool for the job." Adèle E. Clark and Joan H. Fujimura, *The Right Tools for the Job: At Work in Twentieth-Century Life Sciences* (Princeton, NJ: Princeton University Press, 1992).

13. Lily E. Kay, "Virus, enzyme ou gène? Le problème du bactériophage (1917–1947)," in *L'Institut Pasteur: contributions à son histoire*, éd. Michel Morange, 187–197 (Paris: La Découverte, 1991); William C. Summers, *Félix d'Hérelle and the Origins of Molecular Biology* (New Haven, CT: Yale University Press, 1999).

14. Ton Van Helvoort, "The Construction of Bacteriophage as Bacterial Virus: Linking Endogenous and Exogenous Thought Styles," *Journal of the History of Biology* 27 (1994): 91–139.

15. Kay, "Virus, enzyme ou gène?"; Ton Van Helvoort, "The Controversy between John H. Northrop and Max Delbrück on the Formation of Bacteriophage: Bacterial Synthesis or Autonomous Multiplication," *Annals of Science* 49 (1992): 545–575.

16. Emory Ellis and Max Delbrück, "The Growth of Bacteriophage," *Journal of General Physiology* 22 (1939): 365–384.

17. Mullins, "Development of a Scientific Specialty."

18. Franklin W. Stahl, ed., *We Can Sleep Later: Alfred D. Hershey and the Origins of Molecular Biology* (Cold Spring Harbor, NY: Cold Spring Harbor Laboratory Press, 2000).

19. Ernest P. Fischer and Carol Lipson, *Thinking about Science: Max Delbrück and the Origin of Molecular Biology* (New York: W.W. Norton, 1988), 147.

20. Salvador E. Luria and Thomas F. Anderson, "The Identification and Characterization of Bacteriophages with the Electron Microscope," *Proceedings of the National Academy of Sciences of the United States of America* 28 (1942): 127–130.

21. H. Ruska, "Über ein neues bei der bakteriophagen Lyseauftretendes Formelement," *Naturwissenschaften* 29 (1941): 367–368.

22. Luria, *Slot Machine*.

23. Angela N. H. Creager, "The Paradox of the Phage Group: Essay Reviews," *Journal of the History of Biology* 45 (2010): 183–193. Angela Creager proposes two additional reasons for the importance attributed to the phage group despite its scientific failure: the emphasis put on viruses in biomedical research, and the link that it established with radiation biology.

24. Seymour S. Cohen, "The Synthesis of Bacterial Viruses, Synthesis of Nucleic Acid and Protein in *Escherichia coli* Infected with T2r+" Bacteriophage," *Journal of Biological Chemistry* 174 (1948): 295–303.

25. After the war, the phage system became very well known, even outside biological circles. When John von Neumann wrote to Norbert Wiener in 1946 to discuss the direction taken by research in cybernetics and the functioning of the human brain, he quoted the example of the bacteriophage as a model for studies aimed at understanding elementary biological functioning. Pesi R. Masani, *Norbert Wiener, 1894–1964*, Vita Mathematica, vol. 5 (Basel: Birkhaüser Verlag, 1990), 242–249.

26. Gunther S. Stent, "Max Delbrück," *Trends in Biochemical Sciences* 6 (1981): iii–iv.

27. Fischer and Lipson, *Thinking about Science,* 179.

28. An important spin-off of the work of the phage group, in a different branch of science, was the development by Renato Dulbecco, on Delbrück's recommendation, of a new method of quantitatively studying animal viruses. This method was derived from the method used for studying bacteriophages (see Chapter 15).

29. Alfred D. Hershey and Martha Chase, "Independent Functions of Viral Protein and Nucleic Acid in Growth of Bacteriophage," *Journal of General Physiology* 36 (1952): 39–56.

30. Angela N. H. Creager, *Life Atomic: A History of Radioisotopes in Science and Medicine* (Chicago: University of Chicago Press, 2014).

31. Gunther S. Stent, "That Was the Molecular Biology That Was," *Science* 160 (1968): 390–395; Alfred D. Hershey, "Functional Differentiation within Particles of Bacteriophage T2," *Cold Spring Harbor Symposia on Quantitative Biology* 18 (1953):135–140. For more on this, see Matthew Cobb, "Oswald Avery, DNA, and the Transformation of Biology," *Current Biology* 24 (2015): R55–R60.

32. H. V. Wyatt, "How History Has Blended," *Nature* 249 (1974): 803–805.

33. H. V. Wyatt, "Knowledge and Prematurity: The Journey from Transformation to DNA," *Perspectives in Biology and Medicine* 18 (1975): 149–156.

34. André Boivin, Roger Vendrely, and Colette Vendrely, "L'acide désoxyribonucléique du noyau cellulaire, dépositaire des caractères héréditaires; arguments d'ordre analytique," *Comptes rendus hebdomadaires des séances de l'Académie des sciences* 226 (1948): 1061–1063.

35. Roger Vendrely and Colette Vendrely, "La teneur du noyau cellulaire en acide désoxyribonucléique à travers les organes, les individus et les espèces animales," *Experientia* 4 (1948): 434–436.

5. The Birth of Bacterial Genetics

1. William Bulloch, *The History of Bacteriology* (New York: Dover, 1979; first published in 1938).

2. Olga Amsterdamska, "Stabilizing Instability: The Controversy over Cyclogenic Theories of Bacterial Variation during the Interwar Period," *Journal of the History of Biology* 24 (1991): 191–222.

3. Robert E. Kohler, "Innovation in Normal Science: Bacterial Physiology," *Isis* 76 (1985): 162–181.

4. V. B. Smocovitis, "Unifying Biology: The Evolutionary Synthesis and Evolutionary Biology," *Journal of the History of Biology* 25 (1992): 1–65; Theodosius G. Dobzhansky, *Genetics and the Origin of Species* (New York: Columbia University Press, 1937).

5. Cyril Hinshelwood, *The Chemical Kinetics of the Bacterial Cell* (Oxford: Clarendon Press, 1946).

6. William Summers, "From Culture as Organism to Organism as Cell: Historical Origins of Bacterial Genetics," *Journal of the History of Biology* 24 (1991): 171–190.

7. I. M. Lewis, "Bacterial Variation with Special Reference to Behavior of Some Mutabile Strains of Colon Bacteria in Synthetic Media," *Journal of Bacteriology* 26 (1934): 619–639.

8. Salvador E. Luria and Max Delbrück, "Mutations of Bacteria from Virus Sensitivity to Virus Resistance," *Genetics* 28 (1943): 491–511.

9. Max Delbrück and Salvador E. Luria, "Interference between Bacterial Viruses: 1—Interference between Two Bacterial Viruses Acting Upon the Same Host, and the Mechanism of Virus Growth," *Archives of Biochemistry* 1 (1942): 111–141.

10. Salvador E. Luria, *A Slot Machine, a Broken Test Tube: An Autobiography* (New York: Harper and Row, 1984).

11. Luria, *Slot Machine,* 20.

12. Eugene Wollman, Fernand Holweck, and Salvador Luria, "Effect of Radiations on Bacteriophage C16," *Nature* 145 (1940): 935–936.

13. S. E. Luria and T. F. Anderson, "The Identification and Characterization of Bacteriophages with the Electron Microscope," *Proceedings of the National Academy of Sciences of the United States of America* 28 (1942): 127–130.

14. Luria, *Slot Machine,* 75.

15. Ernest P. Fischer and Carol Lipson, *Thinking about Science: Max Delbrück and the Origins of Molecular Biology* (New York: W.W. Norton, 1988), 142–147.

16. Salvador E. Luria, "Recent Advances in Bacterial Genetics," *Bacteriological Reviews* 11 (1947): 1–40. Despite many attempts, Luria and Delbrück's result has never been seriously challenged.

17. René J. Dubos, *The Bacterial Cell in Its Relation to Problems of Virulence, Immunity, and Chemotherapy* (Cambridge, MA: Harvard University Press, 1945), 176 et seq.

18. Harriet Zuckerman and Joshua Lederberg, "Postmature Scientific Discovery?" *Nature* 324 (1986): 629–631. Dobzhansky, one of the key participants in the evolutionary synthesis, played an important role in this unification of biology. He helped circulate information among geneticists, biochemists, and physicists, and in particular he made Avery's findings known to geneticists.

19. Peter Medawar, *The Art of the Soluble* (London: Methuen, 1967).

20. Joshua Lederberg, "A Fortieth Anniversary Reminiscence," *Nature* 324 (1986): 627–628; "Genetic Recombination in Bacteria: A Discovery Account," *Annual Review of Genetics* 21 (1987): 23–46.

21. Joshua Lederberg and Edward L. Tatum, "Novel Genotypes in Mixed Cultures of Biochemical Mutants of Bacteria," *Cold Spring Harbor Symposia on Quantitative Biology* 11 (1946): 113–114. These results were published in a short paper sent to *Nature:* Joshua Lederberg and Edward L. Tatum, "Gene Recombination in *Escherichia coli,*" *Nature* 158 (1946): 558.

22. Thomas D. Brock, *The Emergence of Bacterial Genetics* (Cold Spring Harbor, NY: Cold Spring Harbor Laboratory Press, 1990).

23. Salvador E. Luria, "Mutations of Bacterial Viruses Affecting Their Host Ranges," *Genetics* 30 (1945): 84–99.

24. Alfred D. Hershey, "Mutation of Bacteriophage with Respect to Type of Plaque," *Genetics* 31 (1946): 620–640; Franklin W. Stahl, ed., *We Can Sleep Later: Alfred D. Hershey and the Origins of Molecular Biology* (Cold Spring Harbor, NY: Cold Spring Harbor Laboratory Press, 2000).

25. Alfred D. Hershey, "Spontaneous Mutations in Bacterial Viruses," *Cold Spring Harbor Symposia on Quantitative Biology* 11 (1946): 67–77; M. Delbrück and W. T. Bailey, "Induced Mutations in Bacterial Viruses," *Cold Spring Harbor Symposia on Quantitative Biology* 11 (1946): 33–37.

26. Frederic L. Holmes, *Reconceiving the Gene: Seymour Benzer's Adventures in Phage Genetics,* ed. William Summers (New Haven, CT: Yale University Press, 2006).

27. J. Lederberg, E. M. Lederberg, N. D. Zinder, and E. R. Lively, "Recombination Analysis of Bacterial Heredity," *Cold Spring Harbor Symposia on Quantitative Biology* 16 (1951): 413–443.

28. William Hayes, "Recombination in Bact. coli K12: Unidirectional Transfer of Genetic Material," *Nature* 169 (1952): 118–119.

29. Elie L. Wollman, "Bacterial Conjugation," in *Phage and the Origins of Molecular Biology,* ed. John Cairns, Gunther S. Stent, and James D. Watson, 216–225 (Cold Spring Harbor, NY: Cold Spring Harbor Laboratory Press, 1966; expanded ed., 1992).

30. Elie L. Wollman and François Jacob, "Sur le mécanisme de transfert du matériel génétique au cours de la recombinaison chez *Escherichia coli* Kl2," *Comptes rendus de l'Académie des sciences* 240 (1955): 2449–2451.

31. Lederberg et al., "Recombination Analysis."

32. Lederberg, "Genetic Recombination," 33–34.

33. Norton D. Zinder and Joshua Lederberg, "Genetic Exchange in *Salmonella,*" *Journal of Bacteriology* 64 (1952): 679–699.

34. Milislav Demerec and P. E. Hartman, "Tryptophan Mutants in *Salmonella typhimurium,*" pp. 5–33, and P. E. Hartman, "Linked Loci in the Control of Consecutive Steps in the Primary Pathway of Histidine Synthesis in *Salmonella typhimurium,*" pp. 35–61, in *Genetic Studies in Bacteria,* Publication 612 (Washington, DC: Carnegie Institution of Washington, 1956).

35. M. L. Morse, E. M. Lederberg, and J. Lederberg, "Transduction in *Escherichia coli* K12," *Genetics* 41 (1956): 142–156.

6. The Crystallization of the Tobacco Mosaic Virus

1. Hermann J. Muller, "Variations Due to Change in the Individual Gene," *American Naturalist* 22 (1922): 32–50.

2. For an excellent description of the scientific and social context, see Angela N. H. Creager, *The Life of a Virus: Tobacco Mosaic Virus as an Experimental Model, 1930–1965* (Chicago: University of Chicago Press, 2002).

3. Ton Van Helvoort, "What Is a Virus? The Case of Tobacco Mosaic Disease," *Studies in History and Philosophy of Science: Part A* 22 (1991): 557–588; John G. Shaw and Milton Zaitlin, eds., *Tobacco Mosaic Virus: One Hundred Years of Contributions to Virology* (St. Paul, MN: American Physiopathological Society Press, 1999); Creager, *Life of a Virus*.

4. Robert Olby, *The Path to the Double Helix* (London: Macmillan, 1974), chap. 10; Lily E. Kay, "W. M. Stanley's Crystallization of the Tobacco Mosaic Virus, *1930–1940*," *Isis* 77 (1986): 450–472.

5. Seymour S. Cohen, "Finally the Beginnings of Molecular Biology," *Trends in Biochemical Sciences* 11 (1986): 92–93.

6. Philip J. Pauly, *Controlling Life: Jacques Loeb and the Engineering Ideal in Biology* (Oxford: Oxford University Press, 1987).

7. Wendell M. Stanley, "Isolation of a Crystalline Protein Possessing the Properties of Tobacco Mosaic Virus," *Science* 81 (1935): 644–645.

8. Lily E. Kay, "The Twilight Zone of Life: The Crystallization of the Tobacco Mosaic Virus," 7th Course of the International School of the History of Biological Sciences, Ischia, June 19–28, 1990.

9. Max A. Lauffer, "Contributions of Early Research on Tobacco Mosaic Virus," *Trends in Biochemical Sciences* 9 (1984): 369–371. In 1934, Max Schlesinger had shown that a phage was made of nucleoprotein: Max Schlesinger, "Zur Frage der Chemischen Zusammensetzung des Bakteriophagen," *Biochemische Zeitschrift* 273 (1934): 306–311.

10. Stanley thought the key thing was that his result agreed with an autocatalytic model of life, and of proteins, which are its key constituent. This model was widely accepted by the Rockefeller Institute biologists Northrop, Bergmann, and Niemann (see Chapter 12). It explained the self-activation of proteolytic enzymes such as trypsin, as well as the replication of bacteriophages and of multi-enzyme complexes in protein synthesis (Olby, *Path to the Double Helix*, chap. 9).

11. Kay, "Twilight Zone of Life."

7. Enter the Physicists

1. Bernard T. Feld and Gertrud Weiss Szilard, *The Collected Papers of Leo Szilard: Scientific Papers* (Cambridge, MA: MIT Press, 1972); William Lanouette and Bela Silard, *Genius in the Shadows: A Biography of Leo Szilard, the Man behind the Bomb* (New York: Charles Scribner's Sons, 1992). A list of physicists involved in the biological revolution and a discussion of their contributions can be found in Donald Fleming, "Emigré Physicists and the Biological Revolution," in *The Intellectual Migration: Europe and America, 1930–1960*, ed.

Donald Fleming and Bernard Bailyn, 152–189 (Cambridge, MA: Belknap Press, 1969).

2. Horace F. Judson, *The Eighth Day of Creation: The Makers of the Revolution in Biology* (New York: Simon & Schuster, 1979), 228. Gamow was also a well-known popular science writer who created the character Mr. Tomkins.

3. Szilard and Gamow were among those émigré scientists in the United States who were suspected of having helped the USSR to acquire information on the atomic bomb. See Pavel A. Sudoplatov, Anatoli P. Sudoplatov, Jerrold L. Schecter, and Leona P. Schecter, *Special Tasks: The Memoirs of an Unwanted Witness, a Soviet Spymaster* (Boston: Little, Brown and Company, 1994).

4. Nicholas C. Mullins, "The Development of a Scientific Specialty: The Phage Group and the Origins of Molecular Biology," *Minerva* 10 (1972): 51–82.

5. François Jacob, "La Molécule," in *La Logique du vivant* (Paris: Gallimard, 1970), chap. 5; "The Molecule," in *The Logic of Life: A History of Heredity*, trans. Betty E. Spillmann (New York: Pantheon Press, 1974, chap. 5.

6. Thomas S. Kuhn, *The Structure of Scientific Revolutions* (Chicago: University of Chicago Press, 1970).

7. Andrew Hodges, *Alan Turing: The Enigma of Intelligence* (London: Burnett Books, 1983); Philippe Breton, *Histoire de l'informatique* (Paris: La Découverte, 1987).

8. R. V. Jones, *The Wizard War: British Scientific Intelligence, 1939–1945* (New York: Coward McCann and Geoghan, 1978); F. H. Hinsley and Alan Stripp, eds., *Codebreakers: The Inside Story of Bletchley Park* (Oxford: Oxford University Press, 1993); Stephen Budiansky, *Battle of Wits: Complete Story of Code Breaking in World War II* (New York: Free Press / Viking, 2000).

9. F. Aaserud, *Redirecting Science: Niels Bohr, Philanthropy and the Rise of Nuclear Physics* (Cambridge: Cambridge University Press, 1990); Abraham Pais, *Niels Bohr's Times in Physics: Philosophy and Polity* (Oxford: Clarendon Press, 1991), 388–394.

10. Niels Bohr, "Light and Life," *Nature* 131 (1933): 421–423, 457–459.

11. Lily E. Kay, "Quanta of Life: Atomic Physics and the Reincarnation of Phage," *History and Philosophy of the Life Sciences* 14 (1992): 3–21. Niels Bohr's ideas on complementarity evolved throughout his life. From a complementarity between reductionism and vitalism, he moved to a complementarity between mechanism and finalism. His last article on biology, "Light and Life Revisited" (1962), did not refer to the notion of complementarity (Pais, *Niels Bohr's Times*, 441–444). The outlook of Delbrück, shaped by Bohr's lecture, was more consistent: Daniel J. McKaughan, "The Influence of Niels Bohr on Max Delbrück: Revisiting the Hopes Inspired by 'Light and Life'," *Isis* 96 (2005): 507–529.

12. Erwin Schrödinger, *What Is Life?* (Cambridge: Cambridge University Press, 1944); Robert C. Olby, "Schrodinger's Problem: What Is Life?" *Journal of the*

History of Biology 4 (1971): 119–148. Schrödinger left Austria shortly after its annexation by Germany in 1938.

13. Gunther S. Stent, "That Was the Molecular Biology That Was," *Science* 160 (1968): 390–395.

14. Erwin Schrödinger, *What Is Life?* (Cambridge: Cambridge University Press, 1992; first published in 1944), 21. Schrödinger's position is a direct transposition of the mechanistic and deterministic view of Descartes, expressed in terms of genes and chromosomes: "If all the parts of the seed of some particular animal, for example man, were known exactly, from this alone the face and the form of each of its limbs could be deduced, for entirely mathematical and precise reasons"; René Descartes, *Description du corps humain,* vol. 11 of *Oeuvres complètes,* ed. Charles Adam and Paul Tannery, 11: 277 (Paris: Librairie philosophique J. Vrin, 2000; first published in 1897).

15. Max F. Perutz, "Erwin Schrödinger's 'What Is Life?' and Molecular Biology," in *Schrodinger: Centenary Celebration of a Polymath,* ed. C. V. Kilmister, 234–251 (Cambridge: Cambridge University Press, 1987). For a critique of many interpretations of Schrödinger's use of the term "code," see A. E. Walsby and M. J. S. Hodge, "Schrödinger's Code-Script: Not a Genetic Cipher but a Code of Development," *Studies in History and Philosophy of Science: Part C* 63 (2017): 45–54.

16. Pnina Abir-Am, "Themes, Genres and Orders of Legitimation in the Consolidation of New Scientific Disciplines: Reconstructing the Historiography of Molecular Biology," *History of Science* 23 (1985): 75–117. This general statement does not mean that Schrödinger's book had no direct influence on some scientists: Nicola Williams, "Irene Manton, Erwin Schrödinger and the Puzzle of Chromosome Structure," *Journal of the History of Biology* 49 (2016): 425–459.

17. Theodosius Dobzhansky, *Genetics and the Origin of Species* (New York: Columbia University Press).

18. Michel Morange, "Schrödinger et la biologie moléculaire," *Fundamenta Scientiae* 4 (1983): 219–234.

19. Schrödinger, *What Is Life?* p. 21.

20. Edward Yoxen, "Where Does Schrödinger's 'What Is Life?' Belong in the History of Molecular Biology?" *History of Science* 17 (1979): 17–52.

21. Yoxen, "Where Does Schrödinger's."

22. Walter Moore, *Schrödinger: Life and Thought* (Cambridge: Cambridge University Press, 1989).

23. Erwin Schrödinger, *My View of the World, Part I: Seek for the Road* (Cambridge: Cambridge University Press, 1964).

24. Morange, "Schrödinger et la biologie moléculaire."

25. William M. Johnston, *The Austrian Mind: An Intellectual and Social History, 1848–1938* (Berkeley: University of California Press, 1972).

8. The Influence of the Rockefeller Foundation

1. Robert E. Kohler, "The Management of Science: The Experience of Warren Weaver and the Rockefeller Foundation Programme in Molecular Biology," *Minerva* 14 (1976): 279–306; Pnina Abir-Am, "The Discourse of Physical Power and Biological Knowledge in the 1930s: A Reappraisal of the Rockefeller Foundation's 'Policy' in Molecular Biology," *Social Studies of Science* 12 (1982): 341–382; John A. Fuerst, Ditta Bartels, Robert Olby, Edward J. Yoxen, and Pnina Abir-Am, "Responses and Replies to P. Abir-Am: Final Response of P. Abir-Am," *Social Studies of Science* 14 (1984): 225–263; Lily E. Kay, *The Molecular Vision of Life: Caltech, the Rockefeller Foundation and the Rise of the New Biology* (Oxford: Oxford University Press, 1993). In a later article, Abir-Am's criticisms of the Rockefeller Foundation were much more muted: Pnina G. Abir-Am, "The Rockefeller Foundation and the Rise of Molecular Biology," *Nature Reviews Molecular Biology* 3 (2002): 65–70. Other founda- tions also played an important role in the birth of molecular biology, such as those that financed research on viruses (see Chapters 6 and 15).

2. Robert E. Kohler, *Partners in Science: Foundations and Natural Scientists, 1900–1945* (Chicago: University of Chicago Press, 1991).

3. Kohler, "Management of Science," 290.

4. Kohler, "Management of Science," 291.

5. Warren Weaver, "Molecular Biology: Origins of the Term," *Science* 170 (1970): 591–592.

6. Doris T. Zallen, "The Rockefeller Foundation and French Research," *Les Cahiers pour l'histoire du CNRS* 5 (1989): 35–58.

7. Jean-François Picard, *La République des savants. La recherche française et le CNRS* (Paris: Flammarion, 1990).

8. Abir-Am, "Discourse of Physical Power."

9. Quoted in Horace F. Judson, *The Eighth Day of Creation: The Makers of the Revolution in Biology* (New York: Simon & Schuster, 1979), 41.

10. Judson, *Eighth Day of Creation*, 361–367; Pnina Abir-Am, "The Assessment of Interdisciplinary Research in the 1930s: The Rockefeller Foundation and Physico-Chemical Morphology," *Minerva* 26 (1988): 153–176.

11. All of them were founder members of a theoretical biology club. Pnina G. Abir-Am, "The Biotheoretical Gathering, Trans-Disciplinary Authority and the Incipient Legitimation of Molecular Biology in the 1930s: New Perspective on the Historical Sociology of Science," *History of Science* 25 (1987): 1–70; Pnina G. Abir-Am and Dorinda Outram, eds., *Uneasy Careers and Intimate Lives: Women in Science, 1789–1979* (New Brunswick, NJ: Rutgers University Press, 1987). Chapter 12 is devoted to Dorothy Wrinch. Joseph Needham later turned to the study of Chinese science. On J. H. Woodger and his critique of methodological reductionism, see Nils Roll-Hansen, "E. S. Russell and J. H. Woodger: The Failure of Two Twentieth-Century Opponents of Mechanistic

Biology," *Journal of the History of Biology* 17 (1984): 399–428. Bernal was one of the first protein crystallographers, and a thinker keen to address the most fundamental biological issues; see Andrew Brown, *J. D. Bernal: The Sage of Science* (Oxford: Oxford University Press, 2005).

12. Tim J. Horder and Paul Weindling, "Hans Spemann and the Organizer," in *A History of Embryology,* ed. Tim Horder, Jan Witkowski, and C. C. Wylie, 183–242 (Cambridge: Cambridge University Press, 1986). For an overall survey of his research, see Jan Witkowski's very clear study "The Hunting of the Organizer: An Episode in Biochemical Embryology," *Trends in Biochemical Sciences* 10 (1985): 379–381; Scott F. Gilbert, "Induction and the Origins of Developmental Genetics," in *Developmental Biology,* vol. 7, *A Conceptual History of Modern Embryology,* ed. Scott F. Gilbert, 181–206 (New York: Plenum Press, 1991); Rony Armon, "Between Biochemists and Embryologists: The Biochemical Study of Embryonic Induction in the 1930s," *Journal of the History of Biology* 45 (2012): 65–108.

13. J. M. W. Slack, *From Egg to Embryo: Determinative Events in Early Development* (Cambridge: Cambridge University Press, 1983), cited by Witkowski, "Hunting of the Organizer."

14. Abir-Am and Outram, *Uneasy Careers,* 267.

15. Similarly, the patrons of Caltech were not the creators of molecular biology, contrary to Kay's suggestion, even though the research groups from this institute perhaps contributed more than others. See Kay, *Molecular Vision of Life.*

16. Robert Bud, *The Uses of Life: A History of Biotechnology* (Cambridge: Cambridge University Press, 1993).

17. Antoine Danchin, *Physique, chimie, biologie, un demi-siècle d'interactions: 1927–1977,* text written for the fiftieth anniversary of Institut de Biologie Physico-Chimique, Paris, 1977; Michel Morange, "L'Institut de biologie physico-chimique: de sa fondation à l'entrée dans l'ère moléculaire," *Revue pour l'histoire du CNRS* 7 (2002): 32–40.

9. Physical Techniques in Molecular Biology

1. William Bulloch, *The History of Bacteriology* (Oxford: Oxford University Press, 1938; reprint New York: Dover, 1979); Thomas D. Brock, *Robert Koch: A Life in Medicine and Bacteriology* (Madison, WI: Science Tech Publishers, 1988).

2. A. J. P. Martin and R. L. M. Synge, "A New Form of Chromatogram Employing Two Liquid Phases: 1. A Theory of Chromatography; 2. Application to the Micro-Determination of the Higher Monoamino-Acids in Proteins," *Biochemical Journal* 35 (1941): 1358–1368; A. H. Gordon, A. J. P. Martin, and R. L. M. Synge, "Partition Chromatography in the Study of Protein Constituents," *Biochemical Journal* 37 (1943): 79–86.

3. R. Consden, A. H. Gordon, and A. J. P. Martin, "Qualitative Analysis of Proteins: A Partition Chromatographic Method Using Paper," *Biochemical Journal* 38 (1944): 224–232. Paper chromatography had already been used by the German chemical industries (in particular the dye and oil industry) in the nineteenth century; this method had later been forgotten (reported by A. H. Gordon in P. R. Srinivasan, Joseph S. Fruton, and John T. Edsall, *The Origins of Modern Biochemistry: A Retrospect on Proteins* (New York: New York Academy of Science, 1979).

4. Erwin Chargaff, *Hericlatean Fire: Sketches from a Life before Nature* (New York: Rockefeller University Press, 1978), 90–93.

5. Ion-exchange resins, long used in organic chemistry, played an important role in the Manhattan Project in separating the products of nuclear fission; see Arthur Kornberg, *For the Love of Enzymes: The Odyssey of a Biochemist* (Cambridge: Harvard University Press, 1989). The first resins of this kind to be used in molecular biology were polymethacrylic acid resins (Amberlite IRC 50) and sulfonated polystyrene resins. Many other resins (Dowex, DEAE-cellulose) appeared in the next few years, some of which were better suited to protein and polypeptide separation. Stanford Moore and William Stein used this technique to develop automatic amino acid analyzers. In 1948 Moore and Stein were also the first to use automatic fraction collectors, which simplified column chromatography considerably; see Stanley Moore and William H. Stein, "Chemical Structures of Pancreatic Ribonuclease and Deoxyribonuclease," *Science* 180 (1973): 458–464. For dextran beads, see Jerker Porath and P. Flodin, "Gel Filtration: A Method for Desalting and Group Separation," *Nature* 183 (1959): 1657–1659.

6. Rolf Axen, Jerker Porath, and Sverker Emback, "Chemical Coupling of Peptides and Proteins to Polysaccharides by Means of Cyanogen Halides," *Nature* 214 (1967): 1302–1304; Pedro Cuatrecasas, Meir Wilchek, and Christian Anfinsen, "Selective Enzyme Purification by Affinity Chromatography," *Proceedings of the National Academy of Sciences of the United States of America* 61 (1968): 636–643.

7. Robert Olby, *The Path to the Double Helix* (London: Macmillan, 1974), 11–21 et seq; Lily E. Kay, *The Molecular Vision of Life: Caltech, the Rockefeller Foundation and the Rise of the New Biology* (Oxford: Oxford University Press, 1993), 112 et seq.; Boelie Elzen, "Two Ultracentrifuges: A Comparative Study of the Social Construction of Artefacts," *Social Studies of Science* 16 (1986): 621–662.

8. Milton Kerker, "The Svedberg and Molecular Reality," *Isis* 67 (1976): 190–216.

9. The Svedberg, "The Ultra-Centrifuge and the Study of High-Molecular Compounds," *Nature* 139 (1937): 1051–1062.

10. Dorothy M. Wrinch, "The Pattern of Proteins," *Nature* 137 (1936): 411–412.

11. Linus Pauling and Carl Niemann, "The Structure of Proteins," *Science* 61 (1939): 1860–1867.

12. Milton Kerker, "The Svedberg and Molecular Reality: An Autobiographical Postscript," *Isis* 77 (1986): 278–282.
13. Lily E. Kay, "W. M. Stanley's Crystallization of the Tobacco Mosaic Virus, 1930–1940," *Isis* 77 (1986): 450–472.
14. Albert Claude, "The Coming of Age of the Cell: The Inventory of Cells by Fractionation, Biochemistry and Electron Microscopy Has Affected Our Status and Thinking," *Science* 189 (1975): 433–435; William Bechtel, *Discovering Cell Mechanisms: The Creation of Modern Cell Biology* (Cambridge: Cambridge University Press, 2008).
15. Elzen, "Two Ultracentrifuges."
16. Lily E. Kay, "Laboratory Technology and Biological Knowledge: The Tiselius Electrophoresis Apparatus, 1930–1945," *History and Philosophy of the Life Sciences* 10 (1988): 51–72.
17. R. Consden, A. H. Gordon, and A. J. P. Martin, "Ionophoresis in Silica Jelly: A Method for the Separation of Amino-Acids and Peptides," *Biochemical Journal* 40 (1946): 33–41.
18. Samuel Raymond and Lewis Weintraub, "Acrylamide Gel as a Supporting Medium for Zone Electrophoresis," *Science* 130 (1959): 711.
19. Ulrich K. Laemmli, "Cleavage of Structural Proteins During the Assembly of the Head of Bacteriophage T4," *Nature* 227 (1970): 680–685; Howard Hsueh-Hao Chang, "The Laboratory Technology of Discrete Molecular Separation: The Historical Development of Gel Electrophoresis and the Material Epistemology of Biomolecular Science," *Journal of the History of Biology* 42 (2009): 495–527.
20. Kay, "Laboratory Technology," 51. Arne Tiselius also made significant contributions to chromatography: see the review by A. J. P. Martin and R. L. M. Synge, "Analytical Chemistry of the Proteins," *Advances in Protein Chemistry* 2 (1945): 1–83; and the autobiographical account of Arne Tiselius, "Reflections from Both Sides of the Counter," *Annual Review of Biochemistry* 37 (1968): 1–24.
21. Ronald Bentley, "The Use of Stable Isotopes at Columbia University's College of Physicians and Surgeons," *Trends in Biochemical Sciences* 10 (1985): 171–174.
22. Rudolph Schonheimer, *The Dynamic State of Body Constituents* (Cambridge, MA: Harvard University Press, 1942).
23. Jacques Monod, "From Enzymatic Adaptation to Allosteric Transitions," *Science* 154 (1966): 477.
24. Robert E. Kohler Jr., "Rudolf Schonheimer, Isotopic Tracers and Biochemistry in the 1930s," *Historical Studies in the Physical Sciences* 8 (1977): 257–298; on the history of radioisotopes, see Angela N. H. Creager, *Life Atomic: A History of Radioisotopes in Science and Medicine* (Chicago: University of Chicago Press, 2013).
25. Doris T. Zallen, "The Rockefeller Foundation and Spectroscopy Research: The Programs at Chicago and Utrecht," *Journal of the History of Biology* 25 (1992): 67–89.

26. In particular, by providing quantitative measures of nucleic acids, as in Chargaff's work.

27. V. E. Cosslet, "The Early Years of Electron Microscopy in Biology," *Trends in Biochemical Sciences* 10 (1985): 361–363; Nicolas Rasmussen, "Making a Machine Instrumental: RCA and the Wartime Origins of Biological Electron Microscopy in America, 1940–1945," *Studies in History and Philosophy of Science* 28 (1996): 311–349; Nicolas Rasmussen, *Picture Control: The Electronic Microscope and the Transformation of Biology in America, 1940–1960* (Palo Alto, CA: Stanford University Press, 1999).

28. For example, mesosomes. Nicolas Rasmussen, "Facts, Artifacts and Mesosomes: Practicing Epistemology with the Electron Microscope," *Studies in History and Philosophy of Science* 24 (1993): 227–265. The limitations of electron microscopy are such that the work of specialists in this field is not representative of biological research.

10. The Role of Physics

1. Horace F. Judson, *The Eighth Day of Creation: The Makers of the Revolution in Biology* (New York: Simon & Schuster, 1979), and see in particular pp. 606–607. For a very different point of view, see Matthew Cobb, "1953: When Genes Became 'Information,'" *Cell* 153 (2013): 503–506, and Matthew Cobb, *Life's Greatest Secret: The Race to Crack the Genetic Code* (London: Profile Books, 2015).

2. Arthur Kornberg, "Molecular Origins," *Nature* 214 (1967): 538.

3. Alain Prochiantz, "L'illusion physicaliste dans les sciences de la vie," *Revue internationale de psychopathologie* 8 (1992): 553–569. Even if that were the case, Schrödinger's role would still have been significant. Analogies and metaphors have an important place in scientific research.

4. Nils Roll-Hansen, "The Meaning of Reductionism," in *International Conference on Philosophy of Science* (Vigo, Spain: University of Vigo, 1996), 125–148.

5. Olga Amsterdamska, "Stabilizing Instability: The Controversy over Cyclogenic Theories of Bacterial Variation during the Interwar Period," *Journal of the History of Biology* 24 (1991): 191–222.

6. Donald Fleming and Bernard Bailyn, eds., *The Intellectual Migration: Europe and America, 1930–1960* (Cambridge, MA: Belknap Press, 1969); Paul K. Hoch, "Migration and the Generation of New Scientific Ideas," *Minerva* 25 (1987): 209–237. Molecular biology developed in an international space. See Pnina Abir-Am, "From Multidisciplinary Collaboration to Transnational Objectivity: International Space as Constitutive of Molecular Biology, 1930–1970," in *Denationalizing Science: The Contexts of International Scientific Practice,* ed. Elisabeth Crawford, Terry Shin, and Sverker Sorlin, 153–186 (Dordrecht, the Netherlands: Kluwer Academic, 1993).

7. Evelyn Fox Keller, "Physics and the Emergence of Molecular Biology: A History of Cognitive and Political Synergy," *Journal of the History of Biology* 23 (1990): 389–409.

8. "Everything suggests that the logic of heredity can be compared to that of a calculator. Seldom has the model of an epoch found such a faithful application." François Jacob, *La Logique du vivant* (Paris: Gallimard, 1970); *The Logic of Life: A History of Heredity,* trans. Betty E. Spillmann (New York: Pantheon Press, 1974). The chronologies are in fact remarkably parallel—1936: the Turing machine, the first "theoretical" concept of a computer; 1944: Avery's experiment; 1945: conception of the first digital computer (EDVAC) by John von Neumann; 1948: creation of cybernetics by Norbert Wiener, and publication of information theory by Claude Shannon; 1953: discovery of the double-helix structure of DNA and early reflections on the genetic code. See Norbert Wiener, *Cybernetics: Or Control and Communication in the Animal and the Machine* (Paris: Hermann; Cambridge, MA: MIT Press; New York: Wiley, 1948); Claude E. Shannon, "A Mathematical Theory of Communication," *Bell System Technical Journal* 27 (1948): 379–423, 623–656. In the years after World War II, John von Neumann made many contacts with the founders of molecular biology. William Aspray, *John von Neumann and the Origins of Modern Computing* (Cambridge, MA: MIT Press, 1990), 181 et seq.

9. Andrew Hodges, *Alan Turing: The Enigma of Intelligence* (London: Burnett, 1983). Alan Turing was himself interested in biology; one of his last publications was devoted to morphogenesis: A. M. Turing, "The Chemical Basis of Morphogenesis," *Philosophical Transactions of the Royal Society B* 237 (1952): 37–72. The relations between cyberneticists and molecular biologists, such as Delbrück, were at first difficult: Steve J. Heims, *Constructing a Social Science for Postwar America: The Cybernetics Group, 1946–1953* (Cambridge, MA: MIT Press, 1993), 93–96. Although at the end of the 1940s and at the beginning of the 1950s information theory did not have a practical influence on molecular biology, the vocabulary of information theory became widely used at the beginning of the 1960s: Lily E. Kay, *Who Wrote the Book of Life? A History of the Genetic Code* (Palo Alto, CA: Stanford University Press, 2000); Matthew Cobb, *Life's Greatest Secret: The Race to Crack the Genetic Code* (London: Profile Books, 2015).

11. The Discovery of the Double Helix

1. Watson has written a lively account of this period; see James D. Watson, *The Double Helix: A Personal Account of the Discovery of the Structure of DNA* (London: Weidenfeld and Nicholson, 1968; rev. ed. 1981); see also the biography of Watson: Victor K. McElheny, *Watson and DNA: Making a Scientific Revolution* (Hoboken, NJ: Perseus / John Wiley, 2003). The British historian Robert Olby has written *The Path to the Double Helix* (London: Macmillan,

1974), and more recently a biography of Francis Crick: *Francis Crick, Hunter of Life's Secrets* (Cold Spring Harbor, NY: Cold Spring Harbor Laboratory Press, 2009). Horace F. Judson has reconstituted the periods before and after this discovery, using a large number of interviews; see *The Eighth Day of Creation: The Makers of the Revolution in Biology* (New York: Simon & Schuster, 1979). The path to the double helix has been described anew by Michael Fry, who has included a lot of historical work and oral accounts from participants: Michael Fry, *Landmark Experiments in Molecular Biology* (London: Academic Press, 2016). These are the main sources for this chapter.

2. Francis Crick, *What Mad Pursuit: A Personal View of Scientific Discovery* (New York: Basic Books, 1988).

3. On the history of the Cambridge molecular biology laboratory, see Soraya De Chadarevian, *Designs for Life: Molecular Biology after World War II* (Cambridge: Cambridge University Press, 2002).

4. John M. Thomas and David Phillips, eds., *Selections and Reflections: The Legacy of Sir Lawrence Bragg* (London: Science Reviews, 1991).

5. Crick, *What Mad Pursuit.*

6. Anne Sayre, *Rosalind Franklin and DNA* (New York: W. W. Norton, 1975); in the more recent biography written by Brenda Maddox, Rosalind Franklin is not presented as a symbol of the difficulties met by female scientists but as a rich and complex personality and an active participant in the discovery of the double helix: Brenda Maddox, *Rosalind Franklin: The Dark Lady of DNA* (London: Harper Collins, 2002).

7. Information about the situation in the London laboratory and on the difficult relations between London and Cambridge can be found in: Maurice Wilkins, *The Third Man of the Double Helix; The Autobiography* (Oxford: Oxford University Press, 2003); Alexander Gann and Jan Witkowski, "The Lost Correspondence of Francis Crick," *Nature* 467 (2010): 519–524, and in Olby, *Francis Crick.*

8. Linus Pauling and Robert B. Corey, "Two Hydrogen-Bonded Spiral Configurations of the Polypeptide Chain," *Journal of the American Chemical Society* 72 (1950): 5349; Linus Pauling, Robert B. Corey, and H. R. Branson, "Two Hydrogen-Bonded Helical Configurations of the Polypeptide Chain," *Proceedings of the National Academy of Sciences of the United States of America* 37 (1951): 205–211; Linus Pauling and Robert B. Corey, seven successive papers in *Proceedings of the National Academy of Sciences of the United States of America* 37 (1951): 235–285.

9. Olby, *Path to the Double Helix,* chap. 4.

10. Sir Lawrence Bragg, John C. Kendrew, and Max F. Perutz, "Polypeptide Chain Configuration in Crystalline Proteins," *Proceedings of the Royal Society of London: Series A, Mathematical, Physical and Engineering Sciences* 203 (1950): 321–357.

11. Alexander Rich and Norman Davidson, *Structural Chemistry and Molecular Biology* (San Francisco: W. H. Freeman, 1968).

12. Linus Pauling and Robert B. Corey, "A Proposed Structure for the Nucleic Acids," *Proceedings of the National Academy of Sciences of the United States of America* 39 (1953): 84–97.

13. Keith L. Manchester, "Did a Tragic Accident Delay the Discovery of the Double Helical Structure of DNA?" *Trends in Biochemical Sciences* 20 (1995): 126–128.

14. James D. Watson and Francis H. C. Crick, "A Structure for Deoxyribose Nucleic Acid," *Nature* 171 (1953): 737–738; Maurice H. F. Wilkins, Alexander R. Stokes, and H. R. Wilson, "Molecular Structure of Deoxy Pentose Nucleic Acid," *Nature* 171 (1953): 738–740; Rosalind E. Franklin and Raymond G. Gosling, "Molecular Configuration in Sodium Thymonucleate," *Nature* 171 (1953): 740–741.

15. James D. Watson and Francis H. C. Crick, "Genetical Implications of the Structure of Deoxyribonucleic Acid," *Nature* 171 (1953): 964–967.

16. Bruno J. Strasser, "Who Cares About the Double Helix?" *Nature* 422 (2003): 803–804; Yves Gingras, "Revisiting the 'Quiet Debut' of the Double Helix: A Bibliometric and Methodological Note on the 'Impact' of Scientific Publications," *Journal of the History of Biology* 43 (2010): 159–181.

17. Graeme K. Hunter, *Light as a Messenger: The Life and Science of William Lawrence Bragg* (Oxford: Oxford University Press, 2004); Kersten T. Hall, *The Man in the Monkeynut Coat: William Astbury and the Forgotten Road to the Double Helix* (Oxford: Oxford University Press, 2014).

18. J. D. Bernal, "Structure of Proteins," *Nature* 143 (1939): 663–667.

19. Ernest P. Fischer and Carol Lipson, *Thinking about Science: Max Delbrück and the Origins of Molecular Biology* (New York: W.W. Norton, 1988).

20. The concept of tinkering *(bricolage)* was introduced in 1977 by François Jacob to describe evolution's action on life (see Chapter 23). See François Jacob, "Evolution and Tinkering," *Science* 196 (1977): 1161–1166. As Jacob points out in his article, this concept also applies to scientific research. A slightly different interpretation of the usefulness of this idea in describing the behavior of scientists can be found in Karin D. Knorr-Cetina, *The Manufacture of Knowledge: An Essay on the Constructivist and Contextual Nature of Science* (Oxford: Pergamon Press, 1981), 34–35.

21. This was how, in 1954, Watson and Crick tried to rationally reconstruct the discovery of the double helix structure of DNA. Francis H. C. Crick and James D. Watson, "The Complementary Structure of Deoxyribonucleic Acid," *Proceedings of the Royal Society of London: Series A, Mathematical, Physical and Engineering Sciences* 223 (1954): 80–96.

22. Gunther S. Stent, "Prematurity and Uniqueness in Scientific Discovery," *Scientific American* 227 (1972): 84–93.

23. Crick, *What Mad Pursuit.*

24. Sayre, *Rosalind Franklin and DNA.*

25. Maddox, *Rosalind Franklin.*

26. Angela N. H. Creager and Gregory J. Morgan, "After the Double Helix: Rosalind Franklin's Research on Tobacco Mosaic Virus," *Isis* 99 (2008): 239–272.

27. Rosalind Franklin had single-handedly made some important steps toward the correct structure in the early weeks of 1953, which explains why she immediately accepted Watson and Crick's results. See Aaron Klug, "The Discovery of the DNA Double Helix," *Journal of Molecular Biology* 335 (2004): 3–26.

28. Bernadette Bensaude-Vincent, "Une mythologie révolutionnaire dans la chimie française," *Annals of Science* 40 (1983): 189–196.

29. Pnina Abir-Am, "From Biochemistry to Molecular Biology: DNA and the Acculturated Journey of the Critic of Science Erwin Chargaff," *History and Philosophy of the Life Sciences* 2 (1980): 3–60; Pnina Abir-Am, "How Scientists View Their Heroes: Some Remarks on the Mechanism of Myth Construction," *Journal of the History of Biology* 15 (1982): 281–315.

30. To borrow the title of Robert Olby's book, *Path to the Double Helix.*

31. Judson, *Eighth Day of Creation,* 261 et seq.

32. Crick, *What Mad Pursuit.*

33. Monica Winstanley, "Assimilation into the Literature of a Critical Advance in Molecular Biology," *Social Studies of Science* 6 (1976): 545–549; Barak Gaster, "Assimilation of Scientific Change: The Introduction of Molecular Genetics into Biology Textbooks," *Social Studies of Science* 20 (1990): 431–454.

34. A. H. J. Wang, G. J. Quigley, F. J. Kolpak, et al., "Molecular Structure of a Left-Handed Double Helical DNA Fragment at Atomic Resolution," *Nature* 282 (1979): 680–686. Z-DNA was the first DNA structure to be proved, over 25 years after Watson and Crick's discovery.

35. Max Delbrück and Gunther S. Stent, "On the Mechanism of DNA Replication," in *The Chemical Basis of Heredity,* ed. W. D. McElroy and B. Glass, 699–736 (Baltimore: Johns Hopkins Press, 1957).

36. J. B. S. Haldane, *New Paths in Genetics* (London: Allen and Unwin, 1941), 44.

37. Judson, *Eighth Day of Creation,* 188 et seq.

38. Matthew Meselson and Franklin W. Stahl, "The Replication of DNA in *Escherichia coli,*" *Proceedings of the National Academy of Sciences of the United States of America* 44 (1958): 671–682.

39. Frederic L. Holmes, *Meselson, Stahl, and the Replication of DNA: A History of "The Most Beautiful Experiment in Biology"* (New Haven, CT: Yale University Press, 2001).

12. Deciphering the Genetic Code

1. James D. Watson and Francis H. C. Crick, "Genetical Implications of the Structure of Deoxyribonucleic Acid," *Nature* 171 (1953): 964–967. The best history of the discovery of the genetic code is Lily E. Kay, *Who Wrote the Book of Life? A History of the Genetic Code* (Palo Alto, CA: Stanford University Press,

2000). For a more popular account, see Matthew Cobb, *Life's Greatest Secret: The Race to Crack the Genetic Code* (London, Profile Books, 2015).

2. Horace F. Judson, *The Eighth Day of Creation: The Makers of the Revolution in Biology* (New York: Simon & Schuster, 1979), 261 et seq.

3. By an "unlucky" chance the distance separating amino acids in an extended protein chain is 3.3–3.4 Å (this had been noted by William Astbury as early as 1938). This value is identical to that separating two successive nucleotides in a DNA molecule and led to a number of mistaken hypotheses. It also explains why a hypothesis analogous to Gamow's had previously been proposed by Pauling and Corey when they published their erroneous model of the structure of DNA. See Linus Pauling and Robert B. Corey, "A Proposed Structure for the Nucleic Acids," *Proceedings of the National Academy of Sciences of the United States of America* 39 (1953): 84–97.

4. Francis H. C. Crick, John S. Griffith, and Leslie E. Orgel, "Codes without Commas," *Proceedings of the National Academy of Sciences of the United States of America* 43 (1957): 416–421.

5. Carl R. Woese, *The Genetic Code: The Molecular Basis for Genetic Expression* (New York: Harper and Row, 1967), chap. 2; Martynas Yčas, *The Biological Code* (Amsterdam: North Holland, 1969); Jan A. Witkowski, "The 'Magic' of Numbers," *Trends in Biochemical Sciences* 10 (1985): 139–141.

6. Francis H. C. Crick, "The Present Position of the Coding Problem," *Brookhaven Symposia* 12 (1959): 35–39.

7. Max Bergmann and Carl Niemann, "Newer Biological Aspects of Protein Chemistry," *Science* 86 (1937): 187–190.

8. Archer J. P. Martin and Richard L. M. Synge, "A New Form of Chromatogram Employing Two Liquid Phases: 1. A Theory of Chromatography; 2. Application to the Micro-Determination of the Higher Monoamino-Acids in Proteins," *Biochemical Journal* 35 (1941): 1358–1361; A. H. Gordon, Archer J. P. Martin, and Richard L. M. Synge, "Partition Chromatography in the Study of Protein Constituents," *Biochemical Journal* 37 (1943): 79–86; R. Consden, A. H. Gordon, and Archer J. P. Martin, "Qualitative Analysis of Proteins: A Partition Chromatographic Method Using Paper," *Biochemical Journal* 38 (1944): 224–232.

9. Frederick Sanger, "Sequences, Sequences and Sequences," *Annual Review of Biochemistry* 57 (1988): 1–28. Sanger profited from the fact that proteases had become relatively well understood, in particular as a result of the work of Bergmann's group. To carry out this work, he developed a reagent that was specific to the N-terminal part of amino acids, subsequently known as Sanger's reagent. On Sanger's relations with the Cambridge molecular biologists, see Soraya de Chadarevian, "Sequences, Conformation, Information: Biochemists and Molecular Biologists in the 1950s," *Journal of the History of Biology* 29 (1996): 361–386; see also Miguel Garcia-Sancho, *Biology, Computing, and the History of Molecular Sequencing: from Proteins to DNA, 1945–2000* (New York: Palgrave MacMillan, 2012).

10. A. P. Ryle, Frederick Sanger, L. F. Smith, and Ruth Kital, "The Disulphide Bonds of Insulin," *Biochemical Journal* 60 (1955): 541–556; Frederick Sanger, "Chemistry of Insulin: Determination of the Structure of Insulin Opens the Way to Greater Understanding of Life Processes," *Science* 129 (1959): 1340–1344. The structure of a small antibiotic polypeptide, gramicidin, had been determined slightly earlier, along with—thanks to du Vigneaud—the sequence of two small pituitary gland hormones, ocytocin and vasopressin.

11. Linus Pauling, Harvey A. Itano, Seymour J. Singer, and Ibert C. Wells, "Sickle Cell Anemia, a Molecular Disease," *Science* 110 (1949): 543–548.

12. James V. Neel, "The Inheritance of Sickle Cell Anemia," *Science* 110 (1949): 64–66.

13. Vernon M. Ingram, "A Specific Chemical Difference between the Globins of Normal Human and Sickle-Cell Anaemia Haemoglobin," *Nature* 178 (1956): 792–794; Vernon M. Ingram, "Gene Mutations in Human Haemoglobin: The Chemical Difference between Normal and Sickle Cell Haemoglobin," *Nature* 180 (1957): 326–328.

14. Joseph S. Fruton, "Proteolytic Enzymes as Specific Agents in the Formation and Breakdown of Proteins," *Cold Spring Harbor Symposia on Quantitative Biology* 9 (1941): 211–217; Ditta Bartels, "The Multi-Enzyme Programme of Protein Synthesis: Its Neglect in the History of Biochemistry and Its Current Role in Biotechnology," *History and Philosophy of the Life Sciences* 5 (1983): 187–219.

15. Judson, *Eighth Day of Creation,* 247.

16. The earliest data were obtained by Joachim Hammerling in 1934 on a giant unicellular alga, *Acetabularia mediterranea.* Joachim Hammerling, "Über formbildende Substanzen bei Acetabularia mediterranea, Räumliche und zeitliche Verteilung und ihre Herkunft," *Wilhelm Roux' Archiv für Entwicklungsmechanik der Organismen: Organ für d. gesamte kausale Morphologie* 131 (1934): 1–81.

17. Torbjörn Caspersson and Jack Schultz, "Pentose Nucleotides in the Cytoplasm of Growing Tissues," *Nature* 143 (1939): 602–603; Jean Brachet, "La localisation des acides pentose nucléiques dans les tissus animaux et les oeufs d'amphibiens en voie de développement," *Archives de biologie* 53 (1942): 207–257.

18. Alexander L. Dounce, "Duplicating Mechanism for Peptide Chain and Nucleic Acid Synthesis," *Enzymologia* 15 (1952): 251–258. The term "template" was introduced into biochemistry in 1904 by H. E. Armstrong to denote the "surface" on which enzymatic catalysis takes place; H. E. Armstrong, "Enzyme Action as Bearing on the Validity of the Ionic-Dissociation Hypothesis and on the Phenomena of Vital Change," *Journal of the Chemical Society* 73 (1904): 537. The term thus suggested stereospecific recognition (see Chapter 1).

19. Alexander Dounce, "Nucleic and Template Hypotheses," *Nature* 172 (1953): 541–542. This model was again supported by the similarity of the distance both between amino acids in an unfolded polypeptide chain and between nucleotides in the DNA molecule. (See note 3 in this chapter.)

20. André Boivin and Roger Vendrely, "Sur le rôle possible des deux acides nucléiques dans la cellule vivante," *Experientia* 3 (1947): 32–34. This publication came with a handy English summary that would have made the results available to non-Francophones.

21. Linus Pauling, "A Theory of the Structure and Process of Formation of Antibodies," *Journal of the American Chemical Society* 62 (1940): 2643–2657.

22. For a more complete presentation of theories of antibody biosynthesis, see Arthur M. Silverstein, *A History of Immunology* (San Diego, CA: Academic Press, 1988).

23. Karl Landsteiner, *The Specificity of Serological Reactions* (Springfield, IL: Charles C. Thomas, 1936); Linus Pauling, "A Theory of the Structure and Process of Formation of Antibodies," *Journal of the American Chemical Society* 62 (1940): 2643–2657.

24. F. Breinl and Felix Haurowitz, "Chemische Untersuchung des Präzipitates aus Hämoglobin und Anti-Hämoglobin Serum und Bemerkungen über die Natur der Antikörper," *Hoppe-Seyler's Zeitschrift für physiologische chemie* 192 (1930): 45–57; Jerome Alexander, "Some Intracellular Aspects of Life and Disease," *Protoplasma* 14 (1931): 296–306; Stuart Mudd, "A Hypothetical Mechanism of Antibody Formation," *Journal of Immunology* 23 (1932): 423–427.

25. According to Karl Popper, what distinguishes a scientific theory from a nonscientific one is that the former can be refuted or falsified by experimentation. See Karl Popper, *The Logic of Scientific Discovery* (London: Hutchinson, 1959).

26. Experiments reported by Burnet: F. M. Burnet and F. Fenner, *The Production of Antibodies* (Melbourne: Macmillan, 1949).

27. Linus Pauling and Dan H. Campbell, "The Production of Antibodies in Vitro," *Science* 95 (1942): 440–441; Linus Pauling and Dan H. Campbell, "The Manufacture of Antibodies in Vitro," *Journal of Experimental Medicine* 76 (1942): 211–220.

28. Felix Haurowitz, Paula Schwerin, and Saide Tunç, "The Mutual Precipitation of Proteins and Azoproteins," *Archives of Biochemistry and Biophysics* 11 (1946): 515–520; Felix Haurowitz was one of the Jewish refugees from Central Europe (Prague) to whom the Turkish government generously offered academic positions: Arnold Reisman, *Turkey's Modernization: Refugees from Nazism and Ataturk's Vision* (Washington, DC: New Academia, 2006).

29. Lily E. Kay, "Molecular Biology and Pauling's Immunochemistry: A Neglected Dimension," *History and Philosophy of the Life Sciences* 11 (1989): 211–219; Lily E. Kay, *The Molecular Vision of Life: Caltech, the Rockefeller Foundation*

and the Rise of the New Biology (Oxford: Oxford University Press, 1993), chap. 6.

30. See, for example, MacFarlane Burnet, *The Clonal Selection Theory of Acquired Immunity* (Nashville: Vanderbilt University Press, 1959); Niels K. Jerne, "The Natural Selection Theory of Antibody Formation," *Proceedings of the National Academy of Sciences of the United States of America* 41 (1955): 849–857.

31. Linus Pauling, "Molecular Basis of Biological Specificity," *Nature* 248 (1974): 769–771.

32. Felix Haurowitz, *Chemistry and Biology of Proteins* (New York: Academic Press, 1950); Felix Haurowitz, "The Mechanism of the Immunological Response," *Biological Reviews* 27 (1952): 247–280.

33. Jacques Monod, "The Phenomenon of Enzymatic Adaptation and Its Bearings on Problems of Genetics and Cellular Differentiation," *Growth Symposium* 11 (1947): 223–289.

34. Jacques Monod and Melvin Cohn, "La biosynthèse induite des enzymes (adaptation enzymatique)," *Advances in Enzymology* 13 (1952): 67–119.

35. Kay, "Molecular Biology," 211–219.

36. George W. Gray, "Pauling and Beadle," *Scientific American* 180 (1949): 16–21, quotation pp. 19–20; mentioned in Kay, "Molecular Biology."

37. Alfred H. Sturtevant, "Can Specific Mutations Be Induced by Serological Methods?" *Proceedings of the National Academy of Sciences of the United States of America* 30 (1944): 176–178.

38. Anne-Marie Moulin, *Le Dernier langage de la médecine: histoire de l'immunologie de Pasteur au SIDA* (Paris: Presses Universitaires de France, 1991), 176.

39. Sterling Emerson, "The Induction of Mutations by Antibodies," *Proceedings of the National Academy of Sciences of the United States of America* 30 (1944): 179–183.

40. Seymour S. Cohen, "The Biochemical Origins of Molecular Biology," *Trends in Biochemical Sciences* 9 (1984): 334–336; Robert C. Olby, "Biochemical Origins of Molecular Biology: A Discussion," *Trends in Biochemical Sciences* 11 (1986): 303–305.

41. John Cairns, Gunther S. Stent, and James D. Watson, eds., *Phage and the Origins of Molecular Biology* (Cold Spring Harbor, NY: Cold Spring Harbor Laboratory Press, 1966; expanded ed., 1992); John C. Kendrew, "How Molecular Biology Started," *Scientific American* 216 (1967): 141–144. This idea was taken up and developed by Gunther S. Stent, "That Was the Molecular Biology That Was," *Science* 160 (1968): 390–395.

42. Paul Zamecnik, "The Machinery of Protein Synthesis," *Trends in Biochemical Sciences* 9 (1984): 464–466; "Historical Aspects of Protein Synthesis," in *The Origins of Modern Biochemistry: A Retrospect on Proteins,* ed. P. R. Srinivasan, Joseph S. Fruton, and John T. Edsall, 269–301 (New York: New York Academy of Science, 1979); Hans-Jörg Rheinberger, "Experiment, Difference and Writing:

I. Tracing Protein Synthesis," *Studies in History and Philosophy of Science* 23 (1992): 305–331, and II. "The Laboratory Production of Transfer RNA," *Studies in History and Philosophy of Science* 23 (1992): 389–422; Hans-Jörg Rheinberger, "Experiment and Orientation: Early Systems of in Vitro Protein Synthesis," *Journal of the History of Biology* 26 (1993): 443–471; Hans-Jörg Rheinberger, *Toward a History of Epistemic Things: Synthesizing Proteins in the Test Tube* (Palo Alto, CA: Stanford University Press, 1997). The very detailed studies by Rheinberger clearly show how the strategies chosen by scientists are constrained and directed by experimental "resistance." For example, one of the most difficult steps was the development of methods for rupturing cells that did not lead to the inactivation of cellular proteins.

43. Mahlon Hoagland, *Towards the Habit of Truth: A Life in Science* (New York: W.W. Norton, 1990).

44. Judson, *Eighth Day of Creation,* chap. 6.

45. Marvin R. Lamborg and Paul C. Zamecnik, "Amine Acid Incorporation into Proteins by Extracts of *E. coli,*" *Biochimica et Biophysica Acta* 42 (1960): 206–211; Alfred Tissières, David Schlessinger, and François Gros, "Amine Acid Incorporation into Proteins by *Escherichia coli* Ribosomes," *Proceedings of the National Academy of Sciences of the United States of America* 46 (1960): 1450–1463.

46. These experiments are well described in Judson, *Eighth Day of Creation,* 470–489, and in Michael Fry, *Landmark Experiments in Molecular Biology* (London: Academic Press, 2016).

47. Johann H. Matthaei and Marshall W. Nirenberg, "Characteristics and Stabilization of DNAase Sensitive Protein Synthesis in *E. coli* Extracts," *Proceedings of the National Academy of Sciences of the United States of America* 47 (1961): 1580–1588; Marshall W. Nirenberg and Johann H. Matthaei, "The Dependence of Cell-Free Protein Synthesis in *E. coli* upon Naturally Occurring or Synthetic Polyribonucleotides," *Proceedings of the National Academy of Sciences of the United States of America* 47 (1961): 1588–1602.

48. Judson, *Eighth Day of Creation,* 478, 482; Robert C. Olby, "And on the Eighth Day . . . ," *Trends in Biochemical Sciences* 4 (1979): N215–N216; see also Franklin H. C. Portugal, *The Least Likely Man: Marshall Nirenberg and the Discovery of the Genetic Code* (Boston: MIT Press, 2015).

49. Francis H. C. Crick, "The Recent Excitement in the Coding Problem," *Progress in Nucleic Acid Research* (New York: Academic Press, 1963), 1:163–217.

50. Har G. Khorana, "Polynucleotide Synthesis and the Genetic Code," *Harvey Lectures* 1966–1967, series 62 (New York: Academic Press, 1968), 79–105.

51. Marshall W. Nirenberg and Philip Leder, "RNA Codewords and Protein Synthesis: The Effect of Trinucleotides upon the Binding of sRNA to Ribosomes," *Science* 145 (1964): 1399–1407.

52. Francis H. C. Crick, Leslie Barnett, Sydney Brenner, and R. J. Watts-Tobin, "General Nature of the Genetic Code for Proteins," *Nature* 192 (1961):

1227–1232; Sydney Brenner, *My Life in Science,* ed. Errol C. Friedberg and Eleanor Lawrence (London: PubMed Central, 2001); Errol C. Friedberg, *Sydney Brenner: A Biography* (Cold Spring Harbor, NY: Cold Spring Harbor Laboratory Press, 2010).

53. Francis Crick, *What Mad Pursuit: A Personal View of Scientific Discovery* (New York: Basic Books, 1988).

13. The Discovery of Messenger RNA

1. Francis Crick, *What Mad Pursuit: A Personal View of Scientific Discovery* (New York: Basic Books, 1988).
2. Francis H. C. Crick, "On Protein Synthesis," *Symposia of the Society for Experimental Biology* 12 (1958): 138–163. The spontaneous folding hypothesis was based on a large number of experiments on protein denaturation–renaturation that were begun in the 1930s and brilliantly developed on ribonuclease by Christian Anfinsen in the late 1950s. Francis Crick knew of Anfinsen's recent results when he gave his lecture. See Bruno J. Strasser, "A World in One Dimension: Pauling, Francis Crick, and the Central Dogma of Molecular Biology," *History and Philosophy of the Life Sciences* 28 (2006): 491–512.
3. Crick, "On Protein Synthesis," 153.
4. Johann H. Matthaei and Marshall W. Nirenberg, "Characteristics and Stabilization of DNAase Sensitive Protein Synthesis in *E. coli* Extracts," *Proceedings of the National Academy of Sciences of the United States of America* 47 (1961): 1580–1588.
5. Torbjörn Caspersson and Jack Schultz, "Pentose Nucleotides in the Cytoplasm of Growing Tissues," *Nature* 143 (1939): 602–603; Torbjörn Caspersson, "Nukleinsaureketten und Genvermehrung," *Chromosoma* 1 (1940): 605–619.
6. Jean Brachet, "La localisation des acides pentosenucléiques dans les tissus animaux et les oeufs d'amphibiens en voie de développement," *Archives de biologie* 53 (1942): 207–257.
7. Jean Brachet, Hubert Chantrenne and F. Vanderhaeghe, "Biochemical Interaction of the Nucleus and Cytoplasm of Unicellular Organisms. II. *Acetabulatia mediterranea,*" *Biochimica et Biophysica Acta* 18 (1955): 544–563. The early experiments on anucleation had been done by Joachim Hammerling. For a reevaluation of the role of the "Belgian" molecular biology group at Rouge-Cloître near Brussels, see Denis Thieffry and Richard M. Burian, "Jean Brachet's Alternative Scheme for Protein Synthesis," *Trends in Biochemical Sciences* 21 (1996): 114–117.
8. Ilana Löwy, "Variances in Meaning in Discovery Accounts: The Case of Contemporary Biology," *Historical Studies in the Physical and Biological Sciences* 21 (1990): 87–121.
9. This work was described by Palade in 1974, in his Nobel lecture; George Palade, "Intracellular Aspects of the Process of Protein Synthesis," *Science* 189

(1975): 347–358. See also Hans-Jörg Rheinberger, "From Microsomes to Ribosomes: 'Strategies' of 'Representation'," *Journal of the History of Biology* 28 (1995): 49–89.

10. Heinz Fraenkel-Conrat, "The Role of the Nucleic Acid in the Reconstitution of Active Tobacco Mosaic Virus," *Journal of the American Chemical Society* 78 (1956): 882–883.

11. Arthur B. Pardee, François Jacob, and Jacques Monod, "The Genetic Control and Cytoplasmic Expression of 'Inducibility' in the Synthesis of β-galactosidase by *E. coli*," *Journal of Molecular Biology* 1 (1959): 165–178.

12. Monica Riley, Arthur Pardee, François Jacob, and Jacques Monod, "On the Expression of a Structural Gene," *Journal of Molecular Biology* 2 (1960): 216–225; this experiment is one example of the "suicide experiments" made possible by the use of phosphorus 32: Angela N. H. Creager, "Phosphorus-32 in the Phage Group: Radioisotopes as Historical Tracers of Molecular Biology," *Studies in History and Philosophy of Science Part C: Studies in History and Philosophy of Biological and Biomedical Sciences* 40 (2009): 29–42.

13. François Jacob, *La Statue intérieure* (Paris: Odile Jacob, 1986), chap. 7; *The Statue Within: An Autobiography*, trans. Franklin Philip (New York: Basic Books, 1988).

14. François Gros, *Les Secrets du gène* (Paris: Odile Jacob, 1986), 126–127.

15. Horace F. Judson, *The Eighth Day of Creation: The Makers of the Revolution in Biology* (New York: Simon & Schuster, 1979), 428–436; Jacob, *Statue Within*, chap. 7; Crick, *What Mad Pursuit;* Sydney Brenner, *My Life in Science*, ed. Errol C. Friedberg and Eleanor Lawrence (London: PubMed Central, 2001); Errol C. Friedberg, *Sydney Brenner: A Biography* (Cold Spring Harbor, NY: Cold Spring Harbor Laboratory Press, 2010), chaps. 11 and 12; Michael Fry, *Landmark Experiments in Molecular Biology* (London: Academic Press, 2016).

16. Elliot Volkin and L. Astrachan, "Phosphorus Incorporation in *Escherichia coli* Ribonucleic Acid after Infection with Bacteriophage T2," *Virology* 2 (1956): 149–161.

17. Jacob, *La Statue intérieure*, 415–427; Sydney Brenner, François Jacob, and Matthew Meselson, "An Unstable Intermediate Carrying Information from Genes to Ribosomes for Protein Synthesis," *Nature* 190 (1961): 576–581.

18. François Gros, H. Hiatt, Walter Gilbert, Chuck G. Kurland, R. W. Risebrough, and James D. Watson, "Unstable Ribonucleic Acid Revealed by Pulse Labelling of *Escherichia coli*," *Nature* 190 (1961): 581–585; Gros, *Les Secrets du gène*, 131–135.

19. Crick, *What Mad Pursuit*. Another example of a postmature discovery was discussed in Chapter 2—Beadle and Tatum's one gene-one enzyme hypothesis.

20. Andrei N. Belozersky and A. S. Spirin, "A Correlation between the Composition of Deoxyribonucleic Acid and Ribonucleic Acids," *Nature* 182 (1958): 111–112.

21. Thieffry and Burian, "Jean Brachet's Alternative Scheme"; a good illustration of this "alternative scheme" can be found in a review written by Hubert Chantrenne in 1958: Hubert Chantrenne, "Newer Developments in Relation to Protein Biosynthesis," *Annual Review of Biochemistry* 27 (1958): 35–56.

22. There were several competing theories of cytoplasmic inheritance. The plasmon theory was opposed to the plasmagene theory and rejected the particulate nature of cytoplasmic inheritance.

23. The only study devoted to these theories is Jan Sapp, *Beyond the Gene: Cytoplasmic Inheritance and the Struggle for Authority in Genetics* (New York: Oxford University Press, 1987). See also Jan Sapp, "Hérédité cytoplasmique et histoire de la génétique," in Jean-Louis Fischer and William H. Schneider, *Histoire de la génétique: pratique, techniques et théories* (Paris: ARPEM et Sciences en situation, 1990), 231–246; Jan Sapp, "The Struggle for Authority in the Field of Heredity, 1900–1932: New Perspectives on the Rise of Genetics," *Journal of the History of Biology* 16 (1983): 311–342; Jonathan Harwood, "Genetics and the Evolutionary Synthesis in Interwar Germany," *Annals of Science* 42 (1985): 279–301, and *Styles of Scientific Thought in the German Genetics Community* (Chicago: University of Chicago Press, 1993). Among the many original publications on this subject, in particular see Sol Spiegelman and M. D. Kamen, "Genes and Nucleoproteins in the Synthesis of Enzymes," *Science* 104 (1946): 581–584. The question of whether the nucleus or the cytoplasm determines inheritance and guides embryonic development had been posed at the end of the nineteenth century. The supporters of the cytoplasm also generally believed that development could be influenced by the external medium. The form of the cytoplasm does in fact depend on the surrounding medium: Garland E. Allen, "Morgan's Background and the Revolt from Descriptive and Speculative Biology," in *A History of Embryology,* ed. T. J. Horder, J. A. Witkowski, and C. C. Wylie, 116–146 (Cambridge: Cambridge University Press, 1986).

24. Francis H. C. Crick, "The Present Position of the Coding Problem," *Brookhaven Symposia* 12 (1959): 35–39; cited in Judson, *Eighth Day of Creation,* 346.

25. See, for example, Jerard Hurwitz, Ann Bresler, and Renata Diringer, "The Enzymic Incorporation of Ribonucleotides into Polyribonucleotides and the Effect of DNA," *Biochemical and Biophysical Research Communications* 3 (1960): 15–19.

26. James D. Watson, *The Double Helix: A Personal Account of the Discovery of the Structure of DNA* (London: Weidenfeld and Nicolson, 1968; rev. ed., 1981), pp. 67–69.

27. Alexander Rich, "An Analysis of the Relation between DNA and RNA," *Annals of the New York Academy of Sciences* 81 (1959): 709–72; and "A Hybrid Helix Containing Both Deoxyribose and Ribose Polynucleotides and its Relation to the Transfer of Information between the Nucleic Acids," *Proceedings of the*

National Academy of Sciences of the United States of America 46 (1960): 1044–1053.

28. Elliot Volkin, "The Function of RNA in T2 Infected Bacteria," *Proceedings of the National Academy of Sciences of the United States of America* 46 (1960): 1336–1349.

29. Mahlon B. Hoagland, "Nucleic Acids and Proteins," *Scientific American* 201 (1959): 55–61.

30. Masayasu Nomura, Benjamin D. Hall, and Sol Spiegelman, "Characterization of RNA, Synthesized in *Escherichia coli* after Bacteriophage T2 Infection," *Journal of Molecular Biology* 2 (1960): 306–326; Dario Giacomoni, "The Origin of DNA: RNA Hybridization," *Journal of the History of Biology* 26 (1993): 89–107; Benjamin D. Hall and S. Spiegelman, "Sequence Complementarity of T2-DNA and T2-specific RNA," *Proceedings of the National Academy of Sciences of the United States of America* 47 (1961): 137–146.

31. J. Marmur and D. Lane, "Strand Separation and Specific Recombination in Deoxyribonucleic Acids: Biological Studies," *Proceedings of the National Academy of Sciences of the United States of America* 46 (1960): 453–461; Paul Doty, J. Marmur, J. Eigner, and C. Schildkraut, "Strand Separation and Specific Recombination in Deoxyribonucleic Acids: Physical Chemical Studies," *Proceedings of the National Academy of Sciences of the United States of America* 46 (1960): 461–476.

32. Matthew Cobb, "Who Discovered Messenger RNA?" *Current Biology* 25 (2015): R526–R532.

14. The French School

1. Gunther S. Stent, "That Was the Molecular Biology That Was," *Science* 160 (1968): 390–395.

2. Mirko D. Grmek and Bernardino Fantini, "Le rôle du hasard dans la naissance du modèle de l'opéron," *Revue d'histoire des sciences* 35 (1982): 193–215; Laurent Loison recently showed that, far from being a surprise, a possible convergence had been anticipated by André Lwoff: Laurent Loison, "Le rôle de la physiologie microbienne d'André Lwoff dans la constitution du modèle de l'opéron," in *L'Invention de la régulation génétique: les Nobel 1965 (Jacob, Lwoff, Monod) et le modèle de l'opéron dans l'histoire de la biologie*," éd. Laurent Loison and Michel Morange, 67–84 (Paris: Editions Rue d'Ulm, 2017).

3. Horace F. Judson, *The Eighth Day of Creation: The Makers of the Revolution in Biology* (New York: Simon & Schuster, 1979), chap. 7; Bernardino Fantini, ed., *Jacques Monod: Pour une éthique de la connaissance* (Paris: La Découverte, 1988); Patrice Debré, *Jacques Monod* (Paris: Flammarion, 1996); see also the thematic issue of the *Comptes rendus de l'Académie des Sciences / Biologies* on "Jacques Monod: A Theorist in the Era of Molecular Biology," 338 (2015): 369–423, Jean Gayon, Michel Morange and François Gros eds., with contributions

from Henri Buc, Soraya de Chadarevian, Jean Gayon, Evelyn Fox Keller, Laurent Loison, Francesca Merlin, Michel Morange, and Maxime Schwartz.

4. Jacques Monod, "From Enzymatic Adaptation to Allosteric Transitions," *Science* 154 (1966): 475–483; Jean-Paul Gaudillière, "J. Monod, S. Spiegelman et l'adaptation enzymatique. Programmes de recherche, cultures locales et traditions disciplinaires," *History and Philosophy of the Life Sciences* 14 (1992): 23–71.

5. As already suggested before the war by Marjory Stephenson. See Robert E. Kohler, "Innovation in Normal Science: Bacterial Physiology," *Isis* 76 (1985): 162–181.

6. Jacques Monod, "The Phenomenon of Enzymatic Adaptation and Its Bearings on Problems of Genetics and Cellular Differentiation," *Growth Symposium* 11 (1947): 223–289.

7. Benno Müller-Hill, *The "lac" Operon: A Short History of a Genetic Paradigm* (New York: Walter De Gruyter, 1996).

8. Michel Morange, "L'oeuvre scientifique de Jacques Monod," *Fundamenta Scientiae* 3 (1982): 396–404.

9. Alvin M. Pappenheimer Jr., "Qu'est donc devenue Pz?" in *Les Origines de la biologie moléculaire: Un hommage à Jacques Monod,* ed. André Lwoff and Agnès Ullmann, 55–60 (Paris: "Études Vivantes," 1980); "Whatever Happened to Pz?" in *Origins of Molecular Biology: A Tribute to Jacques Monod,* ed. Agnès Ullmann, 69–75 (Washington, DC: ASM Press, 2003); Monod, "Enzymatic Adaptation to Allosteric Transitions."

10. Melvin Cohn, Jacques Monod, Martin R. Pollock, Sol Spiegelman, and Roger Y. Stanier, "Terminology of Enzyme Formation," *Nature* 172 (1953): 1096–1097.

11. Monod, who had been a communist during and after the war, broke with the party after the Lysenko affair. For five years the attitude of the Communist Party did not change at all. The Mendelian theory of heredity and the respective merits of neo-Darwinism and neo-Lamarckism continued to fuel discussions among French Left intellectuals. The question was not settled, and Monod felt the need to reaffirm publicly his rejection of all neo-Lamarckian ideas. See Joël and Dan Kottek, *L'Affaire Lyssenko* (Brussels: Éditions Complexe, 1986); Dominique Lecourt, *Proletarian Science?* (London: New Left Books, 1976); Z. A. Medvedev, *The Rise and Fall of T. D. Lysenko* (New York: Columbia University Press, 1969); Nils Roll-Hansen, *The Lysenko Effect: The Politics of Science* (Amherst, NY: Humanity Books, 2004).

12. Thomas D. Brock, *The Emergence of Bacterial Genetics* (Cold Spring Harbor, NY: Cold Spring Harbor Laboratory Press, 1990), chap. 7.

13. Charles Galperin, "Le bactériophage, la lysogénie et son déterminisme génétique," *History and Philosophy of the Life Sciences* 9 (1987): 175–224; Laurent Loison, Jean Gayon, and Richard M. Burian, "The Contributions—and Collapse—of Lamarckian Heredity in Pasteurian Molecular

Biology: I. Lysogeny, 1900–1960," *Journal of the History of Biology* 50 (2017): 5–52.

14. François Jacob, *The Statue Within: An Autobiography* (New York: Basic Books, 1988); see also the special issue of *Research in Microbiology* published in honor of Jacob: Julian Davies, ed., Memorial Issue Devoted to François Jacob (1920–2013), *Research in Microbiology* 165 (2014): 311–398; Jacob was less ignorant of research than he made out. After the war, he worked on the production of antibiotics, and attended meetings on microbiology: Michel Morange, "François Jacob: 17 June 1920–19 April 2013," *Biographical Memoirs of Fellows of the Royal Society* 63 (2017): 345–361.

15. Charles Galperin, "La lysogénie et les promesses de la génétique bactérienne," in *L'Institut Pasteur: contributions à son histoire* (Paris: La Découverte, 1991), 198–206; Brock, *Emergence of Bacterial Genetics*.

16. Grmek and Fantini, "Le rôle du hasard."

17. Arthur B. Pardee, François Jacob, and Jacques Monod, "The Genetic Control and Cytoplasmic Expression of 'Inducibility' in the Synthesis of β-galactosidase by *E. coli*," *Journal of Molecular Biology* 1 (1959): 165–178.

18. François Jacob and Jacques Monod, "Genetic Regulatory Mechanisms in the Synthesis of Proteins," *Journal of Molecular Biology* 3 (1961): 318–356. For a description of this research, see Judson, *Eighth Day of Creation*, chap. 7. See also: Jean-Paul Gaudillière, "J. Monod, S. Spiegelman et l'adaptation enzymatique: programmes de recherche, cultures locales et traditions disciplinaires," *History and Philosophy of the Life Sciences* 14 (1992): 23–27; Grmek and Fantini, "Le rôle du hasard"; Kenneth Schaffner, "Logic of Discovery and Justification in Regulatory Genetics," *Studies in History and Philosophy of Science* 4 (1974): 349–385.

19. Jacques Monod, "An Outline of Enzyme Induction," *Recueil des travaux chimiques des Pays-Bas* 77 (1958): 569.

20. François Jacob, "Genetic Control of Viral Function," in *Harvey Lectures, 1958–1959* (New York: Academic Press, 1960), 1–39.

21. All these results were published in a single article that was dense, rich, and enormously influential: F. Jacob and J. Monod, "Genetic Regulatory Mechanisms in the Synthesis of Proteins", *Journal of Molecular Biology* 3 (1061): 318–356.

22. Monod, "Enzymatic Adaptation to Allosteric Transitions"; Jacob, *Statue Within*.

23. H. Edwin Umbarger, "Evidence for a Negative-Feedback Mechanism in the Biosynthesis of Isoleucine," *Science* 123 (1956): 848.

24. Bernard T. Feld and Gertrud Weiss Szilard, *The Collected Papers of Leo Szilard: Scientific Papers* (Cambridge, MA: MIT Press, 1972); William Lanouette and Bela Szilard, *Genius in the Shadows: A Biography of Leo Szilard, the Man Behind the Bomb* (New York: Charles Scribner's Sons, 1992).

25. Michel Morange, "Le concept de gène régulateur," in Fischer and Schneider, *Histoire de la génétique,* 271–291.

26. François Jacob and Jacques Monod, "Gènes de structure et gènes de régulation dans la biosynthèse des protéines," *Comptes rendus de l'Académie des Sciences* 249 (1959): 1282–1284; François Jacob and Jacques Monod, "On the Regulation of Gene Activity," *Cold Spring Harbor Symposia on Quantitative Biology* 26 (1961): 193–211.

27. Jacques Monod, "Remarques conclusives du colloque: Basic Problems in Neoplastic Disease," in *Jacques Monod: pour une éthique de la connaissance,* ed. Bernardino Fantini, 79–96 (Paris: La Découverte, 1988).

28. Evelyn Fox Keller, *A Feeling for the Organism: The Life and Work of Barbara McClintock* (New York: W. H. Freeman, 1983); Nina Fedoroff and David Botstein, eds., *The Dynamic Genome: Barbara McClintock's Ideas in the Century of Genetics* (Cold Spring Harbor, NY: Cold Spring Harbor Laboratory Press, 1991). Nathaniel Comfort has shown that it was not the results of Barbara McClintock on gene mobility that were not accepted, but rather the role that she gave them in the control of gene expression during development: Nathaniel Comfort, *The Tangled Field: Barbara McClintock's Search for Patterns of Genetic Control* (Cambridge, MA: Harvard University Press, 2001).

29. François Jacob, "Comments," *Cancer Research* 20 (1960): 695–697.

30. Elie L. Wollman and François Jacob, *La Sexualité des bactéries* (Paris: Masson, 1959); *Sexuality and the Genetics of Bacteria* (New York: Academic Press, 1961). The change undoubtedly took place in 1961. In the English version of the book, Jacob and Wollman were much more cautious, as shown by the change in the title of the corresponding section: "Les épisomes et la différenciation cellulaire" ("Episomes and Cellular Differentiation") became "Episomes and Cellular Regulation."

31. Jacques Monod and François Jacob, "General Conclusions: Teleonomic Mechanisms in Cellular Metabolism, Growth and Differentiation," *Cold Spring Harbor Symposia on Quantitative Biology* 26 (1961): 389–401.

32. François Jacob and Jacques Monod, "Genetic Repression, Allosteric Inhibition and Cellular Differentiation," in *Cytodifferential and Macromolecular Synthesis,* ed. M. Locke, 30–64 (New York: Academic Press, 1963), quote from p. 31.

33. Edward Yoxen, "Where Does Schrödinger's 'What Is Life?' Belong in the History of Molecular Biology?" *History of Science* 17 (1979): 17–52.

34. Jacob did not recall there being any such direct influence (François Jacob, personal communication). See also Michel Morange, "Le concept de gène régulateur," in Fischer and Schneider, *Histoire de la génétique,* 271–291.

35. Jacques Monod, Jeffries Wyman, and Jean-Pierre Changeux, "On the Nature of Allosteric Transitions: A Plausible Model," *Journal of Molecular Biology* 12 (1965): 88–118.

36. Angela N. H. Creager and Jean-Paul Gaudillière, "Meanings in Search of Experiments or *Vice-Versa:* The Invention of Allosteric Regulation in Paris and Berkeley (1959–1969)," *Historical Studies in the Physical and Biological Sciences* 27 (1996): 1–89.

37. Henri Buc, "Mother Nature and the Design of a Regulatory Enzyme," in Ullmann, *Origins of Molecular Biology,* 255–262; Daniel E. Koshland Jr., "Memories of Jacques Monod," in Ullmann, *Origins of Molecular Biology,* 249–253. The scale and intensity of these controversies might seem surprising. They partly continued the old debates between Darwinians and neo-Lamarckians. Koshland's model was given a neo-Lamarckian interpretation by Carl C. Lindegren, who played a key role in the genetic analysis of *Neurospora* (see Chapter 2) and yeast, but who also became a fervent opponent of "Morganian" genetics. See Carl C. Lindegren, "Lamarckian Proteins," *Nature* 198 (1963): 1224.

38. Jean-Pierre Changeux subsequently turned to the study of the acetylcholine receptor, which he interpreted according to allosteric theory: Thierry Heidmann and Jean-Pierre Changeux, "Structural and Functional Properties of the Acetylcholine Receptor Protein in Its Purified and Membrane-Bound States," *Annual Review of Biochemistry* 47 (1978): 317–335. Jean-Pierre Changeux, "Allosteric Proteins: From Regulatory Enzymes to Receptors—Personal Recollections," *Bioessays* 15 (1993): 634.

39. Jacques Monod wrote the preface to the first French edition of Popper's *The Logic of Scientific Discovery* (London: Hutchinson, 1959), translated as *La Logique de la découverte scientifique* (Paris: Payot, 1973).

40. François Jacob, Sydney Brenner, and François Cuzin, "On the Regulation of DNA Replication in Bacteria," *Cold Spring Harbor Symposia on Quantitative Biology* 28 (1963): 329–348.

41. François Jacob, *Of Flies, Mice, and Men* (Cambridge, MA: Harvard University Press, 1999), chap. 3. Other scientists in his laboratory adopted a more cautious strategy and began to study cellular differentiation in an amoeba, *Dictyostelium discoideum.*

42. Jean-Paul Gaudillière, "Catalyse enzymatique et oxydations cellulaires: l'œuvre de Gabriel Bertrand et son héritage," in *L'Institut Pasteur,* 118–136.

43. Yvette Conry, *L'Introduction du darwinisme en France au XIXᵉ siècle* (Paris: Vrin, 1974).

44. Jan Sapp, *Beyond the Gene: Cytoplasmic Inheritance and the Struggle for Authority in Genetics* (New York: Oxford University Press, 1987), chap. 5; Laurent Loison, "French Roots of French Neo-Lamarckism, 1879–1985," *Journal of the History of Biology* 44 (2011): 713–744.

45. The remarkable work by Philippe L'Héritier and Georges Teissier on experimental populations of *Drosophila* in the 1930s was highly significant. For a description, see Jean Gayon, *Darwin et l'après-Darwin: une histoire de l'hypothèse de sélection naturelle* (Paris: Kimé, 1992), 379–384; *Darwinism's Struggle for Survival: Heredity and the Hypothesis of Natural Selection* (Cambridge: Cambridge University Press, 1998).

46. Richard M. Burian, Jean Gayon, and Doris Zallen, "The Singular Fate of Genetics in the History of French Biology, 1900–1940," *Journal of the History of Biology* 21 (1988): 357–402.

47. Richard M. Burian and Jean Gayon, "Genetics after World War II: The Laboratories at Gif," *Cahiers pour l'histoire du CNRS* 7 (1990): 25–48; see also Richard M. Burian and Jean Gayon, "The French School of Genetics: From Physiological and Population Genetics to Regulatory Molecular Genetics," *Annual Review of Genetics* 33 (1999): 313–349; Richard M. Burian and Jean Gayon, "National Traditions and the Emergence of Genetics: the French Example," *Nature Reviews Genetics* 5 (2004): 150–156.

48. Richard M. Burian and Jean Gayon, "Un évolutionniste bernardien à l'Institut Pasteur? Morphologie des ciliés et évolution physiologique dans l'œuvre d'André Lwoff," in *L'Institut Pasteur,* 165–186.

49. Burian and Gayon, "Un évolutionniste bernardien"; Laurent Loison, "Monod before Monod: Enzymatic Adaptation, Lwoff, and the Legacy of General Biology," *History and Philosophy of the Life Sciences* 35 (2013): 167–192.

50. Sapp, *Beyond the Gene,* chap. 5.

51. For a vivid description of "Monod before Monod," portraying the man in his intellectual, political and societal context, and focusing on his friendship with the writer Albert Camus, see Sean B. Carroll, *Brave Genius: A Scientist, a Philosopher, and Their Daring Adventures from the French Resistance to the Nobel Prize* (New York: Crown, 2013).

52. Jacques Monod, *Chance and Necessity* (London: Collins, 1972).

53. Jacob, *Statue Within.*

54. François Jacob, *The Logic of Life: A History of Heredity* (Princeton, NJ: Princeton University Press, 1973).

15. Normal Science

1. Thomas S. Kuhn, *The Structure of Scientific Revolutions* (Chicago: University of Chicago Press, 1970).

2. Gunther S. Stent, *The Coming of the Golden Age: A View of the End of Progress* (Garden City, NY: Natural History Press, 1969).

3. Ernest P. Fischer and Carol Lipson, *Thinking about Science: Max Delbrück and the Origins of Molecular Biology* (New York: W.W. Norton, 1988).

4. Yoshiki Hotta and Seymour Benzer, "Mapping of Behaviour in *Drosophila* Mosaics," *Nature* 240 (1972): 527–535; Jonathan Weiner, *Time, Love, Memory: A Great Biologist and His Quest for the Origins of Behavior* (New York: Alfred A. Knopf, 1999).

5. Robert Olby, *Francis Crick, Hunter of Life's Secrets* (Cold Spring Harbor, NY: Cold Spring Harbor Laboratory Press, 2009).

6. Stent, *Coming of the Golden Age;* François Jacob, *The Logic of Life: A History of Heredity* (Princeton, NJ: Princeton University Press, 1973).

7. Jacques Monod, *Chance and Necessity* (London: Collins, 1972), 12.

8. See, for example, Pierre-Henri Simon, *Questions aux savants* (Paris: Le Seuil, 1969); Marc Beigbeder, *Le Contre-Monod* (Paris: Grasset, 1972); Madeleine

Barthélémy-Madaule, *L'Idéologie du hasard et de la nécessité* (Paris: Le Seuil, 1972). The intensity of the debate that accompanied the publication of *Le Hasard et la nécessité* is explained more by Monod's personality and his place in French society than by the book's content. Monod was the only scientist among the intellectuals who played an important part in French life between World War II and the end of the 1970s. Very active in the Communist resistance during the war, but having spectacularly broken with the Communist Party over the Lysenko affair, Monod remained a leftist, close to Albert Camus, whose work he admired profoundly. Monod was one of the scientists close to the circle of Pierre Mendès France, who tried to renovate the French university and research systems in the late 1950s, without much success. Basking in the official recognition that followed the Nobel Prize, and as the director of the Pasteur Institute from 1971 on, Monod often intervened in public affairs. He participated in the French Family Planning Movement in the campaign for birth control and the liberalization of contraception and was also active in protests demanding the legalization of abortion, the right to euthanasia, and support for imprisoned Soviet scientists. His views were always clear, but his opponents sometimes thought they were marked by contempt. This explains why many of the debates he took part in were particularly bitter. See Jacques Julliard and Michel Winock, eds., *Les Intellectuels français* (Paris: Le Seuil, 1996), 800–801; Patrice Debré, *Jacques Monod* (Paris: Flammarion, 1996); Sean B. Carroll, *Brave Genius: A Scientist, a Philosopher, and Their Daring Adventures from the French Resistance to the Nobel Prize* (New York: Crown, 2013). A rich archive relating to be Monod can be found at the Service des Archives de l'Institut Pasteur, fond Monod. Monod was also a musician—he was torn for a long time between scientific research and a career in music. He was also interested in literature and even wrote a play, *Le Puits de Syène (1964).*

9. Maxime Schwartz, "Une autre voie?" in *Les Origines de la biologie moléculaire: Un hommage à Jacques Monod,* ed. André Lwoff and Agnès Ullmann, 177–184 (Paris: Études Vivantes, 1980); "Another Route," in *Origins of Molecular Biology: A Tribute to Jacques Monod,* ed. Agnès Ullmann, 207–215 (Washington, DC: ASM Press, 2003).

10. David Baltimore, "Viral RNA-Dependent DNA Polymerase," *Nature* 226 (1970): 1209–1211; Howard M. Temin and S. Mizutani, "RNA-Dependent DNA Polymerase in Virions of Rous Sarcoma Virus," *Nature* 226 (1970): 1211–1213; Lindley Darden, "Exemplars, Abstractions, and Anomalies: Representations and Theory Change in Mendelian and Molecular Genetics," in *Philosophy of Biology,* ed. James G. Lennox and Gereon Walters, 137–158 (Pittsburgh: University of Pittsburgh Press, 1995).

11. Francis H. C. Crick, "On Protein Synthesis," *Symposia of the Society for Experimental Biology* 12 (1958): 138–163.

12. Denis Thieffry and Richard M. Burian, "Jean Brachet's Alternative Scheme for Protein Synthesis," *Trends in Biochemical Sciences* 21 (1996): 114–117.

13. Francis Crick, *What Mad Pursuit: A Personal View of Scientific Discovery* (New York: Basic Books, 1988).

14. James D. Watson, *Molecular Biology of the Gene* (New York: W. A. Benjamin, 1965).

15. Renato Dulbecco, "Production of Plaques in Monolayer Tissue Cultures by Single Particles of an Animal Virus," *Proceedings of the National Academy of Sciences of the United States of America* 38 (1952): 747–752. For a review, see Renato Dulbecco, "From the Molecular Biology of Oncogenic DNA Viruses to Cancer," *Science* 192 (1976): 437–440; Daniel J. Kevles, "Renato Dulbecco and the New Animal Virology: Medicine, Methods and Molecules," *Journal of the History of Biology* 26 (1993): 409–442.

16. Hannah Landecker, *Culturing Life: How Cells Became Technologies* (Cambridge, MA: Harvard University Press, 2007).

17. "Central Dogma Reversed," *Nature* 226 (1970): 1198–1199.

18. James A. Marcum, "From Heresy to Dogma in Accounts of Opposition to Howard Temin's DNA Protovirus Hypothesis," *History and Philosophy of the Life Sciences* 24 (2002): 165–192.

19. Howard M. Temin, "The Protovirus Hypothesis," *Journal of the National Cancer Institute* 46 (1971): iii–viii. Temin thus proposed a Lamarckian view of cell function. The genetic content of a cell reflected its state; because cells could modify the expression of their genes as a function of external signals—as Jacob and Monod had shown—the genetic makeup of a cell depended on this external medium.

20. Soraya de Chadarevian and Bruno Strasser, eds., "Molecular Biology in Postwar Europe," *Studies in History and Philosophy of Science Part C: Studies in History and Philosophy of Biological and Biomedical Sciences* 33 (2002): 361–565.

21. S. J. Singer and Garth L. Nicholson, "The Fluid Mosaic Model of the Structure of Cell Membranes," *Science* 175 (1972): 720–721. This revolution had been prepared by the work done in previous years on the protein pumps present in bacterial membranes; see Mathias Grote, "Purple Matter, Membranes and 'Molecular Pump' in Rhodopsin Research (1960s–1980s)," *Journal of the History of Biology* 46 (2013): 331–368; Mathias Grote, *Membranes to Molecular Machines: Active Matter and the Remaking of Life* (Chicago: University of Chicago Press, 2019).

22. The first journal of molecular biology (appropriately named the *Journal of Molecular Biology*) was started in 1959 by John Kendrew. A selection of papers by Brenner from the journal gives an idea of its importance in the publication of key results in molecular biology: Sydney Brenner, *Molecular Biology: A Selection of Papers* (London: Academic Press, 1989). The initiative for this journal, however, did not come from molecular biologists; Robert C. Olby, "The Molecular Revolution in Biology," in *Companion to the History of Modern Science,* ed. R. C. Olby, G. N. Cantor, J. R. R. Christie, and M. J. S. Hodge,

503–520 (London: Routledge, 1990), p. 507. The rapid expansion of molecular biology undoubtedly explains why it did not have time to crystallize into a genuinely new discipline but fused with preexisting disciplines.

23. Jean-Paul Gaudillière, *Inventer la biomédecine: la France, l'Amérique, et la production des savoirs du vivant* (Paris: La Découverte, 2002); Jean-Paul Gaudillière, "Molecular Biology in the French Tradition? Redefining Local Traditions and Disciplinary Patterns," *Journal of the History of Biology* 26 (1993): 473–498; Jean-Paul Gaudillière, "Chimie biologique ou chimie moléculaire? La biochimie au CNRS dans les années soixante," *Cahiers pour l'histoire du CNRS* 7 (1990): 91–147; Jean-Paul Gaudillière, "Molecular Biologists, Biochemists and Messenger RNA: The Birth of a Scientific Network," *Journal of the History of Biology* 29 (1996): 417–445; Xavier Polanco, "La mise en place d'un réseau scientifique, les rôles du CNRS et de la DGRST dans l'institutionnalisation de la biologie moléculaire en France (1960–1970)," *Cahiers pour l'histoire du CNRS* 7 (1990): 49–90. For an international view of this "seizure of power" by the molecular biologists and their confrontation with the biochemists, see Pnina G. Abir-Am, "The Politics of Macromolecules: Molecular Biologists, Biochemists, and Rhetoric," *Osiris* 7 (1992): 164–191.

24. Bernd Gutte and R. B. Merrifield, "The Synthesis of Ribonuclease A," *Journal of Biological Chemistry* 246 (1971): 1922–1941.

25. The first protein structures (myoglobin and hemoglobin, respectively) had been determined earlier: John C. Kendrew, R. E. Dickerson, B. E. Strandberg, R. G. Hart, D. R. Davies, et al., "A Three-Dimensional Model of the Myoglobin Molecule Obtained by X-Ray Analysis," *Nature* 181 (1958): 662–666; Max F. Perutz, M. G. Rossmann, Ann F. Cullis, Hillary Muirhead, George Will, and A. C. T. North, "Structure of Hemoglobin: A Three-Dimensional Fourier Synthesis at 5.5Å Resolution Obtained by X-Ray Analysis," *Nature* 185 (1960): 416–422. Over the next few years, both an improved definition of these structures and the determination of new three-dimensional protein structures (in particular of enzymes) were produced. The determination of these structures was rendered possible by Perutz's 1951 solution to the phase problem (see Chapter 11), by continuous advances in the quantitative measurement of diffraction patterns, and by computer processing of the experimental data; see Soraya de Chadarevian, *Design for Life: Molecular Biology after World War II* (Cambridge: Cambridge University Press, 2002).

26. Lysozyme was the first enzyme for which the catalytic mechanism was determined using crystallography (1967): C. C. F. Blake, L. N. Johnson, G. A. Mair, A. C. T. North, D. C. Philipps, and V. R. Sarma, "Crystallographic Studies of the Activity of Hen Egg-White Lysozyme," *Proceedings of the Royal Society of London: Series B, Biological Sciences* 167 (1967): 378–388. The structure and mechanism of action of chymotrypsin and carboxypeptidase were determined shortly afterward.

27. Okazaki also showed that DNA replication was a discontinuous process: Reiji
Okazaki, Tuneko Okazaki, Kiwako Sakabe, Kazunori Sugimoto, and Akio
Sugino, "Mechanism of DNA Chain Growth 1. Possible Discontinuity and
Unusual Secondary Structure of Newly Synthesized Chains," *Proceedings of the
National Academy of Sciences of the United States of America* 59 (1968):
598–605. DNA replication turned out to be an extremely complicated process,
requiring the involvement of many protein factors: Arthur Kornberg, *For the
Love of Enzymes: The Odyssey of a Biochemist* (Cambridge, MA: Harvard
University Press, 1989).

28. Richard R. Burgess, Andrew A. Travers, John J. Dunn, and Ekkehard K. F.
Bautz, "Factors Stimulating Transcription by RNA Polymerase," *Nature* 221
(1969): 43–46.

29. Masayasu Nomura, "Reflections on the Days of Ribosome Reconstitution
Research," *Trends in Biochemical Sciences* 22 (1997): 275–279.

30. Sung Hou Kim, Gary Quigley, F. L. Suddath, and Alexander Rich, "High
Resolution X-Ray Diffraction Patterns of Crystalline Transfer RNA That Show
Helical Regions," *Proceedings of the National Academy of Sciences of the United
States of America* 68 (1971): 841–845.

31. For the first nucleic acid sequences, see Robert W. Holley, Jean Apgar,
George A. Everett, James T. Madison, Mark Marquisee, et al., "Structure of a
Ribonucleic Acid," *Science* 147 (1965): 1462–1465. For the first isolated and
purified genes, see Jim Shapiro, Lorne Machattie, Larry Eron, Garrett Ihler,
Karin Ippen, and Jon Beckwith, "Isolation of Pure *Lac* Operon DNA," *Nature*
224 (1969): 768–774. For the isolation of the first genes from eukaryotes
(ribosomal genes), see the review by Max L. Birnstiel, "Gene Isolation is
25 Years Old This Month," *Trends in Genetics* 6 (1990): 380–381. For the
chemically synthesized gene, see K. L. Agarwal, H. Buchi, M. H. Caruthers, N.
Gupta, H. G. Khorana, et al., "Total Synthesis of the Gene for an Alanine
Transfer Ribonucleic Acid from Yeast," *Nature* 227 (1970): 27–34.

32. Walter Gilbert and Benno Müller-Hill, "Isolation of the *Lac* Repressor,"
Proceedings of the National Academy of Sciences of the United States of America
56 (1966): 1891–1898; Mark Ptashne, "Isolation of the Lambda Phage Repressor," *Proceedings of the National Academy of Sciences of the United States of
America* 57 (1967): 306–313.

33. The crisis felt by molecular biologists has been well described by François
Gros: "The history of the development of ideas—in science as in art—shows
that it is dangerous to push the exploitation of a concept or of a methodology to
its limits, because a feeling of saturation will tend to replace the satisfaction
that accompanied the initial phase. Not only did researchers—including the
finest minds in the discipline—start to wonder about the future of molecular
biology, but by the beginning of the 1970s the whole discipline went into crisis.
Of course, impelled by the "dynamic" of its successes, it carried on and
continued to obtain some important results, but originality was not always a

hallmark of the science. One has to admit that research was treading water and the heart had gone out of it. This "low point" was accompanied, if not by a genuine anxiety, at least by a questioning that was not totally unlike anxiety." François Gros, *Les secrets du gène* (Paris: Odile Jacob, 1986), 167.

34. See, for example, Jacqueline Djian, éd., *La Médecine moléculaire* (Paris: Robert Laffont, 1970).

35. Linus Pauling, Harvey A. Itano, S. J. Singer, and Ibert C. Wells, "Sickle Cell Anemia, a Molecular Disease," *Science* 110 (1949): 543–548.

36. Gaudillière, *Inventer la biomédecine.*

37. This retreat can also be justified scientifically. Regulatory mechanisms controlling enzyme activity, such as phosphorylation by protein kinases, do not exist (or rarely exist) in bacteria. It was quite reasonable to imagine that the complexity of multicellular organisms was located as much (if not more) at the biochemical level as at the level of gene expression. For example, a hormone binding to its receptor frequently causes the synthesis of a small molecule called cyclic AMP in the cell, which in turn activates a protein kinase. This pathway of intracellular signaling was particularly well studied. See Earl W. Sutherland, "Studies on the Mechanism of Hormone Action," *Science* 177 (1972): 401–408.

38. M. C. Niu, "Thymus Ribonucleic Acid and Embryonic Differentiation," *Proceedings of the National Academy of Sciences of the United States of America* 44 (1958): 1264–1274; Jean-Paul Gaudillière, "Un code moléculaire pour la différenciation cellulaire: la controverse sur les transferts d'ARN informationnel (1955–1973) et les étapes de diffusion du paradigme de la biologie moléculaire," *Fundamenta Scientiae* 9 (1988): 429–467.

39. Marvin Fishman, R. A. Hammerstrom, and V. P. Bond, "In Vitro Transfer of Macrophage RNA to Lymph Node Cells," *Nature* 198 (1963): 549–551; Gaudillière, "Un code moléculaire."

40. Michel Morange, "La recherche d'un code moléculaire de la mémoire," *Fundamenta Scientiae* 6 (1985): 65–80.

41. The controversy over the possibility of learning in planarian worms was studied by G. D. L. Travis, "Replicating Replication? Aspects of the Social Construction of Learning in Planarian Worms," *Social Studies of Science* 11 (1981): 11–32.

42. W. L. Byrne, D. Samuel, E. L. Bennett, M. R. Rosenzweig, E. Wasserman, et al., "Memory Transfer," *Science* 153 (1966): 658–659.

43. D. F. Tate, L. Galvan, and George Ungar, "Isolation and Identification of Two Learning-Induced Peptides," *Pharmacology Biochemistry and Behavior* 5 (1976): 441–448.

44. In 1972, *Nature* published an article by Ungar and coworkers describing a peptide responsible for the fear reaction to dark—"scotophobin." See George Ungar, D. M. Desiderio, and W. Parr, "Isolation, Identification and Synthesis of a Specific Behaviour-Inducing Brain Peptide," *Nature* 238 (1972): 198–210. In

the same issue the journal published a (much longer) critical review written by one of the article's referees and a response by the authors.

45. For the inside story of memory molecules, see Louis N. Irwin, *Scotophobin: Darkness at the Dawn of the Search for Memory Molecules* (Lanham, MD: Hamilton Books, 2007).

16. Genetic Engineering

1. RNAs were the first nucleic acids to be sequenced, using methods adapted from protein sequencing: ribosomal RNAs by the group of Jean-Pierre Ebel in Strasbourg (France), but also the genomes of RNA bacteriophages. See Jérôme Pierrel, "An RNA Phage Lab: MS2 in Walter Fiers' Laboratory of Molecular Biology in Ghent, from Genetic Code to Gene and Genome, 1963–1976," *Journal of the History of Biology* 45 (2012): 109–138.

2. Oswald T. Avery, Colin M. MacLeod, and Maclyn McCarty, "Studies on the Chemical Nature of the Substance Inducing Transformation of Pneumococcal Types," *Journal of Experimental Medicine* 79 (1944): 137–158.

3. Edward L. Tatum, "A Case History in Biological Research," *Science* 129 (1959): 1711–1715.

4. Joshua Lederberg, "Genetics," in *Encyclopaedia Britannica, Yearbook of Science and the Future* (Chicago: Encyclopaedia Britannica, 1969), 321.

5. Elizabeth H. Szybalska and Waclaw Szybalski, "Genetics of Human Cell Lines, IV. DNA-Mediated Heritable Transformation of a Biochemical Trait," *Proceedings of the National Academy of Sciences of the United States of America* 48 (1962): 2026–2034.

6. Susan Wright, "Recombinant DNA Technology and Its Social Transformation, 1972–1982," *Osiris* 2 (1986): 303–360.

7. For an overview of this work, see Werner Arber, "Promotion and Limitation of Genetic Exchange," *Science* 205 (1979): 361–365.

8. Hamilton O. Smith and K. W. Wilcox, "A Restriction Enzyme from Hemophilus Influenzae 1. Purification and General Properties," *Journal of Molecular Biology* 51 (1970): 379–391; Thomas J. Kelly and Hamilton O. Smith, "A Restriction Enzyme from Hemophilus Influenzae II. Base Sequence of the Recognition Site," *Journal of Molecular Biology* 51 (1970): 393–409. The first restriction enzyme was purified by Matthew Meselson and Robert Yuan, "DNA Restriction Enzyme from *E. coli*," *Nature* 217 (1968): 1110–1114. This enzyme had a low cleavage specificity, which meant that it was not particularly useful.

9. Kathleen Danna and Daniel Nathans, "Specific Cleavage of Simian Virus 40 DNA by Restriction Endonuclease of Hemophilus Influenzae," *Proceedings of the National Academy of Sciences of the United States of America* 68 (1971): 2913–2917.

10. David A. Jackson, Robert H. Symons, and Paul Berg, "Biochemical Method for Inserting New Genetic Information into DNA of Simian Virus 40: Circular

SV40 Molecules Containing Lambda Phage Genes and the Galactose Operon of *Escherichia coli*," *Proceedings of the National Academy of Sciences of the United States of America* 69 (1972): 2904–2909; Doogab Yi, "Cancer, Viruses, and Mass Migration: Paul Berg's Venture into Eukaryotic Biology and the Advent of Recombinant DNA Research and Technology, 1967–1980," *Journal of the History of Biology* 41 (2008): 589–636.

11. Joshua Lederberg, "Genetics of Bacteria," Grant Application to the National Institutes of Health, no. A1 05160–11, December 20, 1967, cited in Wright, "Recombinant DNA Technology," 310.

12. Peter Lobban and A. D. Kaiser, "Enzymatic End to End Joining of DNA Molecules," *Journal of Molecular Biology* 78 (1973): 453–471. Lobban's results were published after a year's delay, which tended to hide the fact that the two experiments were in fact carried out simultaneously. Lobban's thesis advisor, A. D. Kaiser, preferred that Lobban correct his thesis rather than publish his results. Reported by Arthur Kornberg, *For the Love of Enzymes: The Odyssey of a Biochemist* (Cambridge, MA: Harvard University Press, 1989), 275 et seq.

13. Gunther S. Stent, "Prematurity and Uniqueness in Scientific Discovery," *Scientific American* 227 (1972): 84–93.

14. A chronology of the principal discoveries that gave rise to genetic engineering can be found in James D. Watson and John Tooze, *A Documentary History of Gene Cloning* (San Francisco: W. H. Freeman, 1981); Jan Witkowski, "Fifty Years of Molecular Biology's Hall of Fame," *Life Science Job Trends* 2, no. 17 (1988): 1–13. For the development and early applications of genetic engineering, see Stephen S. Hall, *Invisible Frontiers: The Race to Synthesize a Human Gene* (New York: Atlantic Monthly Press, 1987); Nicolas Rasmussen, *Gene Jockeys: Life Science and the Rise of Biotech Enterprise* (Baltimore: Johns Hopkins University Press, 2014); Doogab Yi, *The Recombinant University: Genetic Engineering and the Emergence of Stanford Biotechnology* (Chicago: University of Chicago Press, 2015).

15. Janet E. Mertz and Ronald W. Davis, "Cleavage of DNA by RI Restriction Endonuclease Generates Cohesive Ends," *Proceedings of the National Academy of Sciences of the United States of America* 69 (1972): 3370–3374.

16. Mathias Grote, "Hybridizing Bacteria, Crossing Methods, Cross-checking Arguments: The Transition from Episomes to Plasmids (1961–1969)," *History and Philosophy of the Life Sciences* 30 (2008): 407–430.

17. Stanley N. Cohen, Annie C. Y. Chang, Herbert W. Boyer, and Robert B. Helling, "Construction of Biologically Functional Bacterial Plasmids *in Vitro*," *Proceedings of the National Academy of Sciences of the United States of America* 70 (1973): 3240–3244. To make the plasmids enter the bacteria, these scientists used a calcium chloride–based method developed earlier by Mohrt Mandel, in M. Mandel and A. Higa, "Calcium Dependent Bacteriophage DNA Infection," *Journal of Molecular Biology* 53 (1970): 159–162. Also see Annie C. Y. Chang and Stanley N. Cohen, "Genome Construction between Bacterial Species in

Vitro: Replication and Expression of *Staphylococcus* Plasmid Genes in *Escherichia coli*," *Proceedings of the National Academy of Sciences of the United States of America* 71 (1974): 1030–1034.

18. John F. Morrow, Stanley N. Cohen, Annie C. Y. Chang, Herbert W. Boyer, Howard M. Goodman, and Robert B. Helling, "Replication and Transcription of Eukaryotic DNA in *Escherichia coli*," *Proceedings of the National Academy of Sciences of the United States of America* 71 (1974): 1743–1747.

19. Maxine Singer and Dieter Soll, "Guidelines for DNA Hybrid Molecules," *Science* 181 (1973): 1114.

20. Paul Berg, David Baltimore, Herbert W. Boyer, Stanley N. Cohen, Ronald W. Davis, et al., "Potential Biohazards of Recombinant DNA Molecules," *Proceedings of the National Academy of Sciences of the United States of America* 71 (1974): 2593–2594; Paul Berg, David Baltimore, Herbert W. Boyer, Stanley N. Cohen, Ronald W. Davis, et al., "Potential Biohazards of Recombinant DNA Molecules," *Science* 185 (1974): 303; "NAS Ban on Plasmid Engineering," *Nature* 250 (1974): 175.

21. Paul Berg, David Baltimore, Sydney Brenner, Richard O. Roblin, and Maxine F. Singer, "Asilomar Conference on Recombinant DNA Molecules," *Science* 188 (1975): 44–47.

22. Sir Robert Williams, chair, *Report of the Working Party on the Practice of Genetic Manipulation* (Williams Report), Command 6600 (London: Her Majesty's Stationery Office, 1976).

23. Michael Ruse, "The Recombinant DNA Debate: A Tempest in a Test Tube?" in *Is Science Sexist?* (Dordrecht, the Netherlands: D. Reidel, 1981); Clifford Grobstein, *A Double Image of the Double Helix: The Recombinant DNA Debate* (San Francisco: W. H. Freeman, 1979); Sheldon Krimsky, *Genetic Alchemy: The Social History of the Recombinant DNA Controversy* (Cambridge, MA: MIT Press, 1982); Susan Wright, *Molecular Politics: Developing American and British Regulatory Policy for Genetic Engineering, 1972–1982* (Chicago: University of Chicago Press, 1994); Donald S. Fredrickson, *The Recombinant DNA Controversy, a Memoir: Science, Politics, and the Public Interest, 1974–1981* (Washington, DC: ASM Press, 2001).

24. Hall, *Invisible Frontiers*.

25. Paul Berg, "Dissections and Reconstructions of Genes and Chromosomes," *Science* 213 (1981): 296–303.

26. Susan Wright, "Molecular Biology or Molecular Politics? The Production of Scientific Consensus on the Hazards of Recombinant DNA Technology," *Social Studies of Science* 16 (1986): 593–620.

27. Argiris Efstratiadis, Fotis C. Kafatos, Allan M. Maxam, and Tom Maniatis, "Enzymatic in Vitro Synthesis of Globin Genes," *Cell* 7 (1976): 279–288.

28. Tom Maniatis, Sim Gek Kee, Argiris Efstratiadis, and Fotis C. Kafatos, "Amplification and Characterization of a β-Globin Gene Synthesized in Vitro," *Cell* 8 (1976): 163–182.

29. Pieter Wensink, David J. Finnegan, John E. Donelson, and David S. Hogness, "A System for Mapping DNA Sequences in the Chromosomes of *Drosophila melanogaster*," *Cell* 3 (1974): 315–325.

30. Tom Maniatis, Ross C. Hardison, Elizabeth Lacy, Joyce Lauer, Catherine O'Connel, et al., "The Isolation of Structural Genes from Libraries of Eucaryotic DNA," *Cell* 15 (1978): 687–701.

31. Michael Grunstein and David S. Hogness, "Colony Hybridization: A Method for the Isolation of Cloned DNAs That Contain a Specific Gene," *Proceedings of the National Academy of Sciences of the United States of America* 72 (1975): 3961–3965.

32. In *Drosophila,* the existence of giant chromosomes in the salivary glands makes it easy to localize directly the position of a DNA fragment on the chromosomal map by in situ hybridization. See Wensink et al., "System for Mapping DNA."

33. Hamilton O. Smith, "Nucleotide Sequence Specificity of Restriction Endonucleases," *Science* 205 (1979): 455–462.

34. Phillip A. Sharp, Bill Sugden, and Joe Sambrook, "Detection of Two Restriction Endonuclease Activities in *Haemophilus parainfluenzae* Using Analytical Agarose-Ethidium Bromide Electrophoresis," *Biochemistry* 12 (1973): 3055–3063. The ultracentrifugation method continued to be used in the preparation of plasmids: in the presence of ethidium bromide, plasmids have a density different from that of chromosomal DNA and can easily be separated on a cesium chloride gradient.

35. E. M. Southern, "Detection of Specific Sequences among DNA Fragments Separated by Gel Electrophoresis," *Journal of Molecular Biology* 98 (1975): 503–517; Dario Giacomoni, "The Origin of DNA:RNA Hybridization," *Journal of the History of Biology* 26 (1993): 89–107.

36. James C. Alwine, David J. Kemp, and George R. Stark, "Method for Detection of Specific RNAs in Agarose Gels by Transfer to Diazobenzyloxymethyl-Paper and Hybridization with DNA Probes," *Proceedings of the National Academy of Sciences of the United States of America* 74 (1977): 5350–5354.

37. Francisco Bolivar, Raymond L. Rodriguez, Mary C. Betlach, and Herbert W. Boyer, "Construction and Characterization of New Cloning Vehicles: 1. Ampicillin-Resistant Derivatives of the Plasmid pMB9," *Gene* 2 (1977): 75–93; Francisco Bolivar, Raymond L. Rodriguez, Patricia J. Greene, Mary C. Betlach, Herbert L. Heynecker, et al., "Construction and Characterization of New Cloning Vehicles: II. A Multipurpose Cloning System," *Gene* 2 (1977): 95–113.

38. John Collins and Barbara Hohn, "Cosmids: A Type of Plasmid Gene-Cloning Vector That Is Packageable in Vitro in Bacteriophage λ Heads," *Proceedings of the National Academy of Sciences of the United States of America* 75 (1978): 4242–4246.

39. Allan M. Maxam and Walter Gilbert, "A New Method for Sequencing DNA," *Proceedings of the National Academy of Sciences of the United States of America*

74 (1977): 560–564. The Russian chemist Eugene Sverdlov had proposed a similar method four years earlier, but his work did not attract any attention: E. D. Sverdlov, W. Monastyrskaya, A. V. Chestukhin, and E. I. Budowsky, "The Primary Structure of Oligonucleotides: Partial Apurination as a Method to Determine the Position of Purine and Pyrimidine Residues," *FEBS Letters* 33 (1973): 15–17. The first (partial) sequence of a gene had been obtained more than ten years earlier by combining protein sequencing, frameshift mutations, and a knowledge of the genetic code: Y. Okada, E. Terzaghi, G. Streisinger, J. Emrich, M. Inouye and A. Tsugita, "A Frame-Shift Mutation Involving the Addition of Two Base Pairs in the Lysozyme Gene of Phage T4," *Proceedings of the National Academy of Sciences of the United States of America* 56 (1966): 1692–1698.

40. Frederick Sanger, S. Nicklen, and A. R. Coulson, "DNA Sequencing with Chain-Terminating Inhibitors," *Proceedings of the National Academy of Sciences of the United States of America* 74 (1977): 5463–5467.

41. Clyde A. Hutchison III, Sandra Phillips, Marshall H. Edgell, Shirley Gillam, Patricia Jahnke, and Michael Smith, "Mutagenesis at a Specific Position in a DNA Sequence," *Journal of Biological Chemistry* 253 (1978): 6551–6560.

42. M. J. Gait and R. C. Sheppard, "Rapid Synthesis of Oligodeoxyribonucleotides: A New Solid-Phase Method," *Nucleic Acid Research* 4 (1977): 1135–1158.

43. The authors modified the technique that was already used in bacteria: F. L. Graham and A. J. Van der Erb, "A New Technique for the Assay of Infectivity of Human Adenovirus DNA," *Virology* 52 (1973): 456–467.

44. Michael Wigler, Raymond Sweet, Gek Kee Sim, Barbara Wold, Angel Pellicer, et al., "Transformation of Mammalian Cells with Genes from Prokaryotes and Eukaryotes," *Cell* 16 (1979): 777–785.

45. Tom Maniatis, Ed F. Fritsch, and Joe Sambrook, *Molecular Cloning: A Laboratory Manual* (Cold Spring Harbor, NY: Cold Spring Harbor Laboratory Press, 1982). In laboratories, this book is often called "the recipe book," "the cookbook," or more respectfully, "the Bible." Joan H. Fujimura, "Constructing 'Do-Able' Problems in Cancer Research: Articulating Alignment," *Social Studies of Science* 17 (1987): 257–293.

46. The generalization of data from one organism to the whole of biology was a strategy that was quite reasonable even if it often turned out to be fruitless. Lindley Darden, "Essay Review: Generalizations in Biology," *Studies in History and Philosophy of Science: Part A* 27 (1996): 409–419.

47. β-*globin:* Maniatis et al., "Amplification and Characterization." *Insulin:* Axel Ullrich, John Shine, John Chirgwin, Raymond Pictet, Edmond Tischer, et al., "Rat Insulin Genes: Construction of Plasmids Containing the Coding Sequences," *Science* 196 (1977): 1313–1319. *Rat growth hormone:* Peter H. Seeburg, John Shine, Joseph A. Martial, John D. Baxter, and Howard M. Goodman, "Nucleotide Sequence and Amplification in Bacteria of Structural Gene for Rat Growth Hormone," *Nature* 270 (1977): 486–494. *Human*

placental hormone: John Shine, Peter H. Seeburg, Joseph A. Martial, John D. Baxter, and Howard M. Goodman, "Construction and Analysis of Recombinant DNA for Human Chorionic Somatomammotropin," *Nature* 270 (1977): 494–499.

48. The huge amount of work and "tinkering" that was necessary to obtain results is wonderfully described in Nicolas Rasmussen, *Gene Jockeys: Life Science and the Rise of Biotech Enterprise* (Baltimore: Johns Hopkins University Press, 2014); Sally Smith Hughes, *Genentech: The Beginnings of Biotech* (Chicago: University of Chicago Press, 2011). For those interested in the way the rapid development of Biotech was financially supported, see Cynthia Roberts-Roth, *From Alchemy to IPO: The Business of Biotechnology* (New York: Perseus Books, 2000).

49. Keiichi Itakura, Tadaaki Hirose, Roberto Crea, Arthur D. Riggs, Herbert Heynecker, et al., "Expression in *Escherichia coli* of a Chemically Synthesized Gene for the Hormone Somatostatin," *Science* 198 (1977): 1056–1063.

50. David V. Goeddel, Dennis G. Kleid, Francisco Bolivar, Herbert L. Heynecker, Daniel G. Yansura, et al., "Expression in *Escherichia coli* of Chemically Synthesized Genes for Human Insulin," *Proceedings of the National Academy of Sciences of the United States of America* 76 (1979): 106–110.

51. Lydia Villa-Komaroff, Argiris Efstratiadis, Stephanie Broome, Peter Lomedico, Richard Tizard, et al., "A Bacterial Clone Synthesizing Proinsulin," *Proceedings of the National Academy of Sciences of the United States of America* 75 (1978): 3731; Annie C. Y. Chang, Jack H. Nunberg, Randal J. Kaufman, Henry A. Erlich, Robert T. Schimke, and Stanley N. Cohen, "Phenotypic Expression in *E. coli* of a DNA Sequence Coding for Mouse Dihydrofolate Reductase," *Nature* 275 (1978): 617–624.

52. *Growth hormone:* David V. Goeddel, Herbert L. Heynecker, Toyohara Hozumi, René Arentzen, Keiichi Itakura, et al., "Direct Expression in *Escherichia coli* of a DNA Sequence Coding for Human Growth Hormone," *Nature* 281 (1979): 544–548. *Biologically active interferon:* Shigekazu Nagata, Hideharu Taira, Alan Hall, Lorraine Johnsrud, Michel Streuli, et al., "Synthesis in *E. coli* of a Polypeptide with Human Leukocyte Interferon Activity," *Nature* 284 (1980): 316–320; David V. Goeddel, Elizabeth Yelverton, Axel Ullrich, Herbert L. Heynecker, Giuseppe Miozzari, et al., "Human Leukocyte Interferon Produced by *E. coli* is Biologically Active," *Nature* 287 (1980): 411–416.

53. Jean-Pierre Hemalsteens, Françoise Van Vliet, Marc De Beuckeleer, Ann Depicker, Gilbert Engler, et al., "The Agrobacterium Tumefaciens Ti Plasmid as a Host Vector System for Introducing Foreign DNA in Plant Cells," *Nature* 287 (1980): 654–656.

54. Jon W. Gordon, George A. Scangos, Diane J. Plotkin, James A. Barbosa, and Frank H. Ruddle, "Genetic Transformation of Mouse Embryos by Microinjection of Purified DNA," *Proceedings of the National Academy of Sciences of the United States of America* 77 (1980): 7380–7384.

55. The experiment was a piece of classic molecular hybridization in solution. Its importance derived from the material used. Yuet Wai Kan, Mitchell S. Golbus, and Andrée M. Dozy, "Prenatal Diagnosis of α-Thalassemia: Clinical Application of Molecular Hybridization," *New England Journal of Medicine* 295 (1976): 1165–1167.

56. Michel Morange, *Une lecture du vivant: histoire et épistémologie de la biologie moléculaire* (Louvain-la-Neuve, France: CIACO, 1986).

57. Niels Bohr, "Light and Life," *Nature* 131 (1933): 421–423, 457–459.

58. Lily E. Kay, *Who Wrote the Book of Life? A History of the Genetic Code* (Palo Alto, CA: Stanford University Press, 2000).

17. Split Genes and Splicing

1. C. Yanofsky, B. C. Carlton, J. R. Guest, D. R. Helinski, and U. Henning, "On the Colinearity of Gene Structure and Protein Structure," *Proceedings of the National Academy of Sciences of the United States of America* 51 (1964): 266–272; A. S. Sarabhai, A. O. W. Stetton, Sydney Brenner, and A. Bolle, "Colinearity of the Gene with the Polypeptide Chain," *Nature* 201 (1964): 13–17.

2. Shortly afterward, these results were published in scientific journals. S. M. Berget, A. J. Berk, T. Harrison, and P. A. Sharp, "Spliced Segments at the 5' Termini of Adenovirus-2 Late mRNA: A Role for Heterogeneous Nuclear RNA in Mammalian Cells," *Cold Spring Harbor Symposia on Quantitative Biology* 42 (1978): 523–529; T. R. Broker, L. T. Chow, A. R. Dunn, R. E. Gelinas, J. A. Hassel, et al., "Adenovirus-2 Messengers—An Example of Baroque Molecular Architecture," *Cold Spring Harbor Symposia on Quantitative Biology* 42 (1977): 531–553; H. Westphal and S.-P. Lai, "Displacement Loops in Adenovirus DNA-RNA Hybrids," *Cold Spring Harbor Symposia on Quantitative Biology* 42 (1978): 555–558; Susan M. Berget, Claire Moore, and Phillip A. Sharp, "Spliced Segments at the 5' Terminus of Adenovirus-2 Late mRNA," *Proceedings of the National Academy of Sciences of the United States of America* 74 (1977): 3171–3175; Louise T. Chow, Richard E. Gelinas, Thomas R. Broker, and Richard J. Roberts, "An Amazing Sequence Arrangement at the 5' Ends of Adenovirus 2 Messenger RNA," *Cell* 12 (1977): 1–8; Daniel F. Klessig, "Two Adenovirus mRNAs Have a Common 5' Terminal Leader Sequence Encoded at Least 10 kb Upstream from Their Main Coding Regions," *Cell* 12 (1977): 9–21; Ashley R. Dunn and John A. Hassell, "A Novel Method to Map Transcripts: Evidence for Homology between an Adenovirus mRNA and Discrete Multiple Regions of the Viral Genome," *Cell* 12 (1977): 23–36; J. B. Lewis, C. W. Anderson, and J. F. Atkins, "Further Mapping of Late Adenovirus Genes by Cell-Free Translation of RNA Selected by Hybridization to Specific DNA Fragments," *Cell* 12 (1977): 37–44.

3. B. G. Barrell, G. M. Air, and C. A. Hutchison III, "Overlapping Genes in Bacteriophage ΦX174," *Nature* 264 (1976): 34–41.

4. On SV40: Yosef Aloni, S. Bratosiw, Ravi Dhar, Orgad Laub, Mia Horowitz, and George Khoury, "Splicing of SV40 mRNAs: A Novel Mechanism for the Regulation of Gene Expression in Animal Cells," *Cold Spring Harbor Symposia on Quantitative Biology* 42 (1977): 559–570; M.-T. Hsu and J. Ford, "A Novel Sequence Arrangement of SV40 late RNA," *Cold Spring Harbor Symposia on Quantitative Biology* 42 (1977): 571–576; Yosef Aloni, Ravi Dhar, Orgad Laub, Mia Horowitz, and George Khoury, "Novel Mechanisms for RNA Maturation: The Leader Sequences of Simian Virus 40 mRNA Are Not Transcribed Adjacent to the Coding Sequences," *Proceedings of the National Academy of Sciences of the United States of America* 74 (1977): 3686–3690. On eukaryotes: A. J. Jeffreys and R. A. Flavell, "The Rabbit β-Globin Gene Contains a Large Insert in the Coding Sequence," *Cell* 12 (1977): 1097–1108; R. Breathnach, J. L. Mandel, and P. Chambon, "Ovalbumin Gene Is Split in Chicken DNA," *Nature* 270 (1977): 314–319; Shirley M. Tilghman, David C. Tiemeier, J. G. Seidman, B. Matija Peterlin, Margery Sullivan, et al., "Intervening Sequence of DNA Identified in the Structural Portion of a Mouse β-Globin Gene," *Proceedings of the National Academy of Sciences of the United States of America* 75 (1978): 725–729; Christine Brack and Susumu Tonegawa, "Variable and Constant Parts of the Immunoglobulin Light Chain Gene of a Mouse Myeloma Cell are 1250 Non-Translated Bases Apart," *Proceedings of the National Academy of Sciences of the United States of America* 74 (1977): 5652–5656.
5. Walter Gilbert, "Why Genes in Pieces?" *Nature* 271 (1978): 501.
6. Francis Crick, "Split Genes and RNA Splicing," *Science* 204 (1979): 264–271; R. Breathnach and P. Chambon, "Organization and Expression of Eukaryotic Split Genes Coding for Proteins," *Annual Review of Biochemistry* 50 (1981): 349–383.
7. J. L. Bos, C. Heyting, P. Borst, A. C. Amberg, and E. F. J. Van Bruggen, "An Insert in the Single Gene for the Large Ribosomal RNA in Yeast Mitochondrial DNA," *Nature* 275 (1978): 336–338.
8. Peter J. Curtis, Ned Mantei, Johan Van Den Berg, and Charles Weissmann, "Presence of a Putative 15S Precursor to β-Globin mRNA but not to α-Globin mRNA in Friend Cells," *Proceedings of the National Academy of Sciences of the United States of America* 74 (1977): 3184–3188.
9. Gilbert, "Why Genes in Pieces?" 501; Crick, "Split Genes and RNA Splicing," 264; Jan A. Witkowski, "The Discovery of 'Split' Genes: A Scientific Revolution," *Trends in Biochemical Sciences* 13 (1988): 110–113.
10. J. E. Darnell, L. Philipson, R. Wall, and M. Adesnik, "Polyadenylic Acid Sequences: Role in Conversion of Nuclear RNA into Messenger RNA," *Science* 174 (1971): 507–510.
11. R. P. Perry and D. E. Kelley, "Methylated Constituents of Heterogeneous Nuclear RNA: Presence in Blocked 5′ Terminal Structures," *Cell* 6 (1975): 13–19.
12. O. P. Samarina, "The Distribution and Properties of Cytoplasmic Deoxyribonucleic Acid-Like Ribonucleic Acid (Messenger Ribonucleic Acid)," *Biochimica et Biophysica Acta* 91 (1964): 688–691; G. P. Georgiev, "On the Structural

Organization of Operon and the Regulation of RNA Synthesis in Animal Cells," *Journal of Theoretical Biology* 25 (1969): 473–490; G. P. Georgiev, A. P. Ryskov, C. Coutelle, V. L. Mantieva, and E. R. Avakyan, "On the Structure of Transcriptional Unit in Mammalian Cells," *Biochimica et Biophysica Acta* 259 (1972): 259–283; Darnell et al., "Polyadenylic Acid Sequences," 507–510; Robert A. Weinberg, "Nuclear RNA Metabolism," *Annual Review of Biochemistry* 42 (1973): 329–354.

13. Pierre Chambon, "Split Genes," *Scientific American* 244 (1981): 60–71.

14. Crick, "Split Genes and RNA Splicing," 269–270.

15. Roy J. Britten and Eric H. Davidson, "Gene Regulation for Higher Cells: A Theory," *Science* 165 (1969): 349–357.

16. Claude Jacq, Jaga Lazowska, and Piotr P. Slonimski, "Sur un nouveau mécanisme de la régulation de l'expression génétique," *Comptes rendus de l'Académie de sciences,* series D, 290 (1980): 89–92; Jaga Lazowska, Claude Jacq, and Piotr P. Slonimski, "Sequence of Introns and Flanking Exons in Wild-Type and Box3 Mutants of Cytochrome *b* Reveals an Interlaced Splicing Protein Coded by an Intron," *Cell* 22 (1980): 333–348.

17. Piotr P. Slonimski, "Éléments hypothétiques de l'expression des gènes morcelés: protéines messagères de la membrane nucléaire," *Comptes rendus de l'Académie de sciences,* series D, 290 (1980): 331–334.

18. Antoine Danchin, "Règles de réécriture en biologie moléculaire," *Le Débat,* no. 3 (1980): 111–114.

19. Thomas R. Cech, Arthur J. Zaug, and Paula J. Grabowski, "*In Vitro* Splicing of the Ribosomal RNA Precursor of Tetrahymena: Involvement of a Guanosine Nucleotide in the Excision of the Intervening Sequence," *Cell* 27 (1981): 487–496; Kelly Kruger, Paula J. Grabowski, Arthur J. Zaug, Julie Sands, Daniel E. Gottschling, and Thomas R. Cech, "Self-Splicing RNA: Autoexcision and Autocyclization of the Ribosomal RNA Intervening Sequence of Tetrahymena," *Cell* 31 (1982): 147–157.

20. Cecilia Guerrier-Takada, Kathleen Gardiner, Terry Marsh, Norman Pace, and Sydney Altman, "The RNA Moiety of Ribonuclease P Is the Catalytic Subunit of the Enzyme," *Cell* 35 (1983): 849–857.

21. Gilbert, "Why Genes in Pieces?" 501; James E. Darnell Jr., "Implication of RNA-RNA Splicing in Evolution of Eukaryotic Cells," *Science* 202 (1978): 1257–1260.

22. Gilbert, "Why Genes in Pieces?" 501.

23. Charles S. Craik, Stephen Sprang, Robert Fletterick, and William J. Rutter, "Intron-Exon Splice Junctions Map at Protein Surfaces," *Nature* 299 (1982): 180–182; Charles S. Craik, William J. Rutter, and Robert Fletterick, "Splice Junctions: Association with Variation in Protein Structure," *Science* 220 (1983): 1125–1129.

24. Colin C. F. Blake, "Do Genes-in-Pieces Imply Proteins-in-Pieces?" *Nature* 273 (1978): 267.

25. Gilbert, "Why Genes in Pieces?" 501.

26. Susumu Ohno, *Evolution by Gene Duplication* (New York: Springer-Verlag, 1970).

27. Darnell, "Implication of RNA-RNA Splicing"; Susumu Tonegawa, Allan M. Maxam, Richard Tizard, Ora Bernard, and Walter Gilbert, "Sequence of a Mouse Germ-Line Gene for a Variable Region of an Immunoglobin Light Chain," *Proceedings of the National Academy of Sciences of the United States of America* 75 (1978): 1485–1489; Gilbert, "Why Genes in Pieces?" 501.

28. The concept of tinkering—introduced by Jacob in 1977 before the existence of split genes was known—found an excellent application in this new context: F. Jacob, "Evolution and Tinkering," *Science* 196 (1977): 1161–1166. Also see François Jacob, *Le Jeu des possibles* (Paris: Fayard, 1981); *The Possible and the Actual* (Seattle: University of Washington Press, 1982).

29. Carmen Quinto, Margarita Quiroga, William F. Swain, William C. Nikovits Jr., David N. Standring, et al., "Rat Preprocarboxypeptidase A: cDNA Sequence and Preliminary Characterization of the Gene," *Proceedings of the National Academy of Sciences of the United States of America* 79 (1982): 31–35; Margaret Leicht, George L. Long, T. Chandra, Kotoku Kurachi, Vincent J. Kidd, et al., "Sequence Homology and Structural Comparison between the Chromosomal Human α1-Antitrypsin and Chicken Ovalbumin Genes," *Nature* 297 (1982): 655–659.

30. W. Ford Doolittle, "Genes in Pieces: Were They Ever Together?" *Nature* 272 (1978): 581–582; Darnell, "Implication of RNA-RNA Splicing."

31. Crick, "Split Genes and RNA Splicing," 269.

32. Claude Jacq, J. R. Miller, and G. G. Brownlee, "A Pseudogene Structure in 5S DNA of *Xenopus laevis*," *Cell* 12 (1977): 109–120; Y. Nishioka, A. Leder, and P. Leder, "Unusual α-Globin-Like Gene That Has Cleanly Lost Both Globin Intervening Sequences," *Proceedings of the National Academy of Sciences of the United States of America* 77 (1980): 2806–2809.

33. B. G. Barrell, A. T. Bankier, and J. Drouin, "A Different Genetic Code in Human Mitochondria," *Nature* 282 (1979): 189–194.

34. Stuart Horowitz and Martin A. Gorovsky, "An Unusual Genetic Code in Nuclear Genes of *Tetrahymena*," *Proceedings of the National Academy of Sciences of the United States of America* 82 (1985): 2452–2455. François Caron and Eric Meyer, "Does *Paramecium primaurelia* Use a Different Genetic Code in Its Macronucleus?" *Nature* 314 (1985): 185–188; J. R. Preer, L. B. Preer, B. M. Rudman, and A. J. Barnett, "Deviation from the Universal Code Shown by the Gene for Surface Protein 51.A in *Paramecium*," *Nature* 314 (1985): 188–190.

35. Rob Benne, Janny Van den Burg, Just P. J. Brakenhoff, Paul Sloof, Jacques H. Van Boom, and Marike C. Tromp, "Major Transcript of the Frameshifted Cox II Gene from Trypanosome Mitochondria Contains Four Nucleotides That Are Not Encoded in the DNA," *Cell* 46 (1986): 819–826.

36. Beat Blum, Nancy R. Sturm, Agda M. Simpson, and Larry Simpson, "Chimeric gRNA-mRNA Molecules with Oligo (U) Tails Covalently Linked at Sites of RNA Editing Suggest that U Addition Occurs by Transesterification," *Cell* 65 (1991): 543–550.

37. Scott H. Podolsky and Alfred I. Tauber, *The Generation of Diversity and the Rise of Molecular Biology* (Cambridge, MA: Harvard University Press, 1997).

38. Francis H. C. Crick, "On Protein Synthesis," *Symposia of the Society for Experimental Biology* 12 (1958): 138–163.

39. Joshua Lederberg, "Genes and Antibodies," *Science* 129 (1959): 1649–1653.

40. Niels K. Jerne, "The Natural Selection Theory of Antibody Formation," *Proceedings of the National Academy of Sciences of the United States of America* 41 (1955): 849–857; T. Söderqvist, "Darwinian Overtones: Niels K. Jerne and the Origin of the Selection Theory of Antibody Formation," *Journal of the History of Biology* 27 (1994): 481–529; Michel Morange, "The Complex History of the Selective Model of Antibody Formation," *Journal of Biosciences* 39 (2014): 347–350. David W. Talmage, "Allergy and Immunology," *Annual Review of Medicine* 8 (1957): 239–256. Frank MacFarlane Burnet, *The Clonal Selection Theory of Acquired Immunity* (Cambridge: Cambridge University Press, 1959). This model had first been proposed in 1957 in an Australian journal: Frank MacFarlane Burnet, "A Modification of Jerne's Theory of Antibody Production Using the Concept of Clonal Selection," *Australian Journal of Science* 20 (1957): 67–69.

41. G. J. V. Nossal and Joshua Lederberg, "Antibody Production by Single Cells," *Nature* 181 (1958): 1419–1420.

42. Lederberg, "Genes and Antibodies," 1649.

43. The adoption of Burnet's theory was accompanied by a profound transformation of immunology and led to the concept of the immune system: Arthur M. Silverstein, *A History of Immunology,* 2nd ed. (New York: Elsevier / Academic Press, 2009).

44. Nobumichi Hozumi and Susumu Tonegawa, "Evidence for Somatic Rearrangement of Immunoglobulin Genes Coding for Variable and Constant Regions," *Proceedings of the National Academy of Sciences of the United States of America* 73 (1976): 3628–3632.

45. Stephen M. Hedrick, David I. Cohen, Ellen A. Nielsen, and Mark M. Davis, "Isolation of cDNA Clones Encoding T Cell-Specific Membrane-Associated Proteins," *Nature* 308 (1984): 149–153.

18. The Discovery of Oncogenes

1. Michel Morange, "The Discovery of Cellular Oncogenes," *History and Philosophy of the Life Sciences* 15 (1993): 45–59; Michel Morange, "From the Regulatory Vision of Cancer to the Oncogene Paradigm," *Journal of the History of Biology* 30 (1997): 1–27; Natalie Angier, *Natural Obsessions: The Search for*

the Oncogene (Boston: Houghton Mifflin, 1988). The work of Joan Fujimura casts light on the "strategic" stakes involved in the discovery of oncogenes: Joan Fujimura, "Constructing Doable Problems in Cancer Research: Articulating Alignments," *Social Studies of Science* 17 (1987): 257–293; "The Molecular Biological Bandwagon in Cancer Research: Where Social Worlds Meet," *Social Problems* 35 (1988): 261–283; and *Crafting Science: A Sociohistory of the Quest for the Genetics of Cancer* (Cambridge, MA: Harvard University Press, 1996). A very simplified presentation of this history can be found in Harold Varmus and Robert A. Weinberg, *Genes and the Biology of Cancer* (New York: Scientific American Library, 1993).

2. Robert J. Huebner and George J. Todaro, "Oncogenes of RNA Tumor Viruses as Determinants of Cancer," *Proceedings of the National Academy of Sciences of the United States of America* 64 (1969): 1087–1094. The work of Renato Dulbecco's group on oncogenic DNA viruses, the polyoma virus and SV40, showed that transformation was due to a gene carried by the virus. But the mechanism of action of this gene remained unknown. Furthermore, Dulbecco thought that transformation was the result of other cellular changes that took place after viral invasion, so his findings were not incompatible with the model of Huebner and Todaro. See Renato Dulbecco, "From the Molecular Biology of Oncogenic DNA Viruses to Cancer," *Science* 192 (1976): 437–440. Research on oncogenic viruses received significant funding, enjoying an important position in the crusade against cancer launched in the United States in the late 1960s. It was hoped that these studies would lead to the rapid development of diagnostic tools and therapeutic methods, benefiting both patients and the pharmaceutical industry. In particular, the isolation and characterization of viruses opened the door to the development of vaccines.

3. Howard M. Temin, "The Protovirus Hypothesis," *Journal of the National Cancer Institute* 46 (1971): iii–viii.

4. Edward M. Scolnick, Elaine Rands, David Williams, and Wade P. Parks, "Studies on the Nucleic Acid Sequences of Kirsten Sarcoma Virus: A Model for Formation of a Mammalian RNA-Containing Sarcoma Virus," *Journal of Virology* 12 (1973): 458–463.

5. François Jacob, "Comments," *Cancer Research* 20 (1960): 695–697. The analogy between lysogeny and cancer had already been highlighted by André Lwoff seven years earlier: André Lwoff, "Lysogeny," *Bacteriology Review* 17 (1953): 269–337. See also Charles Galperin, "Virus, provirus et cancer," *Revue d'histoire des sciences* 47 (1994): 7–56.

6. See Chapter 5.

7. Howard M. Temin, "On the Origin of the Genes for Neoplasia: G. H. A. Clowes Memorial Lectures," *Cancer Research* 34 (1974): 2835–2841.

8. Dominique Stehelin, Ramareddy V. Guntaka, Harold E. Varmus, and J. Michael Bishop, "Purification of DNA Complementary to Nucleotide Sequences Required for Neoplastic Transformation of Fibroblasts by Avian Sarcoma

Viruses," *Journal of Molecular Biology* 101 (1975): 349–365; Dominique
Stehelin, Harold E. Varmus, J. Michael Bishop, and Peter K. Vogt, "DNA
Related to the Transforming Gene(s) of Avian Sarcoma Viruses Is Present in
Normal Avian DNA," *Nature* 260 (1976): 170–173.

9. Deborah H. Spector, Harold E. Varmus, and J. Michael Bishop, "Nucleotide
Sequences Related to the Transforming Gene of Avian Sarcoma Virus Are
Present in DNA of Uninfected Vertebrates," *Proceedings of the National
Academy of Sciences of the United States of America* 75 (1978): 4102–4106.

10. See, for example, J. Michael Bishop, "Enemies Within: The Genesis of
Retrovirus Oncogenes," *Cell* 23 (1981): 5–6. Also see Angier, *Natural
Obsessions.*

11. Deborah H. Spector, Karen Smith, Thomas Padgett, Pamela McCombe, Daisy
Roulland-Dussoix, et al., "Uninfected Avian Cells Contain RNA Related to the
Transforming Gene of Avian Sarcoma Viruses," *Cell* 13 (1978): 371–379;
Deborah H. Spector, Barbara Baker, Harold E.Varmus, and J. Michael Bishop,
"Characteristics of Cellular RNA Related to the Transforming Gene of Avian
Sarcoma Viruses," *Cell* 13 (1978): 381–386.

12. Morange, "From the Regulatory Vision," 1–27.

13. In 1973 Graham and Van der Eb had developed a technique for "transfecting"
cells with exogenous DNA. In 1979 this technique became operative through
cotransfection with a resistance gene (see Chapter 16); Weinberg and Cooper
carried out their transfection experiments in the same year. In these experiments, the oncogene is positively selected—transformed cells grow quicker and
can easily be detected and isolated in the culture dish.

14. Joyce McCann, Edmund Choi, Edith Yamasaki, and Bruce N. Ames, "Detection of Carcinogens as Mutagens in the *Salmonella* / Microsome Test: Assay of
300 Chemicals," *Proceedings of the National Academy of Sciences of the United
States of America* 72 (1975): 5135–5139.

15. Chiaho Shih, Ben-Zion Shilo, Mitchell P. Goldfarb, Ann Dannenberg, and
Robert A. Weinberg, "Passages of Phenotypes of Chemically Transformed
Cells via Transfection of DNA and Chromatin," *Proceedings of the National
Academy of Sciences of the United States of America* 76 (1979): 5714–5718;
Geoffrey M. Cooper, Sharon Okenquist, and Lauren Silverman, "Transforming
Activity of DNA of Chemically Transformed and Normal Cells," *Nature* 284
(1980): 418–421.

16. This last result was obtained only by Weinberg's group. Cooper showed that
oncogenes could be extracted from normal, untransformed cells by cleaving
DNA into sufficiently small fragments. Cooper's results agreed with the
then-dominant conception of cancer as deregulation. Morange, "From the
Regulatory Vision," 1–27.

17. Luis F. Parada, Clifford J. Tabin, Chiaho Shih, and Robert A. Weinberg,
"Human EJ Bladder Carcinoma Oncogene Is Homologue of Harvey Sarcoma
Virus *Ras* Gene," *Nature* 297 (1982): 474–478; Channing J. Der, Theodore G.

Krontiris, and Geoffrey M. Cooper, "Transforming Genes of Human Bladder and Lung Carcinoma Cell Lines Are Homologous to the *Ras* Genes of Harvey and Kirsten Sarcoma Viruses," *Proceedings of the National Academy of Sciences of the United States of America* 79 (1982): 3637–3640.

18. Clifford J. Tabin, Scott M. Bradley, Cornelia I. Bargmann, Robert A. Weinberg, Alex G. Papageorge, et al., "Mechanism of Activation of a Human Oncogene," *Nature* 300 (1982): 143–149; E. Premkumar Reddy, Roberta K. Reynolds, Eugenio Santos, and Mariano Barbacid, "A Point Mutation Is Responsible for the Acquisition of Transforming Properties by the T24 Human Bladder Carcinoma Oncogene," *Nature* 300 (1982): 149–152.

19. William S. Hayward, Benjamin G. Neel, and Susan M. Astrin, "Activation of a Cellular *onc* Gene by Promoter Insertion in ALV-Induced Lymphoid Leukosis," *Nature* 290 (1981): 475–480; Steven Collins and Mark Groudine, "Amplification of Endogenous *Myc*-Related DNA Sequences in a Human Myeloid Leukaemia Cell Line," *Nature* 298 (1982): 679–681; Philip Leder, Jim Battey, Gilbert Lenoir, Christopher Moulding, William Murphy, et al., "Translocations among Antibody Genes in Human Cancer," *Science* 222 (1983): 765–771.

20. Russell F. Doolittle, Michael W. Hunkapiller, Leroy E. Hood, Sushilkumar G. Devare, Keith C. Robbins, et al., "Simian Sarcoma Virus *Onc* Gene, v-sis, Is Derived from the Gene (or Genes) Encoding a Platelet-Derived Growth Factor," *Science* 221 (1983): 275–277; Michael D. Waterfield, Geoffrey T. Scrace, Nigel Whittle, Paul Sroobant, Ann Johnsson, et al., "Platelet-Derived Growth Factor Is Structurally Related to the Putative Transforming Protein p28[sis] of Simian Sarcoma Virus," *Nature* 304 (1983): 35–39.

21. J. Downward, Y. Yarden, E. Mayes, G. Scrace, N. Totty, et al., "Close Similarity of Epidermal Growth Factor Receptor and v-*erb*-B Oncogene Protein Sequences," *Nature* 307 (1984): 521–527.

22. James B. Hurley, Melvin I. Simon, David B. Teplow, Janet D. Robishaw, and Alfred G. Gilman, "Homologies between Signal Transducing G Proteins and *ras* Gene Products," *Science* 226 (1984): 860–862.

23. Kathleen Kelly, Brent H. Cochran, Charles D. Stiles, and Philip Leder, "Cell-Specific Regulation of the *c-myc* Gene by Lymphocyte Mitogens and Platelet-Derived Growth Factor," *Cell* 35 (1983): 603–610; Wiebe Kruijer, Jonathan A. Cooper, Tony Hunter, and Inder M. Verma, "Platelet-Derived Growth Factor Induces Rapid but Transient Expression of the *c-fos* Gene and Protein," *Nature* 312 (1984): 711–716; Rolf Müller, Rodrigo Bravo, Jean Burckhardt, and Tom Curran, "Induction of *c-fos* Gene and Protein by Growth Factors Precedes Activation of *c-myc*," *Nature* 312 (1984): 716–720.

24. D. Defeo-Jones, E. M. Scolnick, R. Koller, and R. Dhar, "*Ras*-Related Gene Sequences Identified and Isolated from *Saccharomyces cerevisiae*," *Nature* 306 (1983): 707–709.

25. Michael J. Berridge and Robin F. Irvine, "Inositol Triphosphate, a Novel Second Messenger in Cellular Signal Transduction," *Nature* 312 (1984): 315–321;

Yasutomi Nishizuka, "The Role of Protein Kinase C in Cell Surface Signal Transduction and Tumour Promotion," *Nature* 308 (1984): 693–698.

26. Minoo Rassoulzadegan, Alison Cowie, Antony Carr, Nicolas Glaichenhaus, Robert Kamen, and François Cuzin, "The Roles of Individual Polyoma Virus Early Proteins in Oncogenic Transformation," *Nature* 300 (1982): 713–718.

27. Harmut Land, Luis F. Parada, and Robert A. Weinberg, "Cellular Oncogenes and Multistep Carcinogenesis," *Science* 222 (1983): 771–778.

28. Robert A. Weinberg, "The Action of Oncogenes in the Cytoplasm and Nucleus," *Science* 230 (1985): 770–776.

29. Alfred G. Knudson Jr., "Mutation and Cancer: Statistical Study of Retinoblastoma," *Proceedings of the National Academy of Sciences of the United States of America* 68 (1971): 820–823.

30. Stephen H. Friend, René Bernards, Snezna Rogelj, Robert A. Weinberg, Joyce M. Rapaport, et al., "A Human DNA Segment with Properties of the Gene That Predisposes to Retinoblastoma and Osteosarcoma," *Nature* 323 (1986): 643–646.

31. Fujimura, "Molecular Biological Bandwagon," 261–283.

32. George Klein, "The Role of Gene Dosage and Genetic Transposition in Carcinogenesis," *Nature* 294 (1981): 313–318.

33. Douglas Hanahan and Robert A. Weinberg, "Hallmarks of Cancer: The Next Generation," *Cell* 144 (2011): 646–674.

19. From DNA Polymerase to the Amplification of DNA

1. Kary B. Mullis, "The Unusual Origin of the Polymerase Chain Reaction," *Scientific American* 262 (1990): 36–43; Paul Rabinow, *Making PCR: A Story of Biotechnology* (Chicago: University of Chicago Press, 1996).

2. Henry A. Erlich, David Gelfand, and John J. Sninsky, "Recent Advances in the Polymerase Chain Reaction," *Science* 252 (1991): 1643–1651.

3. Horace F. Judson, *The Eighth Day of Creation: The Makers of the Revolution in Biology* (New York: Simon & Schuster, 1979), 322.

4. Franklin H. Portugal and Jack S. Cohen, *A Century of DNA: A History of the Discovery of the Structure and Function of the Genetic Substance* (Cambridge, MA: MIT Press, 1977), 314–317.

5. Arthur Kornberg, *For the Love of Enzymes: The Odyssey of a Biochemist* (Cambridge, MA: Harvard University Press, 1989); Arthur Kornberg, "Never a Dull Enzyme," *Annual Review of Biochemistry* 58 (1989): 1–30. Two reviews of Kornberg's autobiographical account are also worth reading: Pnina G. Abir-Am, "Noblesse Oblige: Lives of Molecular Biologists," *Isis* 82 (1991): 326–343; Jan Sapp, "Portraying Molecular Biology," *Journal of the History of Biology* 25 (1992): 149–155.

6. Kornberg, "Never a Dull Enzyme," 6.

7. Kornberg, *For the Love of Enzymes,* 121–122.

8. Kornberg, "Never a Dull Enzyme," 11.

9. Marianne Grunberg-Manago and Severo Ochoa, "Enzymatic Synthesis and Breakdown of Polynucleotides: Polynucleotide Phosphorylase," *Journal of the American Chemical Society* 77 (1955): 3165–3166. On the passage of Severo Ochoa from biochemistry to molecular biology, see Maria Jesus Santesmases, "Enzymology at the Core: 'Primers' and 'Templates' in Severo Ochoa's Transition from Biochemistry to Molecular Biology," *History and Philosophy of the Life Sciences* 24 (2002): 193–218.

10. Uriel Z. Littauer and Arthur Kornberg, "Reversible Synthesis of Polyribonucleotides with an Enzyme from *Escherichia coli*," *Journal of Biological Chemistry* 226 (1957): 1077–1092.

11. Arthur Kornberg, "Pathways of Enzymatic Synthesis of Nucleotides and Polynucleotides," in *The Chemical Basis of Heredity*, ed. W. D. McElroy and B. Glass, 579–608 (Baltimore: Johns Hopkins University Press, 1958); I. R. Lehman, Maurice J. Bessman, Ernest S. Simms, and Arthur Kornberg, "Enzymatic Synthesis of Deoxyribonucleic Acid. I. Preparation of Substrates and Partial Purification of an Enzyme from *Escherichia coli*," *Journal of Biological Chemistry* 233 (1958): 163–170; Maurice J. Bessman, I. R. Lehman, Ernest S. Simms, and Arthur Kornberg, "Enzymatic Synthesis of Deoxyribonucleic Acid. II. General Properties of the Reaction," *Journal of Biological Chemistry* 233 (1958): 171–177.

12. I. R. Lehman, Steven R. Zimmerman, Julius Adler, Maurice J. Bessman, Ernest S. Simms, and Arthur Kornberg, "Enzymatic Synthesis of Deoxyribonucleic Acid. V. Chemical Composition of Enzymatically Synthesized Deoxyribonucleic Acid," *Proceedings of the National Academy of Sciences of the United States of America* 44 (1958): 1191–1196.

13. Arthur Kornberg, "Biologic Synthesis of Deoxyribonucleic Acid: An Isolated Enzyme Catalyzes Synthesis of This Nucleic Acid in Response to Directions from Pre-existing DNA," *Science* 131 (1960): 1503–1508.

14. This view of enzymes as demiurges and the ease with which it was accepted by molecular biologists have been discussed by the mathematician René Thom: "I am surprised to see how . . . biologists react to the questions of molecular biology. The behavior of macromolecules is something extraordinarily surprising, and yet in the literature biologists seem to find it quite natural. In DNA replication, in the manner in which the helix splits and the two fragments separate into two distinct cells, they see only the work of enzymes, which they think explains everything"; *Paraboles et catastrophes: entretiens sur les mathématiques, la science et la philosophie* (Paris: Flammarion, 1983), 131. Thom was right to be astonished but wrong to cast doubt on these explanations. The mechanisms behind these extraordinary capacities of proteins were gradually unraveled (Chapter 21).

15. Kornberg, "Never a Dull Enzyme," 13; Kornberg, *For the Love of Enzymes*, 163.

16. Mehran Goulian and Arthur Kornberg, "Enzymatic Synthesis of DNA, XXIII. Synthesis of Circular Replicative Form of Phage ΦX174DNA," *Proceedings of*

the National Academy of Sciences of the United States of America 58 (1967): 1723–1730; Mehran Goulian, Arthur Kornberg, and Robert Sinsheimer, "Enzymatic Synthesis of DNA, XXIV. Synthesis of Infectious Phage ΦX174 DNA," *Proceedings of the National Academy of Sciences of the United States of America* 58 (1967): 2321–2328.

17. Kornberg, "Never a Dull Enzyme," 14. This experiment, however, had been preceded two years earlier by a similar experiment carried out on an RNA virus. S. Spiegelman, T. Haruna, I. B. Holland, G. Beaudreau, and D. Mills, "The Synthesis of a Self-Propagating and Infectious Nucleic Acid with a Purified Enzyme," *Proceedings of the National Academy of Sciences of the United States of America* 54 (1965): 919–927.

18. Paula de Lucia and John Cairns, "Isolation of an *E. coli* Strain with a Mutation Affecting DNA Polymerase," *Nature* 224 (1969): 1164–1166.

19. "How Relevant Is Kornberg Polymerase?" *Nature New Biology* 229 (1971): 65–66; "Is Kornberg Junior Enzyme the True Replicase?" *Nature New Biology* 230 (1971): 258.

20. This "nick translation" activity of DNA Polymerase I, detected by Kornberg's group in 1970, was used by many laboratories to label DNA molecules radioactively. A complete description of this technique can be found in Peter W. J. Rigby, Marianne Dieckmann, Carl Rhodes, and Paul Berg, "Labelling Deoxyribonucleic Acid to High Specific Activity *in Vitro* by Nick Translation with DNA Polymerase I," *Journal of Molecular Biology* 113 (1977): 237–251.

21. Frederick Sanger, "Sequences, Sequences, and Sequences," *Annual Review of Biochemistry* 57 (1988): 1–28; Miguel Garcia-Sancho, *Biology, Computing, and the History of Molecular Sequencing: From Proteins to DNA, 1945–2000* (Newark, NJ: Palgrave MacMillan, 2012); George G. Brownlee, *Fred Sanger, Double Nobel Laureate: A Biography* (Cambridge: Cambridge University Press, 2014).

22. F. Sanger, S. Nicklen, and A. R. Coulson, "DNA Sequencing with Chain Terminating Inhibitors," *Proceedings of the National Academy of Sciences of the United States of America* 74 (1977): 5463–5467.

23. Allan M. Maxam and Walter Gilbert, "A New Method for Sequencing DNA," *Proceedings of the National Academy of Sciences of the United States of America* 74 (1977): 560–564.

24. Mullis, "Unusual Origin," 36–43.

25. Randall K. Saiki, Stephen Scharf, Fred Faloona, Kary B. Mullis, Glenn T. Horn, et al., "Enzymatic Amplification of β-Globin Genomic Sequences and Restriction Site Analysis for Diagnosis of Sickle Cell Anemia," *Science* 230 (1985): 1350–1354.

26. Randall K. Saiki, David H. Gelfand, Susanne Stoffel, Stephen J. Scharf, Russell Higuchi, et al., "Primer-Directed Enzymatic Amplification of DNA with a Thermostable DNA Polymerase," *Science* 239 (1988): 487–491.

27. Erlich et al., "Recent Advances," 1643–1651.

28. According to the definition given by Harriet Zuckerman and Joshua Lederberg, "Post-Mature Scientific Discovery?" *Nature* 324 (1986): 629–631.

29. Mullis, "Unusual Origin," 43. On the basis of these arguments, Du Pont de
Nemours began proceedings to annul the patent held by Cetus: Marcia
Barinaga, "Biotech Nightmare: Does Cetus Own PCR?" *Science* 251 (1991):
739–740.

30. K. Kleppe, E. Ohtsuka, R. Kleppe, I. Molineux, and H. G. Khorana, "Studies
on Polynucleotides XCVI. Repair Replication of Short Synthetic DNAs as
Catalyzed by DNA Polymerase," *Journal of Molecular Biology* 56 (1971):
341–361.

31. Erlich et al., "Recent Advances," 1650.

32. See also the allusion to "deoxyribonucleic bombs," which "exploded" in Kary
Mullis's head after his sudden insight. Mullis, "Unusual Origin," 41.

33. Kimberley Carr, "Nobel Rewards Two Laboratory Revolutions," *Nature* 365
(1993): 685; Tim Appenzeller, "Chemistry: Laurels for a Late-Night Brain-
storm," *Science* 262 (1993): 506–507. The Nobel Prize was jointly awarded to
Michael Smith for developing the technique of directed mutagenesis (see
Chapter 16).

20. The Molecularization of Biology and Medicine

1. Errol C. Friedberg, *Correcting the Blueprint of Life: An Historical Account of the
Discovery of DNA Repair Mechanisms* (Cold Spring Harbor, NY: Cold Spring
Harbor Laboratory Press, 1997).

2. R. Grosschedl and M. L. Birnstiel, "Identification of Regulatory Sequences in
the Prelude Sequences of an H2A Histone Gene by the Study of Specific
Deletion Mutants in Vivo," *Proceedings of the National Academy of Sciences of
the United States of America* 77 (1980): 1432–1436; Rudolf Grosschedl and
Max L. Birnstiel, "Spacer DNA Sequences Upstream of the T-A-T-A-A-A-T-A
Sequence are Essential for Promotion of H2A Histone Gene Transcription in
Vivo," *Proceedings of the National Academy of Sciences of the United States of
America* 77 (1980): 7102–7106; C. Benoist and P. Chambon, "Deletions
Covering the Putative Promoter Region of Early mRNAs of Simian Virus 40
Do Not Abolish T-Antigen Expression," *Proceedings of the National Academy of
Sciences of the United States of America* 77 (1980): 3865–3869; Christophe
Benoist and Pierre Chambon, "*In Vivo* Sequence Requirements of the SV40
Early Promoter Region," *Nature* 290 (1981): 304–310.

3. Peter Gruss, Ravi Dhar, and George Khoury, "Simian Virus 40 Tandem
Repeated Sequences as an Element of the Early Promoter," *Proceedings of the
National Academy of Sciences of the United States of America* 78 (1981): 943–947.

4. Julian Banerji, Sandro Rusconi, and Walter Schaffner, "Expression of a
β-Globin Gene Enhanced by Remote SV40 DNA Sequences", *Cell* 27 (1981):
299–308.

5. Gregory S. Payne, Sara A. Courtneidge, Lyman B. Crittenden, Aly M. Fadly, J. Michael Bishop, and Harold E. Varmus, "Analysis of Avian Leukosis Virus DNA and RNA in Bursal Tumors: Viral Gene Expression Is Not Required for Maintenance of the Tumor State," *Cell* 23 (1981): 311–322; Benjamin G. Neel and William S. Hayward, "Avian Leukosis Virus-Induced Tumors Have Common Proviral Integration Sites and Synthesize Discrete New RNAs: Oncogenesis by Promoter Insertion," *Cell* 23 (1981): 323–334; William S. Hayward, Benjamin G. Neel, and Susan M. Astrin, "Activation of a Cellular *Onc* Gene by Promoter Insertion in ALV-Induced Lymphoid Leukosis," *Nature* 290 (1981): 475–480.

6. Stephen D. Gillies, Sherie L. Morrison, Vernon T. Oi, and Susumu Tonegawa, "A Tissue-Specific Transcription Enhancer Element Is Located at the Major Intron of a Rearranged Immunoglobulin Heavy Chain Gene," *Cell* 33 (1983): 717–728; Julian Banerji, Laura Olson, and Walter Schaffner, "A Lymphocyte-Specific Cellular Enhancer Is Located Downstream of the Joining Region in Immuno-globulin Heavy Chain Genes," *Cell* 33 (1983): 729–740; Cary Queen and David Baltimore, "Immunoglobulin Gene Transcription Is Activated by Downstream Sequence Elements," *Cell* 33 (1983): 741–748; M. S. Neuberger, "Expression and Regulation of Immunoglobulin Heavy Chain Gene Transfected into Lymphoid Cells," *EMBO Journal* 2 (1983): 1373–1378.

7. R. B. Winter, O. G. Berg, and P. H. von Hippel, "Diffusion-Driven Mechanisms of Protein Translocation on Nucleic Acids. 3. The *Escherichia coli* lac Repressor-Operator Interaction: Kinetic Mechanisms and Conclusions," *Biochemistry* 20 (1980): 6961–6977.

8. Harold Weintraub and Mark Groudine, "Chromosomal Subunits in Active Genes Have an Altered Conformation," *Science* 193 (1976): 848–856; Walter A. Scott and Dianne J. Wigmore, "Sites in Simian Virus 40 Chromatin Which Are Preferentially Cleaved by Endonucleases," *Cell* 15 (1978): 1511–1518; A. J. Varshavsky, O. H. Sundin, and M. J. Bohn, "SV40 Viral Minichromosome: Preferential Exposure of the Origin of Replication as Probed by Restriction Endonucleases," *Nucleic Acids Research* 5 (1978): 3469–3477; W. A. Scott and D. J. Wigmore, "Sites in Simian Virus 40 Chromatin Which Are Preferentially Cleaved by Endonucleases," *Cell* 15 (1978): 1511–1518; S. Saragosti, G. Moyne, and M. Yaniv, "Absence of Nucleosomes in a Fraction of SV40 Chromatin between the Origin of Replication and the Region Coding for the Leader RNA," *Cell* 20 (1980): 65–73.

9. James J. Champoux, "Proteins That Affect DNA Conformation," *Annual Review of Biochemistry* 47 (1978): 449–479; Gerald R. Smith, "DNA Supercoiling: Another Level for Regulating Gene Expression," *Cell* 24 (1981): 599–600.

10. M. P. F. Marsden and U. K. Laemmli, "Metaphase Chromosome Structure: Evidence for a Radial Loop Model," *Cell* 17 (1979): 849–858.

11. Laimonis A. Laimins, George Khoury, Cornella Gorman, Bruce Howard, and Peter Gruss, "Host-Specific Activation of Transcription by Tandem Repeats

from Simian Virus 40 and Moloney Murine Sarcoma Virus," *Proceedings of the National Academy of Sciences of the United States of America* 79 (1982): 6453–6457; H. Weiner, M. König, and P. Gruss, "Multiple Point Mutations Affecting the Simian Virus 40 Enhancer," *Science* 219 (1983): 626–631.

12. Vicki L. Chandler, Bonnie A. Maler, and Keith R. Yamamoto, "DNA Sequences Bound Specifically by Glucorticoid Receptor in Vitro Render a Heterologous Promoter Hormone Responsive in Vivo," *Cell* 33 (1983): 489–499.

13. B. F. Luisi, W. X. Xu, Z. Otwinowski, L. P. Friedman, K. R. Yamamoto, and P. B. Sigler, "Crystallographic Analysis of the Interaction of the Glucocorticoid Receptor with DNA," *Nature* 352 (1991): 497–505; Dimitar B. Nikolov, Shu-Hong Hu, Judith Lin, Alexander Gasch, Alexander Hoffmann, et al., "Crystal Structure of TFIID TATA-Box Binding Protein," *Nature* 360 (1992): 40–46.

14. Tom Maniatis, Stephen Goodbourn, and Janice A. Fischer, "Regulation of Inducible and Tissue-Specific Gene Expression," *Science* 236 (1987): 1237–1245; Paula J. Mitchell and Robert Tjian, "Transcriptional Regulation in Mammalian Cells by Sequence-Specific DNA Binding Proteins," *Science* 245 (1989): 371–378; James Darnell, *RNA: Life's Indispensable Molecule* (Cold Spring Harbor, NY: Cold Spring Harbor Laboratory Press, 2011).

15. For a review, see Robert Schleif, "DNA Looping," *Annual Review of Biochemistry* 61 (1992): 199–223.

16. Wouter de Laat and Denis Duboule, "Topology of Mammalian Developmental Enhancers and Their Regulatory Landscapes," *Nature* 502 (2013): 499–506.

17. M. Kidd, "Paired Helical Filaments in Electron Microscopy of Alzheimer's Disease," *Nature* 197 (1963): 192–193.

18. J. Cuillé and P. L. Chelles, "La maladie de la tremblante du mouton est-elle inoculable?," *Comptes rendus hebdomadaires des séances de l'Académie des sciences* 203 (1936): 1552–1554.

19. Tikvah Alper, W. A. Cramp, D. A. Haig, and M. C. Clarke, "Does the Agent of Scrapie Replicate without Nucleic Acid?" *Nature* 214 (1967): 764–766.

20. André Lwoff, "The Concept of Virus," *Journal of General Microbiology* 17 (1957): 239–253.

21. J. B. Griffith, "Self-Replication and Scrapie," *Nature* 215 (1967): 1043–1044.

22. D. Carleton Gajdusek, "Unconventional Viruses and the Origin and Disappearance of Kuru," *Science* 197 (1977): 943–960.

23. Stanley B. Prusiner, "Novel Proteinaceous Particles Cause Scrapie," *Science* 216 (1982): 136–144.

24. David C. Bolton, Michael P. McKinley, and Stanley B. Prusiner, "Identification of a Protein That Purifies with the Scrapie Prion," *Science* 218 (1982): 1308–1311.

25. Prusiner, "Novel Proteinaceous Particles."

26. Stanley B. Prusiner and Maclyn McCarty, "Discovering DNA Encodes Heredity and Prions Are Infectious Proteins," *Annual Review of Genetics* 40 (2006): 25–45.

27. Bruno Oesch, David Westaway, Monika Walchli, Michael P. McKKinley, Stephen B. H. Kent, et al., "A Cellular Gene Encodes Scrapie PrP 27–30 Protein," *Cell* 40 (1985): 735–746; R. Basler, B. Oesch, M. Scott, D. Westaway, M. Walchli, et al., "Scrapie and Cellular PrP Isoforms Are Encoded by the Same Chromosomal Gene," *Cell* 46 (1986): 417–428.

28. George A. Carlson, David T. Kingsbury, Patricia A. Goodman, Shernie Coleman, Susan T. Marshall, et al., "Linkage of Prion Protein and Scrapie Incubation Time Genes," *Cell* 46 (1986): 503–511; David Westaway, Patricia A. Goodman, Carol A. Mirenda, Michael P. McKinley, George A. Carlson, and Stanley B. Prusiner, "Distinct Prion Proteins in Short and Long Scrapie Incubation Period Mice," *Cell* 51 (1987): 651–662.

29. Karen Hsiao, Harry F. Baker, Tim J. Crow, Mark Poulter, Frank Owen, et al., "Linkage of a Prion Protein Missense Variant to Gerstmann-Sträussler Syndrome," *Nature* 338 (1989): 342–344.

30. Jie Kang, Hans-Georg Lemaire, Axel Unterbeck, Michael Salbaum, Colin L. Masters, et al., "The Precursor of Alzheimer's Disease Amyloid A4 Protein Resembles a Cell-Surface Receptor," *Nature* 325 (1987): 733–736.

31. P. H. St George-Hyslop, J. L. Haines, L. A Ferrer, R. Polinsky, C. Van Broeck-hoven, et al., "Genetic Linkage Studies Suggest That Alzheimer's Disease Is Not a Single Homogeneous Disorder," *Nature* 347 (1990): 194–197.

32. Efrat Levy, Mark D. Carman, Ivan J. Fernandez-Madrid, Michael D. Power, Ivan Lieberburg, et al., "Mutation of the Alzheimer's Disease Amyloid Gene in Hereditary Cerebral Hemorrhage, Dutch Type," *Science* 248 (1990): 1124–1126.

33. Gerard D. Schellenberg, Thomas D. Bird, Ellen M. Wijsman, Henry T. Orr, Leojean Anderson, et al., "Genetic Linkage Evidence for a Familial Alzheimer's Disease Locus on Chromosome 14," *Science* 258 (1992): 668–671.

34. Ephrat Levy-Lahad, Ellen M. Wijsman, Ellen Nemens, Leojean Anderson, Kattrina A. B. Goddard, et al., "A Familial Alzheimer's Disease Locus on Chromosome I," *Science* 269 (1995): 970–973; Ephrat Levy-Lahad, Wilma Wasco, Parvoneh Poorkaj, Donna M. Romano, Junko Oshima, et al., "Candidate Gene for the Chromosome 1 Familial Alzheimer's Disease Locus," *Science* 269 (1995): 973–977.

35. Alison Goate, Marie-Christine Chartier-Harlin, Mike Mullan, Jeremy Brown, Fiona Crawford, et al., "Segregation of a Missense Mutation in the Amyloid Precursor Protein Gene with Familiar Alzheimer's Disease," *Nature* 349 (1991): 706; Marie-Christine Chartier-Harlin, Fiona Crawford, Henry Houlden, Andrew Warren, David Hughes, et al., "Early-Onset Alzheimer's Disease Caused by Mutations at Codon 717 of the β-Amyloid Precursor Protein Gene," *Nature* 353 (1991): 844–846.

36. Michael Scott, Dallas Foster, Carol Mirenda, Dan Serban, Frank Coutal, et al., "Transgenic Mice Expressing Hamster Prion Produce Species-Specific Scrapie Infectivity and Amyloid Plaques," *Cell* 59 (1989): 847–857; H. Bueler, A.

Aguzzi, A. Saller, R.-A. Greiner, P. Autenried, et al., "Mice Devoid of PrP Are Resistant to Scrapie," *Cell* 73 (1993): 1339–1347.

37. Susan Lindquist, "Mad Cows Meet Psi-chotic Yeast: The Expansion of the Prion Hypothesis," *Cell* 89 (1997): 495–498.

38. James D. Harper and Peter T. Lansbury Jr., "Models of Amyloid Seeding in Alzheimer's Disease and Scrapie: Mechanistic Truths and Physiological Consequences of the Time-Dependent Solubility of Amyloid Proteins," *Annual Review of Biochemistry* 66 (1997): 385–407.

39. Giuseppe Legname, Ilia V. Baskakov, Hoang-Oanh B. Nguyen, Detlev Riesner, Fred E. Cohen, et al., "Synthetic Mammalian Prions," *Science* 305 (2004): 673–676.

40. Rebecca Nelson, Michael R. Sawaya, Melinda Balbirnie, Anders O Madsen, Christian Riekel, et al., "Structure of the Cross-β Spine of Amyloid-Like Fibrils," *Nature* 435 (2005): 773–778.

41. Pei-Hsien Ren, J. E. Lauckner, Ioulia Kachirskaia, John E. Heuser, Ronald Melki, and Ron R. Kopito, "Cytoplasmic Penetration and Persistent Infection of Mammalian Cells by Polyglutamine Aggregates," *Nature Cell Biology* 11 (2009): 219–225.

42. Martha E. Keyes, "The Prion Challenge to the 'Central Dogma' of Molecular Biology, 1965–1991. Part I: Prelude to Prions," *Studies in History and Philosophy of Science Part C: Studies in History and Philosophy of Biological and Biomedical Sciences* 30 (1999): 1–19; "Part II: The Problem with Prions," *Studies in History and Philosophy of Science Part C: Studies in History and Philosophy of Biological and Biomedical Sciences* 30 (1999): 181–218.

43. Jacques Monod, Jeffries Wyman, and Jean-Pierre Changeux, "On the Nature of Allosteric Transitions: A Plausible Model," *Journal of Molecular Biology* 12 (1965): 88–118.

44. Michel Goedert and Maria Grazia Spillantini, "A Century of Alzheimer's Disease," *Science* 314 (2006): 777–781; Adriano Aguzzi and Magdalini Poly-menidou, "Mammalian Prion Biology: One Century of Evolving Concepts," *Cell* 116 (2004): 313–327.

45. Günter Blobel and David Sabatini, "Ribosome Membrane Interaction in Eukaryotic Cells," in *Biomembranes*, ed. R. A. Manson, 193–195 (New York: Plenum Press, 1971); Michelle Lynne Labonte, "Blobel and Sabatini's Beautiful Idea: Visual Representations of the Conception and Refinement of the Signal Hypothesis," *Journal of the History of Biology* 50 (2017): 797–833.

46. For a description of these early observations, see P. G. H. Clarke and S. Clarke, "Nineteenth Century Research on Naturally Occurring Cell Death and Related Phenomena," *Anatomy and Embryology* 193 (1996): 81–99. Also see A. Glucksmann, "Cell Deaths in Normal Vertebrate Ontogeny," *Biological Reviews of the Cambridge Philosophical Society* 26 (1951): 59–86.

47. Richard A. Lockshin and Carroll M. Williams, "Programmed Cell Death II. Endocrine Potentiation of the Breakdown of the Intersegmental Muscles of Silkmoths," *Journal of Insect Physiology* 10 (1964): 643–649.

48. J. F. Kerr, A. Wyllie, and A. H. Currie, "Apoptosis: A Basic Biological Phenomenon with Wide-Ranging Implications in Tissue Kinetics," *British Journal of Cancer* 26 (1972): 239–257.

49. A. H. Wyllie, "Glucorticoid-Induced Thymocyte Apoptosis Is Associated with Endogenous Endonuclease Activation," *Nature* 284 (1980): 555–556.

50. Sydney Brenner, "The Genetics of *Caenorhabditis elegans*," *Genetics* 77 (1973): 71–94.

51. H. Robert Horvitz and John E. Sulston, "Isolation and Genetic Characterization of Cell-Lineage Mutants of the Nematode *Caenorhabditis elegans*," *Genetics* 96 (1980): 435–464.

52. Victor Ambros and H. Robert Horvitz, "Heterochronic Mutants in the Nematode *Caenorhabditis elegans*," *Science* 226 (1984): 409–416.

53. H. Robert Horvitz, "Worms, Life, and Death (Nobel lecture)," *Chembiochem* 4 (2003): 697–711.

54. Edward M. Hedgecock, John E. Sulston, and J. Nichol Thomson, "Mutations Affecting Programmed Cell Deaths in the Nematode *Caenorhabditis elegans*," *Science* 220 (1983): 1277–1279.

55. Junting Yuan and H. Robert Horvitz, "The *Caenorhabditis elegans* Cell Death Gene *ced-4* Encodes a Novel Protein and Is Expressed during the Period of Extensive Programmed Cell Death," *Development* 116 (1992): 309–320.

56. Junying Yuan, Shai Shaham, Stephane Ledoux, Hilary M. Ellis, and H. Robert Horvitz, "The *C. elegans* Cell Death Gene *ced-3* Encodes a Protein Similar to Mammalian Interleukin-1 β-Converting Enzyme," *Cell* 75 (1993): 841–852.

57. Ding Xue, Shai Shaham, and H. Robert Horvitz, "The *Caenorhabditis elegans* Cell-Death Protein CED-3 Is a Cysteine Protease with Substrate Specificities Similar to Those of the Human CPP32 Protease," *Genes and Development* 10 (1996): 1073–1083; Nancy A. Thornberry and Yuri Lazebnik, "Caspases: Enemies Within," *Science* 281 (1998): 1312–1316.

58. Michael O. Hengartner and H. Robert Horvitz, "*C. elegans* Cell Survival Gene *ced-9* Encodes a Functional Homolog of the Mammalian Proto-Oncogene *bcl-2*," *Cell* 78 (1994): 665–678.

59. J. Michael Bishop, "Enemies Within: The Genesis of Retrovirus Oncogenes," *Cell* 23 (1981): 5–6.

21. Protein Structure

1. Steven R. Jordan and Carl O. Pabo, "Structure of the Lambda Complex at 2.5 Å Resolution: Details of the Repressor-Operator Interactions," *Science* 242 (1988): 893–899.

2. *Lysozyme:* C. C. F. Blake, D. F. Koenig, G. A. Mair, A. C. T. North, D. C. Phillips, and V. R. Sarma, "Structure of Hen Egg-White Lysozyme: A Three-Dimensional Fourier Analysis at 2 Å Resolution," *Nature* 206 (1965): 757–761. *Ribonuclease:* G. Kartha, J. Bello and D. Harker, "Tertiary Structure of

Ribonuclease," *Nature* 213 (1967): 862–865. *Chymotrypsin:* B. W. Matthews, P. B. Sigler, R. Henderson, and D. M. Blow, "Three-Dimensional Structure of Tosyl-α-Chymotrypsin," *Nature* 214 (1967): 652–656. *Carboxypeptidase A:* G. N. Reeke, J. A. Hartsuck, M. L. Ludwig, F. A. Quiocho, T. A. Steitz, and W. N. Lipscomb, "The Structure of Carboxypeptidase A, VI. Some Results at 2.0-Å Resolution and the Complex with Glycyl-Tyrosine at 2.8 Å Resolution," *Proceedings of the National Academy of Sciences of the United States of America* 58 (1967): 2220–2226.

3. S. C. Harrison, A. J. Olson, C. E. Schutt, F. K. Winkler, and G. Bricogne, "Tomato Bushy Stunt Virus at 2.9 Å Resolution," *Nature* 276 (1978): 368–373; Celerino Abad-Zapatero, Sherin S. Abdel-Meguid, John F. Johnson, Andrew G. W. Leslie, Ivan Rayment, et al., "Structure of Southern Bean Mosaic Virus at 2.8 Å Resolution," *Nature* 286 (1980): 33–39.

4. William M. Clemons Jr., Joanna L. C. May, Brian T. Wimberly, John P. McCutcheon, Malcolm S. Capel, and V. Ramakrishnan, "Structure of a Bacterial 30S Ribosomal Subunit at 5.5 Å Resolution," *Nature* 400 (1999): 833–840; Nehad Ban, Poul Nissen, Jeffrey Hansen, Malcolm Capel, Peter B. Moore, and Thomas A. Steitz, "Placement of Protein and RNA Structures into a 5 Å-Resolution Map of the 50S Ribosomal Subunit," *Nature* 400 (1999): 841–847; Jamie H. Cate, Marat M. Yusupov, Gulnara Zh. Yusupova, Thomas N. Earnest, and Harry F. Noller, "X-Ray Crystal Structures of 70S Ribosome Functional Complexes," *Science* 285 (1999): 2095–2104.

5. *Bacterial photosynthetic center:* J. Deisenhofer, O. Epp, K. Miki, R. Huber, and H. Michel, "Structure of the Protein Subunits in the Photosynthetic Reaction Centre of *Rhodopseudomonas viridis* at 3Å Resolution," *Nature* 318 (1985): 618–624. *Histocompatibility antigens:* P. J. Bjorkman, M. A. Saper, B. Samraoui, W. S. Bennett, J. L. Strominger, and D. C. Wiley, "Structure of the Human Class I Histocompatibility Antigen, HLA-A2," *Nature* 329 (1987): 506–512. *Proton pump:* R. Henderson, J. M. Baldwin, T. A. Ceska, F. Zemlin, E. Beckmann, and K. H. Downing, "Model for the Structure of Bacteriorhodopsin Based on High-Resolution Electron Cryo-microscopy, *Journal of Molecular Biology* 213 (1990): 899–929. *Potassium channel:* Declan A. Doyle, Joao Morais Cabral, Richard A. Pfuetzner, Anling Kuo, Jacqueline M. Gulbis, et al., "The Structure of the Potassium Channel: Molecular Basis of K+ Conduction and Seectivity," *Science* 280 (1998): 69–77; Stephen B. Long, Ernest B. Campbell, and Roderick MacKinnon, "Crystal Structure of a Mammalian Voltage-Dependent *Shaker* Family K+ Channel," *Science* 309 (2005): 897–903; Stephen B. Long, Ernest B. Campbell, and Roderick MacKinnon, "Voltage Sensor of Kv1.2: Structural Basis of Electromechanical Coupling," *Science* 309 (2005): 903–908. See also Mathias Grote, *Membranes to Molecular Machines: Active Matter and the Remaking of Life* (Chicago, University of Chicago Press, 2019).

6. Jan Pieter Abrahams, Andrew G. W. Leslie, René Lutter, and John E. Walker, "Structure at 2.8 Å Resolution of F$_1$-ATPase from Bovine Heart Mitochondria,"

Nature 370 (1994): 621–628. Paul D. Boyer, "The ATP Synthase-A Splendid Molecular Motor," *Annual Review of Biochemistry* 66 (1997): 717–749.

7. Hiroyuki Noji, Ryohei Yasuda, Masasuke Yoshida, and Kazuhiko Kinosia Jr, "Direct Observation of the Rotation of F_1-ATPase," *Nature* 386 (1997): 299–302.

8. Special issue of *Cell* on Macromolecular Machines, February 6, 1998.

9. *Proteasomes:* Wolfgang Baumeister, Jochen Walz, Frank Zühl, and Erika Seemüller, "The Proteasome: Paradigm of a Self-Compartmentalizing Protease," *Cell* 92 (1998): 367–380. *Chaperones:* Bernd Bukau and Arthur L. Horwich, "The Hsp70 and Hsp60 Chaperone Machines," *Cell* 92 (1998): 351–366. *Spliceosomes:* Jonathan P. Staley and Christine Guthrie, "Mechanical Devices of the Spliceosome: Motors, Clocks, Springs, and Things," *Cell* 92 (1998): 315–326. *Nucleocytoplasmic transport:* Mutsuhito Ohno, Maarten Fornerod and Iain W. Mattaj, "Nucleocytoplasmic Transport: The Last 200 Nanometers," *Cell* 92 (1998): 327–336.

10. Shimon Weiss, "Fluorescence Spectroscopy of Single Biomolecules," *Science* 283 (1999): 1676–1683; Amit D. Mehta, Matthias Rief, James A. Spudich, David A. Smith, and Robert M. Simmons, "Single-Molecule Biomechanics with Optical Methods," *Science* 283 (1999): 1689–1695; Marcos Sotomayor, and Klaus Schulten, "Single-Molecule Experiments in Vitro and in Silico," *Science* 316 (2007): 1144–1148.

11. Jefferey T. Finer, Robert M. Simmons, and James A. Spudich, "Single Myosin Molecule Mechanics: Piconewton Forces and Nanometer Steps," *Nature* 368 (1994): 113–119.

12. Hong Yin, Michelle D. Wang, Karel Svoboda, Robert Landick, Steven M. Block, and Jeff Gelles, "Transcription against an Applied Force," *Science* 270 (1995): 1653–1657; Jeff Gelles and Robert Landick, "RNA Polymerase as a Molecular Motor," *Cell* 93 (1998): 13–16; Andrey Revyakin, Chenyu Liu, Richard H. Ebright, and Terence R. Strick, "Abortive Initiation and Productive Initiation by RNA Polymerase Involve DNA Scrunching," *Science* 314 (2006): 1139–1143; Achilleis N. Kapanidis, Emmaniel Margeat, Sam On Ho, Ekaterine Kortkhonijia, Shimon Weiss, and Richard H. Ebright, "Initial Transcription by RNA Polymerase Proceeds through a DNA-Scrunching Mechanism," *Science* 314 (2006): 1144–1147.

13. F. Barré-Sinoussi, J. C. Chermann, F. Rey, M. T. Nugeyre, S. Chamaret, et al., "Isolation of a T-Lymphotropic Retrovirus from a Patient at Risk for Acquired Immune Deficiency Syndrome (AIDS)," *Science* 220 (1983): 868–871; Simon Wain-Hobson, Pierre Sonigo, Olivier Danos, Stewart Cole, and Marc Alizon, "Nucleotide Sequence of the AIDS Virus, LAV," *Cell* 40 (1985): 9–17.

14. Hiroaki Mitsuya, Kent J. Weinhold, Phillip A. Hurman, Marty H. St. Clair, Sandra Nusinof, et al., "3'-Azido-3'-Deoxythymidine (BW A509U): An Antiviral Agent That Inhibits the Infectivity and Cytopathic Effect of Human T-Lymphotropic Virus Type 3 / Lymphadenopathy-Associated Virus in Vitro,"

Proceedings of the National Academy of Sciences of the United States of America 82 (1985): 7096–7100.

15. Sigfrid Seelmeier, Holger Schmidt, Vito Turk, and Klaus von der Helm, "Human Immunodeficiency Virus Has an Aspartic-Type Protease That Can Be Inhibited by Pepstatin A," *Proceedings of the National Academy of Sciences of the United States of America* 85 (1988): 6612–6616.

16. Moshe Kotler, Richard A. Katz, Wakeed Danho, Jonathan Leis, and Anna Marie Skalka, "Synthetic Peptides as Substrates and Inhibitors of a Retroviral Protease," *Proceedings of the National Academy of Sciences of the United States of America* 85 (1988): 4185–4189.

17. Maria Miller, Jens Schneider, Bangalore K. Sathyanarayana, Mihaly V. Toth, Garland R. Marshall, et al., "Structure of Complex of Synthetic HIV-1 Protease with a Substrate-Based Inhibitor at 2.3 Å Resolution," *Science* 246 (1989): 1149–1152.

18. T. J. McQuade, A. G. Tomasselli, L. Liu, V. Karacostas, B. Moss, et al., "Synthetic HIV-1 Protease Inhibitor with Antiviral Activity Arrests HIV-Like Particle Maturation," *Science* 247 (1990): 454–456; Alexander Wlodawer and John W. Erickson, "Structure-Based Inhibitors of HIV-1 Protease," *Annual Review of Biochemistry* 62 (1993): 343–385.

19. See, for instance, Steven Rosenberg, Philip J. Barr, Richard C. Najarian, and Robert A. Hallewell, "Synthesis in Yeast of a Functional Oxidation-Resistant Mutant of Human α_1-Antitrypsin," *Nature* 312 (1984): 77–80.

20. Markus G. Grütter, Richard B. Hawkes, and Brian W. Matthews, "Molecular Basis of Thermostability in the Lysozyme from Bacteriophage T4," *Nature* 277 (1979): 667–669.

21. Masazumi Matsumura, Wayne J. Becktel, and Brian W. Matthews, "Hydrophobic Stabilization in T4 Lysozyme Determined Directly by Multiple Substitutions of Ile 3," *Nature* 334 (1988): 406–410; Tom Alber, Dao-pin Sun, Keith Wilson, Joan A. Wozniak, Sean P. Cook, and Brian W. Matthews, "Contributions of Hydrogen Bonds of Thr 157 to the Thermodynamic Stability of Phage T4 Lysozyme," *Nature* 330 (1987): 41–46; H. Nicholson, W. J. Becktel, and B. W. Matthews, "Enhanced Protein Thermostability from Designed Mutations That Interact with α-Helix Dipoles," *Nature* 336 (1988): 651–656; Tom Alber, "Mutational Effects on Protein Stability," *Annual Review of Biochemistry* 58 (1989): 798; Brian W. Matthews, "Structural and Genetic Analysis of Protein Stability," *Annual Review of Biochemistry* 62 (1993): 139–160.

22. Cyrus Chothia, "Structural Invariants in Protein Folding," *Nature* 254 (1975): 304–308.

23. Robin J. Leatherbarrow and Alan R. Fersht, "Protein Engineering," *Protein Engineering* 1 (1986): 7–16.

24. Greg Winter, Alan R. Fersht, Anthony J. Wilkinson, Mark Zoller, and Michael Smith, "Redesigning Enzyme Structure by Site-Directed Mitagenesis: Tyrosyl-tRNA Synthetase and ATP Binding," *Nature* 299 (1982): 756–758;

Robin J. Leatherbarrow, Alan R. Fersht, and Greg Winter, "Transition-State Stabilization in the Mechanism of Tyrosyl-tRNA synthetase Revealed by Protein Engineering," *Proceedings of the National Academy of Sciences of the United States of America* 82 (1985): 7840–7644.

25. Alan J. Russell and Alan R. Fersht, "Rational Modification of Enzyme Catalysis by Engineering Surface Change," *Nature* 328 (1987): 496–500; James T. Kellis Jr., Kerstin Nyberg, Dasa Sali, and Alan R. Fersht, "Contribution of Hydrophobic Interactions to Protein Stability," *Nature* 333 (1988): 784–786.

26. Anthony R. Clarke, Tony Atkinson, and J. John Holbrook, "From Analysis to Synthesis: New Ligand Binding Sites on the Lactate Dehydrogenase Framework. Part I," *Trends in Biochemical Sciences* 14 (1989): 101–105; Anthony R. Clarke, Tony Atkinson, and J. John Holbrook, "From Analysis to Synthesis: New Ligand Binding Sites on the Lactate Dehydrogenase Framework. Part II," *Trends in Biochemical Sciences* 14 (1989): 145–148.

27. Among many possible examples, see Eric Quéméneur, Mireille Moutiez, Jean-Baptiste Charbonnier, and André Menez, "Engineering Cyclophilin into a Proline-Specific Endopeptidase," *Nature* 391 (1998): 301–304.

28. Andrew D. Griffiths and Dan S. Tawfik, "Man-Made Enzymes—from Design to in Vitro Compartmentalisation," *Current Opinion in Biotechnology* 11 (2000): 338–353.

29. Xiaojun Wang, George Minasov, and Brian K. Shoichet, "Evolution of an Antibiotic Resistance Enzyme Constrained by Stability and Activity Trade-Offs," *Journal of Molecular Biology* 320 (2002): 85–95.

30. Elizabeth M. Meiering, Luis Serrano, and Alan R. Fersht, "Effect of Active Site Residues in Barnase on Activity and Stability," *Journal of Molecular Biology* 225 (1992): 585–589; Gideon Schreiber, Ashley M. Buckle, and Alan R. Fersht, "Stability and Function: Two Constraints in the Evolution of Barstar and Other Proteins," *Structure* 2 (1994): 945–951.

31. Antony M. Dean and Joseph W. Thornton, "Mechanistic Approaches to the Study of Evolution: The Functional Synthesis," *Nature Reviews Genetics* 8 (2007): 675–688.

32. Konstantin B. Zeldovich, Peiqiu Chen, and Eugene I. Shaknovich, "Protein Stability Imposes Limits on Organism Complexity and Speed of Molecular Evolution," *Proceedings of the National Academy of Sciences of the United States of America* 104 (2007): 16152–16157; Mark A. DePristo, Daniel M. Weinreich, and Daniel L. Hartl, "Missense Meanderings in Sequence Space: A Biophysical View of Protein Evolution," *Nature Reviews Genetics* 6 (2005): 678–687.

33. Roy A. Jensen, "Enzyme Recruitment in Evolution of New Function," *Annual Review of Microbiology* 30 (1976): 409–425.

34. G. Köhler and C. Milstein, "Continuous Cultures of Fused Cells Secreting Antibody of Predefined Specificity," *Nature* 256 (1975): 495–497.

35. Alfonso Tramontano, Kim D. Janda, and Richard A. Lerner, "Catalytic Antibodies," *Science* 234 (1986): 1566–1569.

36. Andrew D. Napper, Stephen J. Benkovic, Alfonso Tramontano, and Richard A. Lerner, "A Stereospecific Cyclization Catalyzed by an Antibody," *Science* 237 (1987): 1041–1043; Stephen J. Benkovic, Andrew D. Napper, and Richard A. Lerner, "Catalysis of a Stereospecific Bimolecular Amide Synthesis by an Antibody," *Proceedings of the National Academy of Sciences of the United States of America* 85 (1988): 5355–5358.

37. Donald Hilvert, "Critical Analysis of Antibody Catalysis," *Annual Review of Biochemistry* 69 (2000): 751–793.

38. Jacques Monod, Jeffries Wyman, and Jean-Pierre Changeux, "On the Nature of Allosteric Transition: A Plausible Model," *Journal of Molecular Biology* 12 (1965): 88–118; D. E. Koshland, G. Nemethy, and D. Filmer, "Comparison of Experimental Binding Data and Theoretical Models in Proteins Containing Subunits," *Biochemistry* 5 (1966): 365–385.

39. Cyrus Levinthal, "Are There Pathways for Protein Folding?," *Journal of Chemical Physics* 65 (1968): 44–45.

40. Jonathan J. Ewbank and Thomas E. Creighton, "The Molten Globule Protein Conformation Probed by Disulphide Bonds," *Nature* 350 (1991): 618–620.

41. O. B. Ptitsyn, R. H. Pain, G. V. Semisotnov, E. Zerovnik, and O. I. Razgulyaev, "Evidence for a Molten Globule State as a General Intermediate in Protein Folding," *FEBS Letters* 262 (1990): 20–24.

42. Andreas Matouschek, James T. Kellis Jr., Luis Serrano, Mark Bycroft, and Alan R. Fersht, "Transient Folding Intermediates Characterized by Protein Engineering," *Nature* 346 (1990): 440–445.

43. John Ellis, "Proteins as Molecular Chaperones," *Nature* 328 (1987): 378–379.

44. Peter E. Leopold, Mauricio Montal, and José Nelson Onuchic, "Protein Folding Funnels: A Kinetic Approach to the Sequence-Structure Relationship," *Proceedings of the National Academy of Sciences of the United States of America* 89 (1992): 8721–8725; Joseph D. Bryngelson, José Nelson Onuchic, Nicholas D. Socci, and Peter G. Wolynes, "Funnels, Pathways, and the Energy Landscape of Protein Folding: A Synthesis," *Proteins: Structure, Function, and Genetics* 21 (1995): 167–195; Ken A. Dill and Hue Sen Chan, "From Levinthal to Pathways to Funnels," *Nature Structural and Molecular Biology* 4 (1997): 10–19.

45. Vladimir N. Uversky, "Natively Unfolded Proteins: A Point Where Biology Waits for Physics," *Protein Science* 11 (2002): 739–756.

46. J. Andrew McCannon, Bruce R. Gelin, and Martin Karplus, "Dynamics of Folded Proteins," *Nature* 267 (1977): 585–590.

47. Anthony Mittermaier and Lewis E. Kay, "New Tools Provide New Insights in NMR Studies of Protein Dynamics," *Science* 312 (2006): 224–228; Arthur G. Palmer III and Francesca Massi, "Characterization of the Dynamics of Macromolecules Using Rotating-Frame Spin Relaxation NMR Spectroscopy," *Chemical Reviews* 106 (2006): 1700–1719.

48. David E. Shaw, Paul Maragakis, Kresten Lindorff-Larsen, Stefano Piana, Ron O. Dror, et al., "Atomic-Level Characterization of the Structural Dynamics of Proteins," *Science* 330 (2010): 341–346.

49. Elan Z. Elsenmesser, Oscar Millet, Wladimir Labelkovsky, Dimitry M. Korzhnev, Magnus Wolf-Watz, et al., "Intrinsic Dynamics of an Enzyme Underlies Catalysis," *Nature* 438 (2005): 17–121; David D. Boehr, Dan McElheny, H. Jane Dyson, and Peter E. Wright, "The Dynamic Energy Landscape of Dihydrofolate Reductase Catalysis," *Science* 313 (2006): 1638–1642; Katherine Henzler-Wildman and Dorothee Kern, "Dynamic Personalities of Proteins," *Nature* 450 (2007): 964–972; Stephen J. Benkovic, G. G. Hammes, and S. Hammes-Schiffer, "Free-Energy Landscape of Enzyme Catalysis," *Biochemistry* 47 (2008): 3317–3321.

50. Hesam N. Motlagh, James O. Wrabl, Jing Li, and Vincent J. Hilser, "The Ensemble Nature of Allostery," *Nature* 508 (2014): 331–338.

51. Nina M. Goodey and Stephen J. Benkovic, "Allosteric Regulation and Catalysis Emerge via a Common Route," *Nature Chemical Biology* 4 (2008): 474–482.

52. Wade C. Winkler and Charles E. Dann III, "RNA Allostery Glimpsed," *Nature Structural and Molecular Biology* 13 (2006): 569–571; Evgenia N. Nikolova, Runae Kim, Abigail A. Wise, Patrick J. O'Brien, Ioan Andricioaei, and Hashim M. Al-Hashimi, "Transient Hoogsteen Base Pairs in Canonical Duplex DNA," *Nature* 470 (2011): 498–502.

53. Patricia M. Kane, Carl T. Yamashiro, David F. Wolczyk, Norma Neff, Mark Goerl, and Tom H. Stevens, "Protein Splicing Converts the Yeast *TFP1* Gene Product to the 69-kD Subunit of the Vacuolar H$^+$- Adenosine Triphosphatase," *Science* 250 (1990): 651–657; Henry Paulus, "Protein Splicing and Related Forms of Autoprocessing," *Annual Review of Biochemistry* 69 (2000): 447–496.

22. The Rise of Developmental Biology

1. Yoshiki Hotta, and Seymour Benzer, "Mapping of Behaviour in *Drosophila* Mosaics," *Nature* 240 (1972): 527–535.

2. Sydney Brenner, "Nematode Research," *Trends in Biochemical Sciences* 9 (1984): 172; Soraya de Chadarevian, "Of Worms and Programmes: *Caenorhabditis elegans* and the Study of Development," *Studies in History and Philosophy of Science Part C: Studies in History and Philosophy of Biological and Biomedical Sciences* 29 (1998): 81–105; Rachel A. Ankeny, "The Natural History of *Caenorhabditis elegans* Research," *Nature Reviews Genetics* 2 (2001): 474–479.

3. François Jacob, *Of Flies, Mice, and Men* (Cambridge, MA: Harvard University Press, 1998).

4. David Jonah Grunwald and Judith S. Eisen, "Headwaters of the Zebrafish— Emergence of a New Model Vertebrate," *Nature Reviews Genetics* 3 (2002): 717–724; Robert Meunier, "Stages in the Development of a Model Organism as

a Platform for Mechanistic Models in Developmental Biology: Zebrafish, 1970–2000," *Studies in History and Philosophy of Science Part C: Studies in History and Philosophy of Biological and Biomedical Sciences* 43 (2012): 522–531.

5. Mark Ptashne, *A Genetic Switch: Gene Control and Phage Lambda* (Cambridge, MA: Cell Press & Blackwell Scientific, 1986).

6. George Streisinger, Charline Walker, Nancy Dower, Donna Knauber, and Fred Singer, "Production of Clones of Homozygous Diploid Zebra Fish *(Brachydanio rerio),*" *Nature* 291 (1981): 293–296.

7. Denis Thieffry, and Richard M. Burian, "Interview of Jean Brachet by Jan Sapp, Arco Felice, Italy, December 10, 1980," *History and Philosophy of the Life Sciences* 19 (1997): 113–140.

8. Stephen Jay Gould, *Ontogeny and Phylogeny* (Cambridge, MA: Belknap Press, 1977.)

9. Dorothea Bennett, "The T-Locus of the Mouse," *Cell* 6 (1975): 441–454.

10. Michel Morange, "François Jacob's Lab in the Seventies: The *T*-Complex, and the Mouse Developmental Genetic Program," *History and Philosophy of the Life Sciences* 22 (2000): 397–411.

11. Georges Barski, Serge Sorieul, and Francine Cornefert, "Production dans des cultures in vitro de deux souches cellulaires en association, de cellules de caractère "hybride," *Comptes rendus hebdomadaires des séances de l'Académie des sciences* 251 (1960): 1825–1827.

12. Boris Ephrussi, and Serge Sorieul, "Nouvelles observations sur l'hybridation in vitro de cellules de souris," *Comptes rendus hebdomadaires des séances de l'Académie des sciences* 254 (1962): 181–182; Henry Harris and J. F. Watkins, "Hybrid Cells Derived from Mouse and Man: Artificial Heterokaryons of Mammalian Cells from Different Species," *Nature* 205 (1965): 640–646.

13. Richard L. Davidson, Boris Ephrussi, and Kontaro Yamamoto, "Regulation of Pigment Synthesis in Mammalian Cells, as Studied by Somatic Hybridization," *Proceedings of the National Academy of Sciences of the United States of America* 56 (1966): 1437–1440.

14. Robert J. Klebe, Tchaw-ren Chen, and Frank H. Ruddle, "Mapping of a Human Genetic Regulator Element by Somatic Cell Genetic Analysis," *Proceedings of the National Academy of Sciences of the United States of America* 66 (1970): 1920–1927; Mary C. Weiss and Michèle Chaplain, "Expression of Differentiated Functions in Hepatoma Cell Hybrids: Reappearance of Tyrosine Aminotransferase Inducibility after the Loss of Chromosomes," *Proceedings of the National Academy of Sciences of the United States of America* 68 (1971): 3026–3030.

15. S. Gordon, ed., *The Legacy of Cell Fusion* (Oxford: Oxford University Press, 1994).

16. Eric H. Davidson, *Gene Activity in Early Development* (New York: Academic Press, 1968).

17. Michel Morange, "Molecular Hybridization: A Problematic Tool for the Study of Differentiation and Development (1960–1980)," *Journal of Bioscience* 39 (2014): 29–32.

18. R. J. Britten, and D. E. Kohne, "Repeated Sequences in DNA," *Science* 161 (1968): 529–540; Edna Suarez-Diaz, "Satellite-DNA: A Case-Study for the Evolution of Experimental Techniques," *Studies in History and Philosophy of Science Part C: Studies in History and Philosophy of Biological and Biomedical Sciences* 32 (2001): 31–57.

19. Roy J. Britten and Eric H. Davidson, "Gene Regulation for Higher Cells: A Theory," *Science* 165 (1969): 349–357.

20. Roy J. Britten and Eric H. Davidson, "Repetitive and Non-Repetitive DNA Sequences and a Speculation on the Origins of Evolutionary Novelty," *Quarterly Review of Biology* 46 (1971): 111–133.

21. Donald D. Brown and Igor B. Dawid, "Specific Gene Amplification in Oocytes," *Science* 160 (1968): 272–280; Allan C. Spradling and Anthony P. Mahowald, "Amplification of Genes for Chorion Proteins during Oogenesis in *Drosophila melanogaster*," *Proceedings of the National Academy of Sciences of the United States of America* 77 (1980): 1096–1100.

22. Michel Morange, "The Transformation of Molecular Biology on Contact with Higher Organisms, 1960–1980: From a Molecular Description to a Molecular Explanation," *History and Philosophy of the Life Sciences* 19 (1997): 369–393.

23. François Jacob and Jacques Monod, "Gènes de structure et gènes de régulation dans la biosynthèse des protéines," *Comptes rendus hebdomadaires des séances de l'Académie des sciences* 249 (1959): 1282–1284. The origins of Evo-Devo are seen very differently by embryologists and by more generally "traditional" biologists: Manfred D. Laubichler and Jane Maienschein, eds., *From Embryology to Evo-Devo: A History of Developmental Evolution* (Cambridge, MA: MIT Press, 2007).

24. Richard Goldschmidt, *The Material Basis of Evolution* (New Haven, CT: Yale University Press, 1940; reprinted with a preface by S. J. Gould, 1982).

25. François Jacob, and Jacques Monod, "Sur le mode d'action des gènes et leur régulation," *Pontificiae Academiae Scientiarum Scripta Varia* 22 (1962): 85–95.

26. Allan C. Wilson, Steven S. Carlson, and Thomas J. White, "Biochemical Evolution," *Annual Review of Biochemistry* 46 (1977): 573–639.

27. Mary-Claire King and Allan C. Wilson, "Evolution at Two Levels in Humans and Chimpanzees," *Science* 188 (1975): 107–116.

28. Francis H. C. Crick, "On Protein Synthesis," *Symposia of the Society for Experimental Biology* 12 (1958): 138–163.

29. Allan C. Wilson, Linda R. Maxson, and Vincent M. Sarich, "Two Types of Molecular Evolution: Evidence from Studies of Interspecific Hybridization," *Proceedings of the National Academy of Sciences of the United States of America* 71 (1974): 2843–2847; Allan C. Wilson, Vincent M. Sarich, and Linda R. Maxson, "The Importance of Gene Rearrangement in Evolution: Evidence

from Studies on Rates of Chromosomal, Protein, and Anatomical Evolution," *Proceedings of the National Academy of Sciences of the United States of America* 71 (1974): 3028–3030.

30. Antonio Garcia Bellido, P. Ripoll, and Gines Morata, "Developmental Compartmentalisation of the Wing Disk of *Drosophila*," *Nature New Biology* 245 (1973): 251–253.

31. Francis H. C. Crick and Peter A. Lawrence, "Compartments and Polyclones in Insect Development," *Science* 189 (1975): 340–347; Gines Morata and Peter A. Lawrence, "Homeotic Genes, Compartments and Cell Determination in *Drosophila*," *Nature* 265 (1977): 211–216.

32. Gould, *Ontogeny and Phylogeny,* op. cit.

33. William K. Baker, "A Genetic Framework for *Drosophila* Development," *Annual Review of Genetics* 12 (1978): 451–470.

34. Rudolf A. Raff and Thomas C. Kaufman, *Embryos, Genes, and Evolution: The Developmental-Genetic Basis of Evolutionary Change* (London: MacMillan, 1983).

35. Marcel Weber, "Redesigning the Fruitfly: The Molecularization of Drosophila," in *Model Systems, Cases, Exemplary Narratives,* ed. Angela N. H. Creager, E. Lunbeck, and M. Norton Wise, 23–45 (Durham, NC: Duke University Press, 2007). Doogab Yi, *The Recombinant University: Genetic Engineering and the Emergence of Stanford Biotechnology* (Chicago: University of Chicago Press, 2015).

36. Welcome Bender, Pierre Spierer, and David S. Hogness, "Chromosomal Walking and Jumping to Isolate DNA from the *Ace* and *rosy* Loci and the Bithorax Complex in *Drosophila melanogaster,*" *Journal of Molecular Biology* 168 (1983): 17–33; Matthew P. Scott, Amy J. Weiner, Tulle I. Hazelrigg, Barry A. Polisky, Vincenzo Pirrotta, et al., "The Molecular Organization of the *Antennapedia* Locus of *Drosophila*," *Cell* 35 (1983): 763–776.

37. Walter J. Gehring, *Master Control Genes in Development and Evolution: The Homeobox Story* (New Haven, CT: Yale University Press, 1998).

38. Stuart A. Kauffman, "Control Circuits for Determination and Transdetermination," *Science* 181 (1973): 310–318.

39. Christiane Nüsslein-Volhard and Eric Wieschaus, "Mutations Affecting Segment Number and Polarity in *Drosophila*," *Nature* 287 (1980): 795–801.

40. W. McGinnis, M. S. Levine, E. Hafen, A. Kuroiwa, and W. J. Gehring, "A Conserved DNA Sequence in Homeotic Genes of the *Drosophila* Antennapedia and Bithorax Complexes," *Nature* 308 (1984): 428–433.

41. Allen Laughon and Matthew P. Scott, "Sequence of a *Drosophila* Segmentation Gene: Protein Structure Homology with DNA-Binding Proteins," *Nature* 310 (1984): 25–31; John C. W. Sheperd, William McGinnis, Andrés E. Carrasco, Eddy M. De Robertis, and Walter J. Gehring, "Fly and Frog Homeo Domains Show Homologies with Yeast Mating Type Regulatory Proteins," *Nature* 310 (1984): 70–71; William McGinnis, Richard L. Garber, Johannes Wirz, Atsushi Kurowa, and Walter J. Gehring, "A Homologous Protein-Coding Sequence in

Drosophila Homeotic Genes and Its Conservation in Other Metazoans," *Cell* 37 (1984): 403–408.

42. Andrés E. Carrasco, William McGinnis, Walter J. Gehring, and Eddy M. De Robertis, "Cloning of an *X. laevis* Gene Expressed during Early Embryogenesis Coding for a Peptide Region Homologous to *Drosophila* Homeotic Genes," *Cell* 37 (1984): 409–414; William McGinnis, Charles P. Hart, Walter J. Gehring, and Frank H. Ruddle, "Molecular Cloning and Chromosome Mapping of a Mouse DNA Sequence Homologous to Homeotic Genes of *Drosophila,*" *Cell* 38 (1984): 675–680.

43. E. B. Lewis, "A Gene Complex Controlling Segmentation in *Drosophila,*" *Nature* 276 (1978): 565–570.

44. François Jacob, "Evolution and Tinkering," *Science* 196 (1977): 1161–1166; François Jacob, *The Possible and the Actual* (Seattle: University of Washington Press, 1982).

45. François Jacob, "L'irrésistible ascension des gènes *Hox,*" *Revue Médecine / Sciences* 10 (1994): 145–148.

46. Denis Duboule and Pascal Dollé, "The Structural and Functional Organization of the Murine HOX Gene Family Resembles That of *Drosophila* Homeotic Genes," *EMBO Journal* 8 (1989): 1497–1505.

47. Cynthia Kenyon and Bruce Wang, "A Cluster of *Antennapedia*-Class Homeobox Genes in a Nonsegmented Animal," *Science* 253 (1991): 516–517.

48. Michel Morange, "Pseudoalleles and Gene Complexes," *Perspectives in Biology and Medicine* 58 (2016): 196–204.

49. Wouter de Laat and Denis Duboule, "Topology of Mammalian Developmental Enhancers and Their Regulatory Landscapes," *Nature* 502 (2013): 499–506.

50. Spyros Artavanis-Tsakonas, Kenji Matsuno, and Mark E. Fortini, "Notch Signaling," *Science* 268 (1995): 225–232.

51. Gregory R. Dressler and Peter Gruss, "Do Multigene Families Regulate Vertebrate Development?," *Trends in Genetics* 4 (1988): 214–219; Michael Kessel and Peter Gruss, "Murine Developmental Control Genes," *Science* 249 (1990): 374–379.

52. Robert L. Davis, Harold Weintraub, and Andrew B. Lassar, "Expression of a Single Transfected cDNA Converts Fibroblasts to Myoblasts," *Cell* 51 (1987): 987–1000.

53. Michael A. Rudnicki, Patrick N. J. Schnegelsberg, Ronald H. Stead, Thomas Braun, Hans-Henning Arnold, and Rudolf Jaenisch, "*MyoD* or *Myf5* Is Required for the Formation of Skeletal Muscle," *Cell* 75 (1993): 1351–1359.

54. Rebecca Quiring, Uwe Walldorf, Urs Kloter, and Walter J. Gehring, "Homology of the *eyeless* Gene of *Drosophila* to the *Small eye* Gene in Mice and *Aniridia* in Humans," *Science* 265 (1994): 785–789.

55. Georg Haider, Patrick Callaerts, and Walter J. Gehring, "Induction of Ectopic Eyes by Targeted Expression of the *eyeless* Gene in *Drosophila,*" *Science* 267 (1995): 1788–1792.

56. Hans Sommer, José-Pio Beltran, Peter Huijser, Heike Pape, Wolf-Ekkehard Lönnig, et al., "*Deficiens,* a Homeotic Gene Involved in the Control of Flower Morphogenesis *Antirrhinum majus:* The Protein Shows Homology to Transcription Factors," *EMBO Journal* 9 (1990): 605–613. Rosemary Carpenter and Enrico S. Coen, "Floral Homeotic Mutations Produced by Transposon-Mutagenesis in *Antirrhinum majus,*" *Genes and Development* 4 (1990): 1483–1493. Martin F. Yanofsky, Hong Ma, John L. Bowman, Gary N. Drews, Kenneth A. Feldmann, and Elliot M. Meyerowitz, "The Protein Encoded by the *Arabidopsis* Homeotic Gene *Agamous* Resembles Transcription Factors," *Nature* 346 (1990): 35–39.

57. Elliot M. Meyerowitz, John L Bowman, Laura L. Brockman, Gary N. Drews, Thomas Jack, et al., "A Genetic and Molecular Model for Flower Development in *Arabidopsis thaliana,*" *Development* 113, Suppl. 1 (1991): 157–167.

58. Elliot M. Meyerowitz and Robert E. Pruitt, "*Arabidopsis thaliana* and Plant Molecular Genetics," *Science* 229 (1985): 1214–1218; Chris Somerville and Maarten Koorneef, "A Fortunate Choice: The History of *Arabidopsis* as a Model Plant," *Nature Reviews Genetics* 3 (2002): 883–889; Sabina Leonelli, "Growing Weed, Producing Knowledge: An Epistemic History of *Arabidopsis thaliana,*" *History and Philosophy of the Life Sciences* 29 (2007): 193–224.

59. Lewis Wolpert, "Positional Information and Pattern Formation," *Current Topics in Developmental Biology* 6 (1971): 183–224.

60. Wolfgang Driever and Christiane Nüsslein-Volhard, "Gradient of *bicoid* Protein in *Drosophila* Embryos," *Cell* 54 (1988): 83–93; Wolfgang Driever and Christiane Nüsslein-Volhard, "The *bicoid* Protein Determines Position in the Drosophila Embryo in a Concentration-Dependent Manner," *Cell* 54 (1988): 95–104.

61. Ken W. Y. Cho, Elaine A. Morita, Christopher V. E. Wright, and Eddy M. De Robertis, "Overexpression of a Homeodomain Protein Confers Axis-Forming Activity to Uncommitted *Xenopus* Embryonic Cells," *Cell* 65 (1991): 55–64; Ken W. Y. Cho, Bruce Blumberg, Herbert Steinbelsser, and Eddy M. De Robertis, "Molecular Nature of Spemann's Organizer: The Role of the Homeobox Gene *goosecoid,*" *Cell* 67 (1991): 1111–1120; Bruce Blumberg, Christopher V. E. Wright, Eddy M. De Robertis, and Ken W. Y. Cho, "Organism-Specific Homeobox Genes in *Xenopus laevis* Embryos," *Science* 253 (1991): 194–196.

62. Martin Blum, Stephen J. Gaunt, Ken W. Y. Cho, Herbert Steinbelsser, Bruce Blumberg, et al., "Gastrulation in the Mouse: The Role of the Homeobox Gene *goosecoid,*" *Cell* 69 (1992): 1097–1106.

63. Walter J. Gehring, "The Homeobox in Perspective," *Trends in Biochemical Sciences* 17 (1992): 277–280.

23. Molecular Biology and Evolution

1. Salvador E. Luria and Max Delbrück, "Mutations of Bacteria from Virus Sensitivity to Virus Resistance," *Genetics* 28 (1943): 491–511.

2. Ernst Mayr, "Cause and Effect in Biology," *Science* 134 (1961): 1501–1506. For a complementary, but slightly different point of view, see George G. Simpson, "Biology and the Nature of Science: Unification of the Sciences Can Be Most Meaningfully Sought through Study of the Phenomena of Life," *Science* 139 (1963): 81–88.

3. Alexandre E. Peluffo, "The 'Genetic Program': Behind the Genesis of an Influential Metaphor," *Genetics* 200 (2015): 685–696.

4. Emil Zuckerkandl and Linus Pauling, "Evolutionary Divergence and Convergence in Proteins," in *Evolving Genes and Proteins,* ed. V. Bryson and H. J. Vogel, 153–181 (New York: Academic Press, 1965).

5. Francis H. C. Crick, "On Protein Synthesis," *Symposia of the Society for Experimental Biology* 12 (1958): 138–163.

6. Michael R. Dietrich, "Paradox and Persuasion: Negotiating the Place of Molecular Evolution Within Evolutionary Biology," *Journal of the History of Biology* 31 (1998): 85–111.

7. Jack L. King and Thomas H. Jukes, "Non Darwinian Evolution: Random Fixation of Selectively Neutral Mutations," *Science* 164 (1969): 788–798.

8. Motoo Kimura, "Evolutionary Rate at the Molecular Level," *Nature* 217 (1968): 624–626; Motoo Kimura, *The Neutral Theory of Molecular Evolution* (Cambridge: Cambridge University Press, 1983); Michael R. Dietrich, "The Origins of the Neutral Theory of Molecular Evolution," *Journal of the History of Biology* 27 (1994): 21–59; Edna Suarez and Ana Barahona, "The Experimental Roots of the Neutral Theory of Molecular Evolution," *History and Philosophy of the Life Sciences* 18 (1996): 55–81.

9. Willi Hennig, *Grundzüge einer Theorie der phylogenetischen Systematik* (Berlin: Deutscher Zentralverlag, 1950); republished as *Phylogenetic Systematics* (Urbana: University of Illinois Press, 1966).

10. Jan Sapp, "The Iconoclastic Research Program of Carl Woese," in *Rebels, Mavericks, and Heretics in Biology,* ed. Oren Harman and Michael R. Dietrich, 302–320 (New Haven, CT: Yale University Press, 2008).

11. W. Ford Doolittle, "Phylogenetic Classification and the Universal Tree," *Science* 284 (1999): 2124–2128; Yan Boucher and Eric Bapteste, "Revisiting the Concept of Lineage in Prokaryotes: A Phylogenetic Perspective," *Bioessays* 31 (2009): 526–536.

12. Lynn Margulis subsequently argued for a major role of symbiosis at other steps in the evolutionary process, but with limited success: Lynn Margulis, *Symbiosis in Cell Evolution* (San Francisco: W. H. Freeman, 1981).

13. Stephen J. Gould and Niles Eldredge, "Punctuated Equilibria: The Tempo and Mode of Evolution Reconsidered," *Paleobiology* 3 (1977): 115–151.

14. Stephen J. Gould and Richard Lewontin, "The Spandrels of San Marco and the Panglossian Paradigm: A Critique of the Adaptationist Programme," *Proceedings of the Royal Society of London: Series B, Biological Sciences* 205 (1979): 581–598.

15. Pere Alberch, "Ontogenesis and Morphological Diversification," *American Zoologist* 20 (1980): 653–667; Diego Rasskin-Gutman and Miquel de Renzi, eds., *Pere Alberch: The Creative Trajectory of an Evo-Devo Biologist* (Valencia, Spain: Universitat de Valencia, 2009).

16. Richard Goldschmidt, *The Material Basis of Evolution* (New Haven, CT: Yale University Press, 1940; reprinted with a preface by S. J. Gould, 1982). The importance of heterochronic mutations was discussed by Gould in Stephen J. Gould, *Ontogeny and Phylogeny* (Cambridge, MA: Belknap Press, 1977).

17. Luis W. Alvarez, Walter Alvarez, Frank Asano, and Helen V. Michel, "Extraterrestrial Cause for the Cretaceous-Tertiary Extinction," *Science* 208 (1980): 1095–1108.

18. F. Jacob, "Evolution and Tinkering," *Science* 196 (1977): 1161–1166.

19. Joram Piatigorsky and Graeme Wislow, "Enzyme / Crystallins: Gene Sharing as an Evolutionary Strategy," *Cell* 57 (1989): 197–199; Joram Piatigorsky, "Lens Crystallins: Innovation Associated with Changes in Gene Regulation," *Journal of Biological Chemistry* 267 (1992): 4277–4280.

20. Calvin B. Bridges, "The BAR Gene: A Duplication," *Science* 83 (1936): 210–211.

21. Susumu Ohno, *Evolution by Gene Duplication* (Berlin: Springer-Verlag, 1970).

22. Howard M. Temin, "The Protovirus Hypothesis: Speculations on the Significance of the RNA-directed DNA Synthesis for Normal Development and for Carcinogenesis," *Journal of the National Cancer Institute* 46 (1971): III–VII.

23. The hypothesis of a role of reverse transcription in evolution persisted during the 1980s and the beginning of the 1990s: Jürgen Brosius, "Retroposons— Seeds of Evolution," *Science* 251 (1991): 753.

24. John Cairns, J. Overbaugh, and S. Miller, "The Origin of Mutants," *Nature* 335 (1988): 141–145.

25. James A. Shapiro, "Adaptive Mutation: Who's Really in the Garden?" *Science* 268 (1995): 373–374; on the different challenges faced in the last fifty years by the Evolutionary Synthesis, in particular due to molecular data, see the excellent book by Francesca Merlin, *Mutations et aléas: Le hasard dans la théorie de l'évolution* (Paris: Hermann, 2013).

26. Nina Fedoroff, "How Jumping Genes Were Discovered," *Nature Structural Biology* 8 (2001): 300–301.

27. Stanley N. Cohen, "Transposable Genetic Elements and Plasmid Evolution," *Nature* 263 (1976): 731–738.

28. Nathaniel C. Comfort, *The Tangled Field: Barbara McClintock's Search for the Patterns of Genetic Control* (Cambridge MA: Harvard University Press, 2001).

29. Elie L. Wollman and François Jacob, *La sexualité des bactéries* (Paris: Masson, 1959); *Sexuality and the Genetics of Bacteria* (New York: Academic Press, 1961).

30. John B. Gurdon, "Adult Frogs Derived from the Nuclei of Single Somatic Cells," *Developmental Biology* 4 (1962): 256–273.

31. Alan M. Weiner, Prescott L. Deininger and Argiris Efstratiadis, "Nonviral Retroposons: Genes, Pseudogenes, and Transposable Elements Generated by

the Reverse Flow of Genetic Information," *Annual Review of Biochemistry* 55 (1986): 631–661.

32. Charles A. Thomas Jr., "The Genetic Organization of Chromosomes," *Annual Review of Genetics* 5 (1971): 237–256.

33. R. J. Britten and D. E. Kohne, "Repeated Sequences in DNA," *Science* 161 (1968): 529–540.

34. Roy J. Britten and Eric H. Davidson, "Gene Regulation for Higher Cells: A Theory," *Science* 165 (1969): 349–357; Roy J. Britten and Eric H. Davidson, "Repetitive and Non-Repetitive DNA Sequences and a Speculation on the Origins of Evolutionary Novelty," *Quarterly Review of Biology* 46 (1971): 111–133.

35. Susumu Ohno, "So Much 'Junk' DNA in Our Genome," *Brookhaven Symposium in Biology* 23 (1972): 366–370.

36. W. Ford Doolittle and Carmen Sapienza, "Selfish Genes, the Phenotype Paradigm and Genome Evolution," *Nature* 284 (1980): 601–603. L. E. Orgel and F. H. C. Crick, "Selfish DNA: The Ultimate Parasite," *Nature* 284 (1980): 604–607.

37. Richard Dawkins, *The Selfish Gene* (Oxford: Oxford University Press, 1976).

38. Michael Lynch, *The Origins of Genome Architecture* (Sunderland, MA: Sinauer, 2007).

39. ENCODE Project Consortium, "An Integrated Encyclopedia of Genetic Elements in the Human Genome," *Nature* 489 (2012): 57–74.

40. Renaud de Rosa, Jennifer K. Grenier, Tatiana Andreeva, Charles E. Cook, André Adoutte, et al., "*Hox* Genes in Brachiopods and Priapulids and Protostome Evolution," *Nature* 399 (1999): 772–776.

41. Neil Shubin, Cliff Tabin, and Sean Carroll, "Deep Homology and the Origins of Evolutionary Novelty," *Nature* 457 (2009): 818–823.

42. Sean B. Carroll, "Evo-Devo and an Expanding Evolutionary Synthesis: A Genetic Theory of Morphological Evolution," *Cell* 134 (2008): 25–36.

43. Daniel R. Matute, I. A. Butler, and Jerry A. Coyne, "Little Effect of the tan Locus on Pigmentation in Female Hybrids between *Drosophila santomea* and *D. melanogaster*," *Cell* 139 (2009): 1180–1188.

44. Chiou-Hwa Yuh, Hamid Bolouri, and Eric H. Davidson, "*Cis*-Regulatory Logic in the *endo16* Gene: Switching from a Specification to a Differentiation Mode of Control," *Development* 128 (2001): 617–629.

45. Eric H. Davidson, *The Regulatory Genome: Gene Regulatory Networks in Development and Evolution* (Burlington, MA: Academic Press, 2006).

46. Isabelle Peter and Eric H. Davidson, *Genomic Control Process: Development and Evolution* (New York: Academic Press, 2015).

47. Eric H. Davidson and Douglas H. Erwin, "Gene Regulatory Networks and the Evolution of Animal Body Plans," *Science* 311 (2006): 796–800; Douglas H. Erwin and Eric H. Davidson, "The Evolution of Hierarchical Gene Regulatory Networks," *Nature Reviews Genetics* 10 (2009): 141–148.

48. Douglas H. Erwin and James W. Valentine, "'Hopeful Monsters,' Transposons, and Metazoan Radiation," *Proceedings of the National Academy of Sciences of the United States of America* 81 (1984): 5482–5483.

49. Nicolas Di-Poï, Juan I. Montoya-Burgos, Hilary Miller, Olivier Pourquié, Michel C. Milinkovitch, and Denis Duboule, "Changes in *Hox* Genes' Structure and Function during the Evolution of the Squamate Body Plan," *Nature* 464 (2010): 95–103.

50. Arhat Abzhanov, Winston P. Kuo, Christine Hartmann, B. Rosemary Grant, Peter R. Grant, and Clifford J. Tabin, "The Calmodulin Pathway and Evolution of Elongated Beak Morphology in Darwin's Finches," *Nature* 442 (2006): 563–567.

51. Sangeet Lamichhaney, Fan Han, Jonas Berglund, Chao Wang, Markus Sällman Alme, et al., "A Beak Size Locus in Darwin's Finches Facilitated Character Displacement during Drought," *Science* 352 (2016): 470–474.

52. Mary-Claire King and Allan C. Wilson, "Evolution at Two Levels in Humans and Chimpanzees," *Science* 188 (1975): 107–116.

53. David Cyranoski, "Almost Human," *Nature* 418 (2002): 910–912.

54. Wolfgang Enard, Molly Przeworski, Simon E. Fisher, Cecilia S. L. Lai, Victor Wiebe, et al., "Molecular Evolution of *FOXP2*, a Gene Involved in Speech and Language," *Nature* 418 (2002): 869–872.

55. Mehmet Somel, Xiling Liu, Lin Tang, Zheng Yan, Halyang Hu, et al., "MicroRNA-Driven Developmental Remodeling in the Brain Distinguishes Humans from Other Primates," *PLoS Biology* 9 (2011): e1001214.

56. Richard E. Green, Johannes Krause, Adrian W. Briggs, Tonislav Maricic, Udo Stenzel, et al., "A Draft Sequence of the Neandertal Genome," *Science* 328 (2010): 710–722.

57. Marcel Margulies, Michael Egholm, William E. Altman, Said Attiya, Joel S. Bader, et al., "Genome Sequencing in Microfabricated High-Density Picolitre Reactors," *Nature* 437 (2005): 376–380.

58. Svante Pääbo, "Molecular Cloning of Ancient Egyptian Mummy DNA," *Nature* 314 (1985): 644–645; Svante Pääbo, "Ancient DNA: Extraction, Characterization, Molecular Cloning, and Enzymatic Amplification," *Proceedings of the National Academy of Sciences of the United States of America* 86 (1989): 1939–1943; Mathias Röss, Pawel Jaruga, Tomasz H. Zastawny, Miral Dizdaroglu, and Svante Pääbo, "DNA Damage and DNA Sequence Retrieval from Ancient Tissues," *Nucleic Acids Research* 24 (1996): 1304–1307; M. Stiller, R. E. Green, M. Ronan, J. F. Simons, L. Du, et al., "Patterns of Nucleotide Misincorporations during Enzymatic Amplification and Direct Large-Scale Sequencing of Ancient DNA," *Proceedings of the National Academy of Sciences of the United States of America* 103 (2006): 13578–13584.

59. Matthias Meyer, Martin Kircher, Marie-Theres Gansauge, Heng Li, Fernando Racimo, et al., "A High-Coverage Genome Sequence from an Archaic Denisovan Individual," *Science* 338 (2012): 222–226.

60. Emilia Huerta-Sanchez, Xin Jin, Asan, Zhuoma Bianba, Benjamin M. Peter, et al., "Altitude Adaptation in Tibetans Caused by Introgression of Denisovan-Like DNA," *Nature* 512 (2014): 194–197.

61. Kara C. Hoover, Omer Gokcumen, Zoya Qureshy, Elise Bruguera, Aulaphan Savangsuksa, et al., "Global Survey of Variation in a Human Olfactory Receptor Gene Reveals Signatures of Non-Neutral Evolution," *Chemical Senses* 40 (2015): 481–488.

62. George Gaylord Simpson, *Tempo and Mode in Evolution* (New York: Columbia University Press, 1944).

63. Rebecca L. Cann, Mark Stoneking, and Allan C. Wilson, "Mitochondrial DNA and Human Evolution," *Nature* 325 (1987): 31–36.

64. Luigi Luca Cavalli-Sforza, Alberto Piazza, Paolo Menozzi, and Joanna Mountain, "Reconstruction of Human Evolution: Bringing Together Genetic, Archaeological, and Linguistic Data," *Proceedings of the National Academy of Sciences of the United States of America* 85 (1988): 6002–6006.

65. Luca Cavalli-Sforza, "The Human Genome Diversity Project: Past, Present and Future," *Nature Reviews Genetics* 6 (2005): 333–340; Jenny Reardon, *Race to the Finish: Identity and Governance in the Age of Genomics* (Princeton, NJ: Princeton University Press, 2005).

66. Stephen Leslie, Bruce Winney, Garrett Hellenthal, Dan Davison, Abdelhamid Boumertit, et al., "The Fine-Scale Genetic Structure of the British Population," *Nature* 519 (2015): 309–314.

67. Renyi Liu and Howard Ochman, "Stepwise Formation of the Bacterial Flagellar System," *Proceedings of the National Academy of Sciences of the United States of America* 104 (2007): 7116–7121.

68. Michael Behe, *The Edge of Evolution: The Search for the Limits of Darwinism* (New York: Free Press, 2007).

69. Richard E. Lenski and Michael Travisano, "Dynamics of Adaptation and Diversification: A 10,000-Generation Experiment with Bacterial Populations," *Proceedings of the National Academy of Sciences of the United States of America* 91 (1994): 6808–6814; Olivier Tenaillon, Jeffrey E. Barrick, Noah Ribeck, Daniel E. Deatherage, Jeffrey L. Blanchard, et al., "Tempo and Mode of Genome Evolution in a 50,000 Generation Experiment," *Nature* 536 (2016): 165–170.

70. Antony M. Dean and Joseph W. Thornton, "Mechanistic Approaches to the Study of Evolution: The Functional Synthesis," *Nature Reviews Genetics* 8 (2007): 675–688.

71. Michael J. Harms and Joseph W. Thornton, "Historical Contingency and Its Biophysical Basis in Glucocorticoid Receptor Evolution," *Nature* 512 (2014): 203–207.

72. Benjamin Prud'homme, Caroline Minervino, Mélanie Hocine, Jessica D. Cande, Hélène Aicha Aouane, et al., "Body Plan Innovation in Treehoppers through the Evolution of an Extra Wing-Like Appendage," *Nature* 473 (2011): 83–86.

73. Michael J. Kerner, Dean J. Naylor, Yasushi Ishihama, Tobias Maier, Hung-Chun Chang, et al., "Genome-Wide Analysis of Chaperonin-Dependent Protein Folding in *Escherichia coli*," *Cell* 122 (2005): 209–220.

24. Gene Therapy

1. Edward L. Tatum, "A Case History in Biological Research," *Science* 129 (1959): 1711–1715.
2. Oswald T. Avery, Colin MacLeod, and Maclyn McCarty, "Studies on the Chemical Nature of the Substance Inducing Transformation of Pneumococcal Types," *Journal of Experimental Medicine* 79 (1944): 137–158.
3. Elizabeth Hunter Sztbalska and Waclaw Szybalski, "Genetics of Human Cell Lines, IV. DNA-Mediated Heritable Transformation of a Biochemical Trait," *Proceedings of the National Academy of Sciences of the United States of America* 48 (1962): 2026–2034.
4. R. D. Hotchkiss, "Portents for a Genetic Engineering," *Journal of Heredity* 56 (1965): 197–202.
5. Bernard Davis, "Prospects for Genetic Engineering in Man," *Science* 170 (1980): 1279–1283.
6. Jacques Monod, *Chance and Necessity* (London: Collins, 1972).
7. Richard A. Morgan and W. French Anderson, "Human Gene Therapy," *Annual Review of Biochemistry* 62 (1993): 191–217; Ronald G. Crystal, "Transfer of Genes to Humans: Early Lessons and Obstacles to Success," *Science* 270 (1995): 404–410.
8. Joseph Zabner, Larry A. Couture, Richard J. Gregory, Scott M. Graham, Alan E. Smith, and Michael J. Welsh, "Adenovirus-Mediated Gene Transfer Transiently Corrects the Chloride Transport Defect in Nasal Epithelia of Patients with Cystic Fibrosis," *Cell* 75 (1993): 207–218.
9. Joshua R. Sanes, John L. R. Rubenstein, and Jean-François Nicolas, "Use of a Recombinant Retrovirus to Study Post-Implantation Cell Lineage in Mouse Embryos," *EMBO Journal* 5 (1986): 3133–3142.
10. J. G. Izant and H. Weintraub, "Inhibition of Thymidine Kinase Gene Expression by Anti-Sense RNA: A Molecular Approach to Genetic Analysis," *Cell* 36 (1984): 1007–1015; John L. R. Rubenstein, Jean-François Nicolas, and François Jacob, "L'ARN non sens (nsARN): Un outil pour inactiver spécifiquement l'expression d'un gène donné in vivo," *Comptes rendus hebdomadaires des séances de l'Académie des sciences, série III* 299 (1984): 271–274.
11. Barbara J. Culliton, "Politics and Genes," *Nature Medicine* 1 (1995): 181.
12. C. A. Stein and Y. C. Cheng, "Antisense Oligonucleotides as Therapeutic Agents—Is the Bullet Really Magical?" *Science* 261 (1993): 1004–1012; K. W. Wagner, "Gene Inhibition Using Antisense Oligodeoxynucleotides," *Nature* 372 (1994): 333–335.

13. Marina Cavazzana-Calvo, Salima Hacein-Bey, Geneviève de Saint Basile, Fabian Gross, Eric Yvon, et al., "Gene Therapy of Human Severe Combined Immunodeficiency (SCID)-X1 Disease," *Science* 288 (2000): 669–672.

14. Alessandro Aluti, Shlmon Slavin, Mermet Aker, Francesca Ficara, Sara Deola, et al., "Correction of ADA-SCID by Stem Cell Gene Therapy Combined with Nonmyeloablative Conditioning," *Science* 296 (2002): 2410–2413.

15. S. Hacein-Bey-Abina, C. Von Kalle, M. Schmidt, M. P. McCormack, N. Wulffraat, et al., "LMO2-Associated Clonal T Cell Proliferation in Two Patients after Gene Therapy for SCID-X1," *Science* 302 (2003): 415–419.

16. Alain Jacquier and Bernard Dujon, "An Intron-Encoded Protein Is Active in a Gene Conversion Process That Spreads an Intron into a Mitochondrial Gene," *Cell* 41 (1985): 383–394.

17. L. Li, L. P. Wu, and S. Chandrasegaran, "Functional Domains in *Fok* I Restriction Endonuclease," *Proceedings of the National Academy of Sciences of the United States of America* 89 (1992): 4275–4279.

18. Yang-Gyun Kim and Srinivasan Chandrasegaran, "Chimeric Restriction Endonuclease," *Proceedings of the National Academy of Sciences of the United States of America* 91 (1994): 883–667.

19. Nicolas P. Pavletich and Carl O. Pabo, "Crystal Structure of a Five-finger GLI-DNA Complex: New Perspectives on Zinc Fingers," *Science* 261 (1993): 1701–1707; John R. Desjarlais and Jeremy M. Bero, "Toward Rules Relating Zinc Finger Protein Sequences and DNA Binding Site Preferences," *Proceedings of the National Academy of Sciences of the United States of America* 89 (1992): 7345–7349; John R. Desjarlais and Jeremy M. Bero, "Use of a Zinc-Finger Consensus Sequence Framework and Specificity Rules to Design Specific DNA Binding Proteins," *Proceedings of the National Academy of Sciences of the United States of America* 90 (1993): 2256–2260; John R. Desjarlais and Jeremy M. Bero, "Length-Encoding Multiplex Binding Site Determination: Application to Zinc Finger Proteins," *Proceedings of the National Academy of Sciences of the United States of America* 91 (1994): 11099–11103; L. Falrall, John W. R. Schwabe, Lynda Chapman, John T. Finch, and Daniela Rhodes, "The Crystal Structure of a Two Zinc-Finger Peptide Reveals an Extension to the Rules for Zinc-Finger / DNA Recognition," *Nature* 366 (1993): 483–487.

20. Yang-Gyon Kim, Jooyeun Cha, and Srinivasan Chandrasegaran, "Hybrid Restriction Enzymes: Zinc Finger Fusions to *Fok* I Cleavage Domain," *Proceedings of the National Academy of Sciences of the United States of America* 93 (1996): 1156–1160.

21. Marina Bibikova, Dana Carroll, David J. Segal, Jonathan K. Trautman, Jeff Smith, et al., "Stimulation of Homologous Recombination Through Targeted Cleavage by Chimeric Nucleases," *Molecular and Cellular Biology* 21 (2001): 289–297; Marina Bibikova, Kelly Beumer, Jonathan K. Trautman, and Dana Carroll, "Enhancing Gene Targeting with Designed Zinc Finger Nucleases," *Science* 300 (2003): 764.

22. M. H. Porteus and D. Baltimore, "Chimeric Nucleases Stimulate Gene Targeting in Human Cells," *Science* 300 (2003): 763.

23. Mario R. Capecchi, "Altering the Genome by Homologous Recombination," *Science* 244 (1989): 1288–1292; Beverly H. Koller and Oliver Smithies, "Altering Genes in Animals by Gene Targeting," *Annual Review of Immunology* 10 (1992): 705–730.

24. H. Puchta, B. Dujon, and B. Hohn, "Homologous Recombination in Plant Cells Is Enhanced by in Vivo Induction of Double Strand Breaks into DNA by a Site-Specific Endonuclease," *Nucleic Acids Research* 21 (1993): 5034–5040; P. Rouet, F. Smith, and M. Jasin, "Expression of a Site-Specific Endonuclease Stimulates Homologous Recombination in Mammalian Cells," *Proceedings of the National Academy of Sciences of the United States of America* 91 (1994): 6064–6068; A. Choulika, A. Perrin, B. Dujon, and J.-F. Nicolas, "Induction of Homologous Recombination in Mammalian Chromosomes by Using the I-SceI System of *Saccharomyces cerevisiae*," *Molecular and Cellular Biology* 15 (1995): 1968–1973.

25. Sundar Durai, Mala Mani, Karthikeyan Kandavelou, Joy Wu, Matthew H. Porteus, and Srinivasan Chandrasegaran, "The Zinc Finger Nucleases: Custom-Designed Molecular Scissors for Genome Engineering of Plant and Mammalian Cells," *Nucleic Acids Research* 33 (2005): 5978–5990; Dana Carroll, "Progress and Prospects: Zinc-Finger Nucleases as Gene Therapy Agents," *Gene Therapy* 15 (2008): 1463–1468.

26. E. E. Perez, J. Wang, J. C. Miller, Y. Jouvenot, K. A. Kim, et al., "Establishment of HIV-1 Resistance in CD4$^+$ T Cells by Genome Editing Using Zinc-Finger Nucleases, *Nature Biotechnology* 26 (2008): 808–816.

27. M. Christian, T. Cermak, E. L. Doyle, C. Schmidt, F. Zhang, et al., "Targeting DNA Double-Strand Breaks with TAL Effector Nucleases," *Genetics* 186 (2010): 757–761; Jeffrey C. Miller, Siyuan Tan, Guijuan Qiao, Kyle A. Barlow, Jianbin Wang, et al., "A TALE Nuclease Architecture for Efficient Genome Editing, *Nature Biotechnoloy* 29 (2011): 143–148.

28. Fyodor D. Urnov, Edward J. Rebar, Michael C. Holmes, H. Steve Zhang, and Philip D. Gregory, "Genome Editing with Engineered Zinc Finger Nucleases," *Nature Reviews Genetics* 11 (2010): 636–646.

29. C. Pourcel, G. Salvignol, and G. Vergnaud, "CRISPR Elements in *Yersinia pestis* Acquire New Repeats by Preferential Uptake of Bacteriophage DNA, and Provide Additional Tools for Evolutionary Studies," *Microbiology* 151 (2005): 653–663; F. J. M. Mojica, C. Diez-Vollasenor, J. Garcia-Martoinez, and E. Soria, "Intervening Sequences of Regularly Spaced Prokaryotic Repeats Derive from Foreign Genetic Elements," *Journal of Molecular Evolution* 60 (2005): 174–182; A. Bolotin, B. Quinquis, A. Sorokin, and S. D. Ehrlich, "Clustered Regularly Interspaced Short Palindromic Repeats (CRISPRs) Have Spacers of Extrachromosomal Origin," *Microbiology* 151 (2005): 2551–2561; R. Barrangou, C. Frenaux, H. Deveau, M. Richards, P. Boyaval, et al., "CRISPR

Provides Acquired Resistance against Viruses in Prokaryotes," *Science* 315 (2007): 1709–1712.

30. A. Fire, S. Xu, M. K. Montgomery, S. A. Kostas, S. E. Driver, and C. C. Mello, "Potent and Specific Genetic Interference by Double-Stranded RNA in *Cernorhabditis elegans,*" *Nature* 391 (1998): 806–811.

31. E. Deltcheva, K. Chylinski, C. M. Sharma, K. Gonzales, Y. Chao, et al., "CRISPR RNA Maturation by Trans-Encoded Small RNA and Host Factor RNase III," *Nature* 471 (2011): 602–607.

32. M. Jinek, K. Chylinski, I. Fonfara, M. Hauer, J. A. Doudna, and E. Charpentier, "A Programmable Dual-RNA-Guided DNA Endonuclease in Adaptive Bacterial Immunity," *Science* 337 (2012): 816–821.

33. E. Pennisi, "The CRISPR Craze," *Science* 341 (2013): 833–836.

34. François Jacob and Jacques Monod, "Genetic Regulatory Mechanisms in the Synthesis of Proteins," *Journal of Molecular Biology* 3 (1961): 318–356.

25. The Central Place of RNA

1. Phillip A. Sharp, "The Centrality of RNA," *Cell* 136 (2009): 577–580; James Darnell, *RNA: Life's Indispensable Molecule* (Cold Spring Harbor, NY: Cold Spring Harbor Laboratory Press, 2011).

2. A. Rich and D. R. Davies, "A New Two Stranded Helical Structure: Polyadenylic Acid and Polyuridylic Acid," *Journal of the American Chemical Society* 78 (1956): 3548–3549; Gary Felsenfeld and Alexander Rich, "Studies on the Formation of Two- and Three-Stranded Polyribonucleotides," *Biochimica et Biophysica Acta* 26 (1957): 457–468; Alexander Rich, "Formation of Two- and Three-Stranded Helical Molecules by Polyinosinic Acid and Polyadenylic Acid," *Nature* 181 (1958): 521–525; Alexander Rich, "A Hybrid Helix Containing Both Deoxyribose and Ribose Polynucleotides and Its Relation to the Transfer of Information between the Nucleic Acids," *Proceedings of the National Academy of Sciences of the United States of America* 46 (1960): 1044–1053.

3. Benjamin D. Hall and S. Spiegelman, "Sequence Complementarity of T2-DNA and T2-specific RNA," *Proceedings of the National Academy of Sciences of the United States of America* 47 (1961): 137–146.

4. François Jacob and Jacques Monod, "Genetic Regulatory Mechanisms in the Synthesis of Proteins," *Journal of Molecular Biology* 3 (1961): 318–356.

5. Roy J. Britten and Eric H. Davidson, "Gene Regulation for Higher Cells: A Theory," *Science* 165 (1969): 349–357.

6. See, for instance, Jeffrey H. Miller, and Henry M. Sobell, "A Molecular Model for Gene Expression," *Proceedings of the National Academy of Sciences of the United States of America* 55 (1966): 1201–1205.

7. Carl R. Woese, *The Genetic Code: The Molecular Basis for Genetic Expression* (New York: Harper and Row, 1967); F. H. C. Crick, "The Origin of the Genetic

Code," *Journal of Molecular Biology* 38 (1968): 367–37; L. E. Orgel, "Evolution of the Genetic Apparatus," *Journal of Molecular Biology* 38 (1968): 381–393.

8. Hotchkiss presented this idea in the discussion of a talk by Fraenkel-Conrat; for quotes and his later view, see Rollin D. Hotchkiss, "DNA in the Decade before the Double Helix," *Annals of the New York Academy of Sciences* 758 (1995): 55–73.

9. Bruce M. Paterson, Bryan E. Roberts, and Edward L. Kuff, "Structural Gene Identification and Mapping by DNA-mRNA Hybrid-Arrested Cell-Free Translation," *Proceedings of the National Academy of Sciences of the United States of America* 74 (1977): 4370–4374.

10. Mary L. Stephenson and Paul C. Zamecnik, "Inhibition of Rous Sarcoma Viral RNA Translation by a Specific Oligodeoxyribonucleotide," *Proceedings of the National Academy of Sciences of the United States of America* 75 (1978): 285–288; Paul C. Zamecnik and Mary L. Stephenson, "Inhibition of Rous Sarcoma Virus Replication and Cell Transformation by a Specific Oligodeoxy-nucleotide," *Proceedings of the National Academy of Sciences of the United States of America* 75 (1978): 280–284.

11. Takeshi Mizuno, Mei-Yin Chou, and Masayori Inouye, "A Unique Mechanism Regulating Gene Expression: Translational Inhibition by a Complementary RNA Transcript (micRNA)," *Proceedings of the National Academy of Sciences of the United States of America* 81 (1984): 1966–1970; for a general description of these early works, see Michel Morange, "Regulation of Gene Expression by Non-Coding RNAs: The Early Steps," *Journal of Bioscience* 33 (2008): 327–331.

12. Jonathan G. Izant and Harold Weintraub, "Inhibition of Thymidine Kinase Gene Expression by Anti-Sense RNA: A Molecular Approach to Genetic Analysis," *Cell* 36 (1984): 1007–1015.

13. John L. R. Rubenstein, Jean-François Nicolas, and François Jacob, "L'ARN non sens (nsARN): un outil pour inactiver spécifiquement l'expression d'un gène donné in vivo," *Comptes rendus de l'Académie des Sciences, série III* 299 (1984): 271–274.

14. Rosalind C. Lee, Rhonda L. Feinbaum, and Victor Ambros, "The *C. elegans* Heterochronic *Gene lin-4* Encodes Small RNAs With Antisense Complementarity to *lin-14*," *Cell* 75 (1993): 843–854.

15. See, for instance, Pamela J. Green, Ophry Pines, and Masayori Inouye, "The Role of Antisense RNA in Gene Regulation," *Annual Review of Biochemistry* 55 (1986): 569–597; J.-J. Toulmé and C. Hélène, "Antimessenger Oligodeoxyribo-nucleotides: An Alternative to Antisense RNA for Artificial Regulation of Gene Expression—A Review," *Gene* 72 (1988): 51–58; C. A. Stein and Y. C. Cheng, "Antisense Oligonucleotides as Therapeutic Agents—Is the Bullet Really Magical?" *Science* 261 (1993): 1004–1012.

16. Kelly Kruger, Paula J. Grabowski, Arthur J. Zaug, Julie Sands, Daniel E. Gottschling, and Thomas R. Cech, "Self-Splicing RNA: Autoexcision and

Autocyclization of the Ribosomal RNA Intervening Sequence of Tetrahy-mena," *Cell* 31 (1982): 147–157.

17. Cecilia Guerrier-Takeda, Kathleen Gardiner, Terry Marsh, Norman Pace, and Sidney Altman, "The RNA Moiety of Ribonuclease P Is the Catalytic Subunit of the Enzyme," *Cell* 35 (1983): 849–857.

18. C. M. Visser, "Evolution of Biocatalysts 1. Possible Pre-Genetic-Code RNA Catalysts Which Are Their Own Replicase," *Origins of Life* 14 (1984): 391–400; Norman R. Pace and Terry L. Marsh, "RNA Catalysis and the Origin of Life," *Origins of Life* 16 (1985): 97–116.

19. Wally Gilbert, "The RNA World," *Nature* 319 (1986): 618; Thomas R. Cech, "A Model for the RNA-Catalyzed Replication of RNA," *Proceedings of the National Academy of Sciences of the United States of America* 83 (1986): 4360–4363.

20. Huey-Nan Wu, Yu-June Lin, Fu-Pang Lin, Shinji Makino, Ming-Fu Chang, and Michael M. C. Lai, "Human Hepatitis δ Virus RNA Subfragments Contain an Autocleavage Activity," *Proceedings of the National Academy of Sciences of the United States of America* 86 (1989): 1831–1835; Robert H. Symons, "Small Catalytic RNAs," *Annual Review of Biochemistry* 61 (1992): 641–671; Anna Marie Pyle, "Ribozymes: A Distinct Class of Metalloenzymes," *Science* 261 (1993): 709–714; Elizabeth A. Doherty and Jennifer A. Doudna, "Ribozyme Structures and Mechanisms," *Annual Review of Biochemistry* 69 (2000): 597–615.

21. Craig Tuerk and Larry Gold, "Systematic Evolution of Ligands by Exponential Enrichment: RNA Ligands to Bacteriophage T4 DNA Polymerase," *Science* 249 (1990): 505–510; Andrew D. Ellington and Jack W. Szostak, "In Vitro Selection of RNA Molecules That Bind Specific Ligands," *Nature* 346 (1990): 818–822.

22. For a brief history of RNA interference, see Michel Morange, "Transfers from Plant Biology: From Cross Protection to RNA Interference and DNA Vaccination," *Journal of Bioscience* 37 (2012): 949–952. There are very few histories of the discovery of small regulatory RNAs so far. I recommend (although it is written in French): Frédérique Théry, *La Face cachée des cellules: quand le monde des ARN bouscule la biologie* (Paris: Editions Matériologiques, 2016).

23. L. Broadbent, "Epidemiology and Control of Tobacco Mosaic Virus," *Annual Review of Physiopathology* 14 (1976): 75–96.

24. R. W. Fulton, "Practices and Precautions in the use of Cross Protection for Plant Virus Disease Control," *Annual Review of Physiopathology* 24 (1986): 67–81.

25. Luis Sequeira, "Cross Protection and Induced Resistance: Their Potential for Plant Disease Control," *Trends in Biotechnology* 2 (1984): 25–29.

26. Patricia Powell Abel, Richard S. Nelson, Barun De, Nancy Hoffmann, Stephen G. Rogers, et al., "Delay of Disease Development in Transgenic Plants That Express the Tobacco Mosaic Virus Coat Protein Gene," *Science* 232 (1986): 738–743; M. W. Bevan, S. E. Mason, and P. Godelet, "Expression of

Tobacco Mosaic Virus Coat Protein by a Cauliflower Mosaic Virus Promoter in Plants Transformed by *Agrobacterium*," *EMBO Journal* 4 (1985): 1921–1926.

27. R. N. Beachy, S. Loesch-Fries, and N. E. Turner, "Coat Protein-Mediated Resistance against Virus Infection," *Annual Review of Physiopathology* 28 (1990): 351–472.

28. J. C. Sanford and S. A. Johnston, "The Concept of Parasite-Derived Resistance— Deriving Resistance Genes from the Parasite's Own Genome," *Journal of Theoretical Biology* 113 (1985): 395–405; Rebecca Grumet, John C. Sanford, and Stephen A. Johnston, "Pathogen-Derived Resistance to Viral Infection Using a Negative Regulatory Molecule," *Virology* 161 (1987): 561–569.

29. D. Baulcombe, "Strategies for Virus Resistance in Plants," *Trends in Genetics* 5 (1989): 56–60.

30. Marian Longstaff, Gianinna Brigneti, Frédéric Boccard, Sean Chapman, and David Baulcombe, "Extreme Resistance to Potato Virus X Infection in Plants Expressing a Modified Component of the Putative Viral Replicase," *EMBO Journal* 12 (1993): 379–386.

31. Joseph R. Ecker and Ronald W. Davis, "Inhibition of Gene Expression in Plant Cells by Expression of Antisense RNA," *Proceedings of the National Academy of Sciences of the United States of America* 83 (1986): 5372–5376.

32. C. L. Niblett, Elizabeth Dickson, K. H. Fernow, R. K. Horst, and M. Zaitlin, "Cross Protection Among Four Viroids," *Virology* 91 (1978): 198–203.

33. John A. Lindbo and William G. Dougherty, "Untranslatable Transcripts of the Tobacco Etch Virus Coat Protein Gene Sequence Can Interfere with Tobacco Etch Virus Replication in Transgenic Plants and Protoplasts," *Virology* 189 (1992): 725–733.

34. M. A. Matzke, M. Primig, J. Trnovsky, and A. J. M. Matzke, "Reversible Methylation and Inactivation of Marker Genes in Sequentially Transformed Tobacco Plants," *EMBO Journal* 8 (1989): 643–649.

35. Carolyn Napoli, Christine Lemieux, and Richard Jorgensen, "Introduction of a Chimeric Chalcone Synthase Gene into Petunias Results in Reversible Co-Suppression of Homologous Genes *in trans*," *Plant Cell* 2 (1990): 279–289; Alexander R. van der Krol, L. A. Mur, Marcel Beld, Joseph N. M. Mol, and Antoine. R. Stuitje, "Flavonoid Genes in Petunia: Addition of a Limited Number of Gene Copies May Lead to a Suppression of Gene Expression," *Plant Cell* 2 (1990): 291–299.

36. C. J. S. Smith, C. F. Watson, J. Ray, C. R. Bird, P. C. Morris, et al., "Antisense RNA Inhibition of Polygalacturonase Gene Expression in Transgenic Toma- toes," *Nature* 334 (1988): 724–726.

37. C. J. Smith, C. F. Watson, C. R. Bird, J. Ray, W. Schuch, and D. Grierson, "Expression of a Truncated Tomato Polygalacturonase Gene Inhibits Expres- sion of the Endogenous Gene in Transgenic Plants," *Molecular and General Genetics* 224 (1990): 477–481.

38. Michael Wassenegger, Sabine Heimes, Leonhard Riedel, and Heinz L. Sanger, "RNA-Directed De Novo Methylation of Genomic Sequences in Plants," *Cell* 76 (1994): 567–576.

39. Simon N. Covey, Nadia S. Al-Kaff, Amagoia Langara, and David S. Turner, "Plants Combat Infection by Gene Silencing," *Nature* 385 (1997): 781–782; F. Ratcliff, B. D. Harrison, and D. C. Baulcombe, "A Similarity between Viral Defense and Gene Silencing in Plants," *Science* 276 (1997): 1558–1560.

40. Andrew Fire, SiQun Xu, Mary K. Montgomery, Steven A. Kostas, Samuel E. Driver, and Craig C. Mello, "Potent and Specific Genetic Interference by Double-Stranded RNA in *Caenorhabditis elegans,*" *Nature* 391 (1998): 806–811.

41. Phillip D. Zamore, Thomas Tuschl, Phillip A. Sharp, and David P. Bartels, "RNAi: Double-Stranded RNA Directs the ATP-Dependent Cleavage of mRNA at 21 to 23 Nucleotide Intervals," *Cell* 101 (2000): 25–33; Sayda M. Elbashir, Winifried Lendeckel, and Thomas Tuschl, "RNA Interference Is Mediated by 21- and 22-Nucleotide RNAs," *Genes and Development* 15 (2001): 188–200.

42. Amy E. Pasquinelli, Brenda J. Reinhart, Frank Slack, Mark Q. Martindale, Mitzi I. Kuroda, et al., "Conservation of the Sequence and Temporal Expression of *let-7* Heterochronic Regulatory RNA," *Nature* 408 (2000): 86–89.

43. Mariana Lagos-Quintana, Reinhard Rauhut, Wilfried Lendeckel, and Thomas Tuschl, "Identification of Novel Genes Coding for Small Expressed RNAs," *Science* 294 (2001): 853–858; Nelson C. Lau, Lee P. Lim, Earl G. Weinstein, and David P. Bartel, "An Abundant Class of Tiny RNAs with Probable Regulatory Roles in *Caenorhabditis elegans,*" *Science* 294 (2001): 858–862; Rosalind C. Lee and Victor Ambros, "An Extensive Class of Small RNAs in *Caenorhabditis elegans,*" *Science* 294 (2001): 862–864.

44. Maureen O'Malley, Kevin C. Elliott, and Richard M. Burian, "From Genetic to Genomic Regulation: Iterativity in MicroRNA Research," *Studies in History and Philosophy of Science Part C: Studies in History and Philosophy of Biological and Biomedical Sciences* 41 (2010): 407–417.

45. Anamaria Necsulea, Magali Soumillon, Maria Warnefors, Angelica Liechti, Tasman Daish, et al., "The Evolution of lncRNA Repertoires and Expression Patterns in Tetrapods," *Nature* 505 (2014): 635–640.

46. James R. Prudent, Tetsuo Uno, and Peter G. Schultz, "Expanding the Scope of RNA Catalysis," *Science* 264 (1994): 1924–1927.

47. David S. Wilson and Jack W. Szostak, "In Vitro Selection of Functional Nucleic Acids," *Annual Review of Biochemistry* 68 (1999): 611–647.

48. Nehad Ban, Poul Nissen, Jeffrey Hansen, Malcolm Capel, Peter B. Moore, and Thomas A. Steitz, "Placement of Protein and RNA Structures into a 5 A-Resolution Map of the 50S Ribosomal Subunit," *Nature* 400 (1999): 841–847.

49. Poul Nissen, Jeffrey Hansen, Nehad Ban, Peter B. Moore, and Thomas A. Steitz, "The Structural Basis of Ribosome Activity in Peptide Bond Synthesis," *Science* 289 (2000): 920–930.

50. Harry F. Noller, "Ribosomal RNA and Translation," *Annual Review of Biochemistry* 60 (1991): 191–227.
51. Gerald Joyce, "The Antiquity of RNA-Based Evolution," *Nature* 418 (2002): 214–221.
52. E. Dolgin, "A Cellular Puzzle: The Weird and Wonderful Architecture of RNA," *Nature* 523 (2015): 398–399.
53. Jonathan Knight, "Switched on to RNA," *Nature* 425 (2003): 232–233.
54. Ali Nahvi, Narasimhan Sudarsan, Margaret S. Ebert, Xiang Zou, Kenneth L. Brown, and Ronald R. Breaker, "Genetic Control by a Metabolite Binding mRNA," *Chemistry and Biology* 9 (2002): 1043–1049.
55. Wade Winkler, Ali Nahvi, and Ronald R. Breaker, "Thiamine Derivatives Bind Messenger RNAs Directly to Regulate Bacterial Gene Expression," *Nature* 419 (2002): 952–956.
56. Alexander Serganov and Evgeny Nudler, "A Decade of Riboswitches," *Cell* 152 (2013): 17–24.

26. Epigenetics

1. Matthew Cobb, *Generation: The Seventeenth-Century Scientists Who Unraveled the Secrets of Sex, Life, and Growth* (New York: Bloomsbury, 2006).
2. Garland E. Allen, *Thomas Hunt Morgan: The Man and His Science* (Princeton, NJ: Princeton University Press, 1979).
3. Michel Morange, "The Attempt of Nikolaï Koltzoff (Koltsov) to Link Genetics, Embryology and Physical Chemistry," *Journal of Bioscience* 36 (2011): 211–214.
4. Max Delbrück, *Unités biologiques douées de continuité génétique,* Colloques Internationaux, vol. 8 (Paris: CNRS, 1949), 33–34.
5. Janine Beisson, and T. M. Sonneborn, "Cytoplasmic Inheritance of the Organization of the Cell Cortex in *Paramecium aurelia,*" *Proceedings of the National Academy of Sciences of the United States of America* 53 (1965): 275–282.
6. Heather L. True and Susan L. Lindquist, "A Yeast Prion Provides a Mechanism for Genetic Variation and Phenotypic Diversity," *Nature* 407 (2000): 477–483.
7. C. H. Waddington, "The Epigenotype," *Endeavour* 1 (1942): 18–20.
8. C. H. Waddington, "Canalization of Development and the Inheritance of Acquired Characters," *Nature* 150 (1942): 563–565.
9. C. H. Waddington, "Gene Regulation in Higher Cells," *Science* 166 (1969): 639–640.
10. D. L. Nanney, "Epigenetic Control Systems," *Proceedings of the National Academy of Sciences of the United States of America* 44 (1958): 712–717.
11. Edgar Stedman and Ellen Stedman, "Cell Specificity of Histones," *Nature* 166 (1950): 780–781.
12. Ru-Chih C. Huang and James Bonner, "Histone, a Suppressor of Chromosomal RNA Synthesis," *Proceedings of the National Academy of Sciences of the United States of America* 48 (1962): 1216–1222.

13. V. G. Allfrey, R. Faulkner, and A. E. Mirsky, "Acetylation and Methylation of Histones and Their Possible Role in the Regulation of RNA Synthesis," *Proceedings of the National Academy of Sciences of the United States of America* 51 (1964): 786–794.

14. Vincent G. Allfrey and Alfred E. Mirsky, "Structural Modifications of Histones and Their Possible Role in the Regulation of RNA synthesis," *Science* 144 (1964): 559.

15. François Jacob and Jacques Monod, "Genetic Regulatory Mechanisms in the Synthesis of Proteins," *Journal of Molecular Biology* 3 (1061): 318–356.

16. Jacques Monod and François Jacob, "General Conclusions: Teleonomic Mechanisms in Cellular Metabolism, Growth and Differentiation," *Cold Spring Harbor Symposia on Quantitative Biology* 26 (1961): 389–401.

17. Eric H. Davidson, *Gene Activity in Early Development* (New York: Academic Press, 1968).

18. Roy J. Britten and Eric H. Davidson, "Gene Regulation for Higher Cells: A Theory," *Science* 165 (1969): 349–357.

19. Ada L. Olins and Donald E. Olins, "Spheroid Chromatin Units (*v* Bodies)," *Science* 183 (1974): 330–332.

20. Roger D. Kornberg, "Chromatin Structure: A Repeating Unit of Histones and DNA," *Science* 184 (1974): 868–871; T. J. Richmond, J. T. Finch, B. Rushton, D. Rhodes, and A. Klug, "Structure of the Nucleosome Core Particle at 7 Å Resolution," *Nature* 311 (1984): 532–537.

21. Werner Arber, and Stuart Linn, "DNA Modification and Restriction," *Annual Review of Biochemistry* 38 (1969): 467–500.

22. P. R. Srinivasan and Ernest Borek, "Enzymatic Alteration of Nucleic Acid Structure," *Science* 145 (1964): 548–553.

23. E. Scarano, M. Iaccarino, P. Grippo, and E. Parisi, "The Heterogeneity of Thymine Methyl Group Origin in DNA Pyrimidine Isostichs of Developing Sea Urchin Embryos," *Proceedings of the National Academy of Sciences of the United States of America* 57 (1967): 1394–1400.

24. R. Holliday and J. E. Pugh, "DNA Modification Mechanisms and Gene Activity during Development," *Science* 187 (1975): 228–232; A. D. Riggs, "X Inactivation, Differentiation, and DNA Methylation," *Cytogenetics and Cell Genetics* 14 (1975): 9–25.

25. Mary F. Lyon, "Possible Mechanisms of X Chromosome Inactivation," *Nature New Biology* 232 (1971): 229–232.

26. Aharon Razin and Howard Cedar, "Distribution of 5-Methylcytosine in Chromatin," *Proceedings of the National Academy of Sciences of the United States of America* 74 (1977): 2725–2728; Reuven Stein, Yosef Gruenbaum, Yaakov Pollack, Aharon Razin, and Howard Cedar, "Clonal Inheritance of the Pattern of DNA Methylation in Mouse Cells," *Proceedings of the National Academy of Sciences of the United States of America* 79 (1982): 61–65.

27. Gary Felsenfeld and James McGhee, "Methylation and Gene Control," *Nature* 296 (1982): 602–603.

28. Michael Behe and Gary Felsenfeld, "Effects of Methylation on a Synthetic Polynucleotide: The B-Z Transition in Poly(dG-m⁵dC).poly(dG-m⁵dC)," *Proceedings of the National Academy of Sciences of the United States of America* 78 (1981): 1619–1623.

29. Robin Holliday, "The Inheritance of Epigenetic Defects," *Science* 238 (1987): 163–170.

30. Lijing Jiang, "Causes of Aging Are Likely to Be Many: Robin Holliday and Changing Molecular Approaches to Cell Aging, 1963–1988," *Journal of the History of Biology* 47 (2014): 547–584.

31. See, for instance, Jack Taunton, Christian A. Hassig, and Stuart L. Schreiber, "Mammalian Histone Deacetylase Related to the Yeast Transcriptional Regulator Rpd3p," *Science* 272 (1996): 408–411.

32. En Li, Timothy H. Bestor, and Rudolf Jaenisch, "Targeted Mutation of the DNA Methyltransferase Gene Results in Embryonic Lethality," *Cell* 69 (1992): 915–926.

33. Alan P. Wolffe and Dmitry Pruss, "Deviant Nucleosomes: The Functional Specialization of Chromatin," *Trends in Genetics* 12 (1996): 58–62.

34. Craig L. Peterson and Ira Herskowitz, "Characterization of the Yeast *SWI1*, *SWI2*, and *SWI3* Genes, Which Encode a Global Activator of Transcription," *Cell* 68 (1992): 573–583; Jacques Côté, Janet Quinn, Jerry L. Workman, and Craig L. Peterson, "Stimulation of GAL4 Derivative Binding to Nucleosomal DNA by the Yeast SWI / SNF Complex," *Science* 265 (1994): 53–60.

35. See, for instance, Andrew J. Bannister, Phillip Zeggerman, Janet F. Partridge, Eric A. Miska, Jean O. Thomas, et al., "Selective Recognition of Methylated Lysine 9 on Histone H3 by the HP1 Chromo Domain," *Nature* 410 (2001): 120–124.

36. As demonstrated by the publication of a special issue in *Science*: Guy Riddihough and Elizabeth Pennisi, eds., "Epigenetics," *Science* 293 (2001): 1063–1102.

37. Shivl S. Grewal and Sarah C. R. Elgin, "Transcription and RNAi in the Formation of Heterochromatin," *Nature* 447 (2007): 399–406.

38. Matthew D. Anway, Andrea S. Cupp, Mehmet Uzumcu, and Michael K. Skinner, "Epigenetic Transgenerational Actions of Endocrine Disruptors and Male Fertility," *Science* 308 (2005): 1466–1469.

39. Robert A. Martienssen and Vincent Colot, "DNA Methylation and Epigenetic Inheritance in Plants and Filamentous Fungi," *Science* 293 (2001): 1070–1074.

40. D. Haig, "The (Dual) Origin of Epigenetics," *Cold Spring Harbor Symposia on Quantitative Biology* 49 (2004): 67–70.

41. Bryan M. Turner, "Histone Acetylation and an Epigenetic Code," *BioEssays* 22 (2000): 836–845; Thomas Jenuwein and C. David Allis, "Translating the Histone Code," *Science* 293 (2001): 1074–1080.

42. Vivien Marx, "Reading the Second Genomic Code," *Nature* 491 (2012): 143–147.

43. Hongshan Guo, Ping Zhu, LiYing Yan, Rong Li, Boqiang Hu, et al., "The DNA Methylation Landscape of Human Early Embryos," *Nature* 511 (2014): 606–610; Zachary D. Smith, Michelle M. Chan, Kathryn C. Humm, Rahul Karnik, Shila Mekhoubad, et al., "DNA Methylation Dynamics of the Human Preimplantation Embryo," *Nature* 511 (2014): 611–615.

27. Sequencing the Human Genome

1. Joel Davis, *Mapping the Code: The Human Genome Project and the Choices of Modern Science* (New York: John Wiley & Sons, 1990); Robert Shapiro, *The Human Blueprint: The Race to Unlock the Secrets of our Genetic Script* (New York: St. Martin's Press, 1991); Robert Cook-Deegan, *The Gene Wars: Science, Politics, and the Human Genome* (New York: W. W. Norton, 1994); Kevin Davies, *Cracking the Genome: Inside the Race to Unlock Human DNA* (Baltimore: Johns Hopkins University Press, 2001); John Sulston and Georgina Ferry, *Common Thread: A Story of Science, Politics, Ethics, and the Human Genome* (Washington, DC: Joseph Henry Press, 2002); Adam Bostanci, "Sequencing Human Genomes," in *From Molecular Genetics to Genomics: The Mapping Cultures of Twentieth-Century Genetics*, ed. Jean-Paul Gaudillière and Hans-Jörg Rheinberger, 158–179 (New York: Routledge, 2004); Craig Venter, *Life Decoded: My Genome: My Life* (London: Viking, 2007); Victor K. McElheny, *Drawing the Map of Life: Inside the Human Genome Project* (New York: Basic Books, 2010).

2. Renato Dulbecco, "A Turning Point in Cancer Research: Sequencing the Human Genome," *Science* 231 (1986): 1055–1056.

3. Lloyd M. Smith, Jane Z. Sanders, Robert J. Kaiser, Peter Hughes, Chris Dodd, et al., "Fluorescence Detection in Automated DNA Sequence Analysis," *Nature* 321 (1986): 674–679.

4. David Botstein, Raymond L. White, Mark Skolnick, and Ronald W. Davis, "Construction of a Genetic Linkage Map in Man Using Restriction Fragment Length Polymorphisms," *American Journal of Human Genetics* 32 (1980): 314–331.

5. Victor A. McKusick and Frank H. Ruddle, "A New Discipline, a New Name, a New Journal," *Genomics* 1 (1987): 1–2; Alexander Powell, Maureen O'Malley, Staffan Müller-Wille, Jane Calvert, and John Dupré, "Disciplinary Baptisms: A Comparison of the Naming Stories of Genetics, Molecular Biology, Genomics, and Systems Biology," *Historical Studies in the Physical and Biological Sciences* 29 (2007): 5–32.

6. T. D. Yager, D. A. Nickerson, and L. E. Hood, "The Human Genome Project: Creating an Infrastructure for Biology and Medicine," *Trends in Biochemical Sciences* 16 (1991): 454–458.

7. Jean Dausset, Howard Cann, Daniel Cohen, Mark Lathrop, Jean-Marc Lalouel, and Ray White, "Centre d'Etude du Polymorpisme Humain (CEPH): Collaborative Genetic Mapping of the Human Genome," *Genomics* 6 (1990): 575–577.

8. Maynard Olson, Leroy Hood, Charles Cantor, and David Botstein, "A Common Language for Physical Mapping of the Human Genome," *Science* 245 (1989): 1434–1435.

9. Mark D. Adams, Jenny M. Kelley, Jeannine D. Gocayne, Mark Dubnick, Michael H. Polymeropoulos, et al., "Complementary DNA Sequencing: Expressed Sequence Tags and Human Genome Project," *Science* 252 (1991): 1651–1656.

10. G. D. Schuler, M. S. Boguski, E. A. Stewart, L. D. Stein, G. Gyapay, et al., "A Gene Map of the Human Genome," *Science* 274 (1996): 540–546; P. Deloukas, G. D. Schuler, G. Gyapay, E. M. Beasley, C. Siderlund, et al., "A Physical Map of 30,000 Human Genes," *Science* 282 (1998): 744–746; David R. Bentley, Kim D. Prutti, Panigiotis Deloukas, Greg D. Schuler, and Jim Ostell, "Coordination of Human Genome Sequencing via a Consensus Framework Map," *Trends in Genetics* 14 (1998): 381–384.

11. Bernard D. Davis and Colleagues, "The Human Genome and Other Initiatives," *Science* 249 (1990): 342–343; Yager et al., "Human Genome Project"; Martin C. Rechsteiner, "The Human Genome Project: Misguided Science Policy," *Trends in Biochemical Science* 16 (1991): 455–460.

12. Peter Coles, "A Different Approach," *Nature* 347 (1990): 701.

13. Paul R. Billings, "Promotion of the Human Genome Project," *Science* 250 (1990): 1071.

14. Louis M. Kunkel, Anthony P. Monaco, William Middlesworth, Hans D. Ochs, and Samuel A. Latt, "Specific Cloning of DNA Fragments Absent from the DNA of a Male Patient with an X Chromosome Deletion," *Proceedings of the National Academy of Sciences of the United States of America* 82 (1985): 4778–4782; Anthony P. Monaco, Rachael L. Neve, Chris Colletti-Feener, Corlee J. Bertelson, David M. Kurnit, and Louis M. Kunkel, "Isolation of Candidate cDNAs for Portions of the Duchenne Muscular Dystrophy Gene," *Nature* 323 (1986): 646–650.

15. M. Koenig, E. P. Hoffman, C. J. Bertelson, A. P. Monaco, C. Feener, and L. M. Kunkel, "Complete Cloning of the Duchenne Muscular Dystrophy (DMD) cDNA and Preliminary Genomic Organization of the DMD Gene in Normal and Affected Individuals," *Cell* 50 (1987): 509–517; E. P. Hoffman, Robert H. Brown Jr., and Louis M. Kunkel, "Dystrophin: The Protein Product of the Duchenne Muscular Dystrophy Locus," *Cell* 51 (1987): 919–928; Elizabeth E. Zubrzycka-Gaarn, Dennis E. Bulman, George Karpatin, Arthur H. M. Burghes, Bonnie Belfall, et al., "The Duchenne Muscular Dystrophy Gene Product Is Localized in Sarcolemma of Human Skeletal Muscle," *Nature* 333 (1988): 466–469.

16. Robert G. Knowlton, Odile Cohen-Haguenauer, Nguyen Van Cong, Jean Frézal, Valerie A. Brown, et al., "A Polymorphic DNA Marker Linked to Cystic

Fibrosis Is Located on Chromosome 7," *Nature* 318 (1985): 380–382; Ray White, Scott Woodward, Mark Leppert, Peter O'Connell, Mark Holl, et al., "A Closely Linked Marker to Cystic Fibrosis," *Nature* 318 (1985): 382–384; Brandon J. Wainwright, Peter J. Scambler, Jorg Schmidtke, Eila A. Watson, Hai-Yang Law, et al., "Localization of Cystic Fibrosis Locus to Human Chromosome 7cen-q22," *Nature* 318 (1985): 384–385; Lap-Chee Tsui, Manuel Buchwald, David Barker, Jeffrey C. Braman, Robert Knowlton, et al., "Cystic Fibrosis Locus Defined by a Genetically Linked Polymorphic DNA Marker," *Science* 230 (1985): 1054–1057.

17. Leslie Roberts, "The Race for the Cystic Fibrosis Gene," *Science* 240 (1988): 141–144; Leslie Roberts, "Race for Cystic Fibrosis Gene Near End," *Science* 240 (1988): 282–285.

18. Jean L. Marx, "The Cystic Fibrosis Gene Is Found," *Science* 245 (1989): 923–925; Johanna M. Rommens, Michael C. Lannuzzi, Bat-Sheva Kerem, Mitchell M. Drumm, Georg Melmer, et al., "Identification of the Cystic Fibrosis Gene: Chromosome Walking and Jumping," *Science* 245 (1989): 1059–1065; John R. Riordan, J. M. Rommens, Bat-Sheva Kerem, Noa Alon, Richard Rozmahel, et al., "Identification of the Cystic Fibrosis Gene: Cloning and Characterization of Complementary DNA," *Science* 245 (1989): 1066–1073; Bat-Sheva Kerem, Johanna M. Rommens, Janet A. Buchanan, Danuta Markiewicz, Tara K. Cox, et al., "Identification of the Cystic Fibrosis Gene: Genetic Analysis," *Science* 245 (1989): 1073–1080.

19. Yoshio Miki, Jeff Swensen, Donna Shatuck-Eldens, P. Andrew Futreal, Keith Harshman, et al., "A Strong Candidate for the Breast and Ovarian Cancer Susceptibility Gene *BRCA1*," *Science* 266 (1994): 66–71; Mary-Claire King, "The Race to Clone *BRCA1*," *Science* 343 (2014): 1462–1465.

20. Huntington's Disease Collaborative Research Group, "A Novel Gene Containing a Trinucleotide Repeat That Is Expanded and Unstable on Huntington's Disease Chromosomes," *Cell* 72 (1993): 971–983.

21. June L. Davies, Yoshihiko Kawaguchi, Simon T. Bennett, James B. Copeman, Heather J. Cordell, et al., "A Genome-Wide Search for Human Type 1 Diabetes Susceptibility Genes," *Nature* 371 (1994): 130–136; Linda R. Brzustowicz, Kathleen A. Hodgkinson, Eva W. C. Chow, William G. Honer, and Anne S. Bassett, "Location of a Major Susceptibility Locus for Familial Schizophrenia on Chromosome 1q21-q22," *Science* 288 (2000): 678–682.

22. Dean H. Hamer, Stella Hu, Victoria L. Magnuson, Nan Hu and Angela M. L. Pattatucci, "Linkage between DNA Markers on the X Chromosome and Male Sexual Orientation," *Science* 261 (1993): 321–327.

23. Simon LeVay, "A Difference in Hypothalamic Structure between Heterosexual and Homosexual Men," *Science* 253 (1991): 1034–1037.

24. George Rice, Carol Anderson, Neil Risch, and George Ebers, "Male Homosexuality: Absence of Linkage to Microsatellite Markers at Xq28," *Science* 284 (1999): 665–667.

25. Jonathan Weiner, *Time, Love, Memory: A Great Biologist and His Quest for the Origins of Behavior* (New York: Alfred A. Knopf, 1999).

26. Jonathan Flint, Robin Corley, John C. DeFries, David W. Fulker, Jeffrey A. Gray, et al., "A Simple Genetic Basis for a Complex Psychological Trait in Laboratory Mice," *Science* 269 (1995): 1432–1435.

27. John C. Crabbe, Douglas Wahlsten, and Bruce C. Dudek, "Genetics of Mouse Behavior: Interactions with Laboratory environment," *Science* 284 (1999): 1670–1672; Douglas Wahlsten, Alexander Bachmanov, Deborah A. Finn, and John C. Crabbe, "Stability of Inbred Mouse Strain Behavior and Brain Size between Laboratories and across Decades," *Proceedings of the National Academy of Sciences of the United States of America* 103 (2006): 16364–16369.

28. Lon R. Cardon, Shelley D. Smith, David W. Fulker, William J. Kimberling, Bruce F. Pennington, and John C. DeFries, "Quantitative Trait Locus for Reading Disability on Chromosome 6," *Science* 266 (1994): 276–279.

29. Alessio Vassarotti and André Goffeau, "Sequencing the Yeast Genome: The European Effort," *Trends in Biotechnology* 10 (1992): 15–18.

30. S. G. Oliver, Q. J. M. van der Aart, M. L. Agostoni-Carbone, M. Aigle, L. Alberghina, et al., "The Complete DNA Sequence of Yeast Chromosome III," *Nature* 359 (1992): 38–46; B. Dujon, D. Alexandraki, B. André, W. Ansorge, V. Baladron, et al., "Complete DNA Sequence of Yeast Chromosome XI," *Nature* 369 (1994): 371–378; A. Goffeau, B. G. Barrell, H. Bussey, R. W. Davis, B. Dujon, et al., "Life with 6000 Genes," *Science* 274 (1996): 546–567.

31. André Goffeau, "Genomic-Scale Analysis Goes Upstream," *Nature Biotechnology* 16 (1998): 907–908.

32. R. Wilson, R. Ainscough, K. Anderson, C. Baynes, M. Berks, et al., "2.2 Mb of Contiguous Nucleotide Sequence from Chromosome III of *C. elegans*," *Nature* 368 (1994): 32–38.

33. *C. elegans* Sequencing Consortium, "Genome Sequence of the Nematode *C. elegans*: A Platform for Investigating Biology," *Science* 282 (1998): 2012–2046.

34. Robert D. Fleischmann, Mark D. Adams, Owen White, Rebecca A. Clayton, Ewen F. Kirkness, et al., "Whole-Genome Random Sequencing and Assembly of *Haemophilus influenzae* Rd," *Science* 269 (1995): 496–512; Claire M. Fraser, Jeannine D. Gocayne, Owen White, Mark D. Adams, Rebecca A. Clayton, et al., "The Minimal Gene Complement of *Mycoplasma genitalium*," *Science* 270 (1995): 397–403.

35. Stephen Anderson, "Shotgun DNA Sequencing Using Cloned DNase I-Generated Fragments," *Nucleic Acid Research* 9 (1981): 3015–3027.

36. Through sequencing, and in particular but not uniquely through the shotgun strategy, computers have played an increasingly important role in biological research: Hallam Stevens, *Life Out of Sequence: A Data-Driven History of Bioinformatics* (Chicago: University of Chicago Press, 2013).

37. James L. Weber and Eugene W. Myers, "Human Whole-Genome Shotgun Sequencing," *Genome Research* 7 (1997): 401–409; Philip Green, "Against a Whole-Genome Shotgun," *Genome Research* 7 (1997): 410–417.

38. F. Kunst, N. Ogasawara, I. Moszer, A. M. Albertini, G. Alloni, et al., "The Complete Genome Sequence of the Gram-Positive Bacterium *Bacillus subtilis*," *Nature* 390 (1997): 249–256.

39. Donna L. Daniels, Guy Plunkett III, Valerie Burland, and Frederick R. Blattner, "Analysis of the *Escherichia coli* Genome: DNA Sequence of the Region from 84.5 to 86.5 Minutes," *Science* 257 (1992): 771–778; Frederick R. Blattner, Guy Plunkett III, Craig A. Bloch, Nicole T. Perna, Valerie Burland, et al., "The Complete Genome Sequence of *Escherichia coli* K12," *Science* 277 (1997): 1453–1462.

40. Mark D. Adams, Susan E. Celniker, Robert A. Holt, Cheryl A. Evans, Jeannine D. Gocayne, et al., "The Genome Sequence of *Drosophila melanogaster*," *Science* 287 (2000): 2185–2195; Michael Ashburner, *Won for All: How the* Drosophila *Genome Was Sequenced* (Cold Spring Harbor, NY: Cold Spring Harbor Laboratory Press, 2006).

41. Laurie Goodman, "Random Shotgun Fire," *Genome Research* 8 (1998): 567–568.

42. Robert H. Waterston, Eric S. Lander, and John E. Sulston, "On the Sequencing of the Human Genome," *Proceedings of the National Academy of Sciences of the United States of America* 99 (2002): 3712–3716; Phil Green, "Whole Genome Disassembly," *Proceedings of the National Academy of Sciences of the United States of America* 99 (2002): 4143–4144; Eugene W. Myers, Granger G. Sutton, Hamilton O. Smith, Mark D. Adams, and J. Craig Venter, "On the Sequencing and Assembly of the Human Genome," *Proceedings of the National Academy of Sciences of the United States of America* 99 (2002): 4145–4146.

43. *Nature*, February 15, 2001; *Science*, February 16, 2001.

44. Teresa K. Atwood, "The Babel of Bioinformatics," *Science* 290 (2000): 471–473.

45. Brent Ewing and Phil Green, "Analysis of Expressed Sequence Tags Indicates 35,000 Human Genes," *Nature Genetics* 25 (2000): 232–234; Hughes Roest Crollius, Olivier Jaillon, Alain Bernot, Corinne Dasilva, Laurence Bouneau, et al., "Estimate of Human Gene Number Provided by Genome-Wide Analysis Using *Tetraodon nigroviridis* DNA Sequence," *Nature Genetics* 25 (2000): 235–238.

46. M. S. Clark, "Comparative Genomics: The Key to Understanding the Human Genome Project," *BioEssays* 21 (1999): 121–130.

47. Edward M. Marcotte, Matteo Pellegrini, Ho-Leung Ng, Danny W. Rice, Todd O. Yeates, and David Eisenberg, "Detecting Protein Functions and Protein-Protein Interactions from Genome Sequences," *Science* 285 (1999): 751–753.

48. Matteo Pellegrini, Edward M. Marcotte, Michael J. Thompson, David Eisenberg, and Todd O. Yeates, "Assigning Protein Functions by Comparative

Genome Analysis: Protein Phylogenetic Profiles," *Proceedings of the National Academy of Sciences of the United States of America* 96 (1999): 4285–4288.

49. *Yeast:* Manolis Kellis, Bruce W. Birren, and Eric M. Lander, "Proof and Evolutionary Analysis of Ancient Genome Duplication in the Yeast *Saccharomyces cerevisiae,*" *Nature* 428 (2004): 617–624; Fred S. Dietrich, Sylvia Voegeli, Sophie Brachat, Anita Lerch, Krista Gates, et al., "The *Ashbya gossypii* Genome as a Tool for Mapping the Ancient *Saccharomyces cerevisiae* Genome," *Science* 304 (2004): 304–307; Bernard Dujon, David Sherman, Gilles Fischer, Pascal Durrens, Serge Casaregola, et al., "Genome Evolution in Yeasts," *Nature* 430 (2004): 35–44. *Worms:* Lincoln D. Stein, Zhirong Bao, Darin Blasiar, Thomas Blumenthal, Michael R. Brent, et al., "The Genome Sequence of *Caenorhabditis briggsae:* A Platform for Comparative Genomics," *PLoS Biology* 1 (2003): 166–192.

50. Carol J. Bult, Owen White, Gary J. Olsen, Lixin Zhou, Robert D. Fleischmann, et al., "Complete Genome Sequence of the Methanogenic Archaeon *Methanococcus jannaschii,*" *Science* 273 (1996): 1058–1073; Arabidopsis Genome Initiative, "Analysis of the Genome Sequence of the Flowering Plant *Arabidopsis thaliana,*" *Nature* 408 (2000): 796–813.

51. Paramvir Dehal, Yutaka Satou, Robert A. Campbell, Jarrod Chapman, Bernard Degnan, et al., "The Draft Genome of *Ciona intestinalis:* Insights into Chordate and Vertebrate Origins," *Science* 298 (2002): 2157–2167.

52. Rowland H. Davis, *The Microbial Models of Molecular Biology: From Genes to Genomes* (Oxford: Oxford University Press, 2003).

53. Robert E. Service, "The Race for the $1000 Genome," *Science* 311 (2006): 1544–1546.

54. William E. Evans and Mary V. Relling, "Pharmacogenomics: Translating Functional Genomics into Rational Therapeutics," *Science* 286 (1999): 487–491.

55. Robert J. Klein, Caroline Zeiss, Emily Y. Chew, Jen-Yue Tsai, Richard S. Sackler, et al., "Complement Factor H Polymorphism in Age-Related Macular Degeneration," *Science* 308 (2005): 385–489.

56. Michael R. Stratton, "Exploring the Genomes of Cancer Cells: Progress and Promise," *Science* 331 (2011): 1553–1558.

57. Peter C. Nowell, "The Clonal Evolution of Tumor Cell Populations," *Science* 194 (1976): 23–28.

58. S. J. Gamblin, L. F. Haire, R. J. Russell, D. J. Stevens, B. Xiao, et al., "The Structure and Receptor Binding Properties of the 1918 Influenza Hemagglutinin," *Science* 303 (2004): 1838–1842; Nuno R. Faria, Andrew Rambaut, Marc A. Suchard, Guy Baele, Trevor Bedford, et al., "The Early Spread and Epidemic Ignition of HIV-1 in Human Populations," *Science* 346 (2014): 56–61.

59. J. Craig Venter, Karin Remington, John F. Heidelberg, Aaron L. Halpern, Doug Rusch, et al., "Environmental Genome Shotgun Sequencing of the Sargasso Sea," *Science* 304 (2004): 66–74.

60. K. Zaremba-Niedzwiedzka, E. F. Caceres, J. H. Saw, D. Bäckström, L. Juzokaite, et al., "Asgard Archaea Illuminate the Origin of Eukaryotic Cellular Complexity," *Nature* 541 (2017): 353–358.
61. Maureen A. O'Malley, "Exploratory Experimentation and Scientific Practice: Metagenomics and the Proteorhodopsin Case," *History and Philosophy of the Life Sciences* 29 (2007): 337–360.
62. Walter Gilbert, "Towards a Paradigm Shift in Biology," *Nature* 349 (1991): 99.
63. Peter Medawar, *The Art of the Soluble* (London: Methuen, 1967).
64. Sandra Leonelli and Rachel A. Ankeny, "Re-Thinking Organisms: The Impact of Databases on Model Organism Biology," *Studies in History and Philosophy of Science Part C: Studies in History and Philosophy of Biological and Biomedical Sciences* 43 (2012): 29–36.
65. Bruno J. Strasser, "Collecting Nature: Practices, Styles, and Narratives," *Osiris* 27 (2012): 303–320.

28. Systems Biology and Synthetic Biology

1. Leland H. Hartwell, John J. Hopfield, Stanislas Leibler, and Andrew W. Murray, "From Molecular to Modular Cell Biology," *Nature* 402 (1999): C47–C52.
2. Ludwig von Bertalanffy, *General Systems Theory* (New York: George Braziller, 1968).
3. Donald J. Lockhart, Helin Dong, Michael C. Byrne, Maximilian T. Follettie, Michael V. Gallo, et al., "Expression Monitoring by Hybridization to High-Density Oligonucleotide Arrays," *Nature Biotechnology* 14 (1996): 1675–1680.
4. Kevin P. White, Scott A. Rifkin, Patrick Hurban, and David S. Hogness, "Microarray Analysis of *Drosophila* Development during Metamorphosis," *Science* 286 (1999): 2179–2184; Stuart K. Kim, Jim Lund, Moni Kirali, Kyle Duke, Min Jiang, et al., "Gene Expression Map for *Caenorhabditis elegans,*" *Science* 293 (2001): 2087–2092.
5. John Quackenbush, "Microarrays—Guilt by Association," *Science* 302 (2003): 240–241.
6. Ash A. Alizadeh, Michael B. Elsen, R. Eric Davis, Chi Ma, Izidore S. Lossos, et al., "Distinct Types of Diffuse Large B-Cell Lymphoma Identified by Gene Expression Profiling," *Nature* 403 (2000): 503–511.
7. Peter Uetz, L. Giot, Gerard Cagney, Traci A. Mansfield, Richard S. Judson, et al., "A Comprehensive Analysis of Protein–Protein Interactions in *Saccharomyces cerevisiae,*" *Nature* 403 (2000): 623–627; Takashi Ito, Tomoko Chiba, Kitsuko Ozawa, Mikio Yoshida, Masahira Hattori, and Yoshiyuki Sakaki, "A Comprehensive Two-Hybrid Analysis to Explore the Yeast Protein Interactome," *Proceedings of the National Academy of Sciences of the United States of America* 98 (2001): 4569–4574; L. Giot, J. S. Bader, C. Brouwer, A. Chaudhuri, B. Kuang, et al., "A Protein Interaction Map of *Drosophila melanogaster,*" *Science* 302 (2003): 1727–1736.

8. Stanley Fields, and Ok-kyu Song, "A Novel Genetic System to Detect Protein-Protein Interactions," *Nature* 340 (1989): 245–246.

9. Peter O. Brown and David Botstein, "Exploring the New World of the Genome with DNA Microarrays," *Nature Genetics* 21, Suppl. 1 (1999): 33–37.

10. Réka Albert, Hawoong Jeong, and Albert-Laszlo Barabasi, "Diameter of the World-Wide Web," *Nature* 401 (1999): 130–131.

11. Richard Gallagher and Tim Appenzeller, "Beyond Reductionism," *Science* 284 (1999): 79.

12. Gerhard Schlosser and Gunter P. Wagner, *Modularity in Development and Evolution* (Chicago: University of Chicago Press, 2004).

13. R. Milo, S. Shen-Orr, S. Itzkovitz, N. Kashtan, D. Chklovskii, and U. Alon, "Network Motifs: Simple Building Blocks of Complex Networks," *Science* 298 (2002): 824–827.

14. Andreas Wagner, "Robustness against Mutations in Genetic Networks of Yeast," *Nature Genetics* 24 (2000): 355–361.

15. H. Jeong, S. P. Mason, A.-L. Barabasi, and Z. N. Oltvai, "Lethality and Centrality in Protein Networks," *Nature* 411 (2001): 41–42.

16. Daniel E. L. Promislow, "Protein Networks, Pleiotropy, and the Evolution of Senescence," *Proceedings of the Royal Society of London: Series B, Biological Sciences* 271 (2004): 1225–1234.

17. Denis Duboule and Adam S. Wilkins, "The Evolution of 'Bricolage,'" *Trends in Genetics* 14 (1998): 54–59.

18. Michael P. H. Stumpf, Carsten Wiuf, and Robert M. May, "Subnets of Scale-Free Networks Are Not Scale-Free: Sampling Properties of Networks," *Proceedings of the National Academy of Sciences of the United States of America* 102 (2005): 4221–4224; Evelyn Fox Keller, "Revisiting 'Scale-Free' Networks," *Bioessays* 27 (2005): 1060–1068.

19. Bert Vogelstein, David Lane, and Arnold J. Levine, "Surfing the p53 Network," *Nature* 408 (2000): 307–310.

20. Pierre-Alain Braillard, "Systems Biology and the Mechanistic Framework," *History and Philosophy of the Life Sciences* 32 (2010): 43–62.

21. Kunihiko Taneko, *Life: An Introduction to Complex Systems Biology* (New York: Springer, 2006).

22. Avigdor Eldar and Michael B. Elowitz, "Functional Roles for Noise in Genetic Circuits," *Nature* 467 (2010): 167–174.

23. Manuel Porcar and Juli Pereto, *Synthetic Biology: From iGEM to the Artificial Cell* (Dordrecht, the Netherlands: Springer, 2014).

24. Hartwell et al., "From Molecular to Modular Cell Biology," C47–C52.

25. Michael B. Elowitz and Stanislas Leibler, "A Synthetic Oscillatory Network of Transcriptional Regulation," *Nature* 403 (2000): 335–338; Timothy S. Gardner, Charles R. Cantor, and James J. Collins, "Construction of a Genetic Toggle Switch in *Escherichia coli*," *Nature* 403 (2000): 339–342; Anselm Levskaya, Aaron A. Chevalier, Jeffrey J. Tabor, Zachary Booth Simpson,

Laura A. Lavery, et al., "Engineering *Escherichia coli* to See Light," *Nature* 438 (2005): 441–442.

26. Dae-Kyun Ro, Eric M. Paradise, Mario Ouellet, Karl J. Fisher, Karyn L. Newman, et al., "Production of the Antimalarial Drug Precursor Artemisinic Acid in Engineered Yeast," *Nature* 440 (2006): 940–943.

27. Daniel G. Gibson, John I. Glass, Carole Lartigue, Vladimir N. Noskov, Ray-Yuan Chuang, et al., "Creation of a Bacterial Cell Controlled by a Chemically Synthesized Genome," *Science* 329 (2010): 52–56.

28. Jesse Stricker, Scott Cookson, Matthew R. Bennett, William H. Mather, Lev M. Tsimring, and Jeff Hasty, "A Fast, Robust and Tunable Synthetic Gene Oscillator," *Nature* 456 (2008): 516–519; Michael J. Dougherty and Frances H. Arnold, "Directed Evolution: New Parts and Optimized Function," *Current Opinion in Biotechnology* 20 (2009): 486–491.

29. U. Alon, "Biological Networks: The Tinkerer as an Engineer," *Science* 301 (2003): 1866–1867.

30. George von Dassow, Elt Meir, Edwin M. Munro, and Garrett M. Odell, "The Segment Polarity Network Is a Robust Developmental Module," *Nature* 406 (2000): 188–192.

31. S. Benzer, "Induced Synthesis of Enzymes in Bacteria Analyzed at the Cellular Level," *Biochimica et Biophysica Acta* 11 (1953): 383–395; Aravinthan D. T. Samuel and Howard C. Berg, "Fluctuation Analysis of Rotational Speeds of the Bacterial Flagellar Motor," *Proceedings of the National Academy of Sciences of the United States of America* 92 (1995): 3502–3506.

32. William J. Blake, Charles R. Cantor, and J. J. Collins, "Noise in Eukaryotic Gene Expression," *Nature* 422 (2003): 633–637; Jonathan M. Raser and Erin K. O'Shea, "Noise in Gene Expression: Origins, Consequences, and Control," *Science* 309 (2005): 2010–2013; Long Cai, Nir Friedman, and X. Sunney Xie, "Stochastic Protein Expression in Individual Cells at the Single Molecule Level," *Nature* 440 (2006): 358–362.

33. Harley H. McAdams and Adam Arkin, "It's a Noisy Business! Genetic Regulation at the Nanomolar Scale," *Trends in Genetics* 15 (1999): 65–69.

34. Eldar and Elowitz, "Functional Roles for Noise," 167–174.

35. For a general overview of extant postgenomics, see Sarah S. Richardson and Hallam Stevens, eds., *Post-Genomics: Perspectives on Biology after the Genome* (Durham, NC: Duke University Press, 2015).

29. Images, Representations and Metaphors in Molecular Biology

1. Soraya de Chadaverian and Nick Hopwood, eds., *Models: The Third Dimension of Science* (Stanford, CA: Stanford University Press, 2004); Norberto Serpente, "Justifying Molecular Images in Cell Biology Textbooks: From Construction to Primary Data," *Studies in History and Philosophy of Science Part C: Studies in History and Philosophy of Biological and Biomedical Sciences* 55 (2016): 105–116.

2. Linus Pauling, *The Nature of the Chemical Bond* (Ithaca, NY: Cornell University Press, 1939).

3. Thomas H. Morgan, "The Relation of Genetics to Physiology and Medicine," Nobel Lecture, June 4, 1934, www.nobelprize.org/uploads/2018/06/morgan -lecture.pdf. The importance of mapping in twentieth-century biology has been well studied in Jean-Paul Gaudillière and Hans-Jörg Rheinberger, eds., *The Mapping Cultures of Twentieth-Century Genetics* (New York: Routledge, 2004).

4. George W. Beadle and Boris Ephrussi, "Development of Eye Colors in *Drosophila:* Diffusible Substances and Their Interrelations," *Genetics* 22 (1937): 76–86.

5. James D. Watson and Francis H. C. Crick, "A Structure for Deoxyribose Nucleic Acid," *Nature* 171 (1953): 737–738.

6. John Yudkin, "Enzyme Variation in Micro-Organisms," *Biological Reviews* 13 (1938): 93–106.

7. Jacques Monod, Jeffries Wyman, and Jean-Pierre Changeux, "On the Nature of Allosteric Transitions: A Plausible Model," *Journal of Molecular Biology* 12 (1965): 88–118.

8. J. C. Kendrew, G. Bodo, M. Kintzis, R. G. Parrish, H. Wickoff, and D. C. Phillips, "A Three-Dimensional Model of the Myoglobin Molecule Obtained by X-ray Analysis," *Nature* 181 (1958): 662–666.

9. François Jacob and Jacques Monod, "Genetic Regulatory Mechanisms in the Synthesis of Proteins," *Journal of Molecular Biology* 3 (1961): 318–356.

10. Roy J. Britten and Eric H. Davidson, "Gene Regulation for Higher Cells: A Theory," *Science* 165 (1969): 349–357.

11. André Lwoff, "Lysogeny," *Bacteriology Reviews* 17 (1953): 269–337.

12. Maria Trumpler, "Converging Images: Techniques of Intervention and Forms of Representation of Sodium Channel Proteins in Nerve Cell Membranes," *Journal of the History of Biology* 30 (1997): 55–89.

13. Norberto Serpente, "Cells from Icons to Symbols: Molecularizing Cell Biology in the 1980s," *Studies in History and Philosophy of Science Part C: Studies in History and Philosophy of Biological and Biomedical Sciences* 42 (2011): 403–411.

14. J. F. Danielli and H. Davson, "A Contribution to the Theory of Permeability of Thin Films," *Journal of Cellular and Comparative Physiology* 5 (1935): 495–508.

15. S. J. Singer and G. L. Nicolson, "The Fluid Mosaic Model of the Structure of Cell Membranes," *Science* 175 (1972): 720–731.

16. M. Levitt and C. Chothia, "Structural Patterns in Globular Proteins," *Nature* 261 (1976): 552–558.

17. Jane S. Richardson, "The Anatomy and Taxonomy of Protein Structure," *Advances in Protein Chemistry* 34 (1981): 167–339; Jane S. Richardson and David Richardson, "Doing Molecular Biophysics: Finding, Naming, and Picturing Signal within Complexity," *Annual Review of Biochemistry* 42 (2013): 1–28.

18. S. Bahar, "Ribbon Diagrams and Protein Taxonomy: A Profile of Jane S. Richardson," *Newsletter of the Division of Biological Physics of the American Physical Society* 4 (2004): 5–8.

19. Carl-Ivar Branden and John Tooze, *Introduction to Protein Structure* (New York: Garland, 1991).

20. Bruce Alberts, Alexander Johnson, Julian Lewis, David Morgan, Martin Raff, et al., *Molecular Biology of the Cell* (New York: Garland, 1983); Norberto Serpente, "Beyond a Pedagogical Tool: 30 Years of *Molecular Biology of the Cell*," *Nature Reviews / Molecular Biology* 14 (2013): 120–125.

21. There is nonetheless a remarkable exception in this monotonous landscape of representations: David Goodsell's drawings showing how crowded is the interior of a cell. David S. Goodsell, *The Machinery of Life* (New York: Springer-Verlag, 1998).

22. Sarah S. Richardson and Hallam Stevens, eds., *Post-genomics: Perspectives on Biology after the Genome* (Durham, NC: Duke University Press, 2015).

23. Martin Krzywinski, Jacqueline Schein, Inanç Birol, Joseph Connors, Randy Gascoyne, et al., "Circos: An Information Aesthetic for Comparative Genomics," *Genome Research* 19 (2011): 1639–1645.

24. Ken A. Dill and Hue Sun Chan, "From Levinthal to Pathways to Funnels," *Nature Structural and Molecular Biology* 4 (1997): 10–19.

25. Evelyn Fox Keller, *Refiguring Life: Metaphors of Twentieth Century Biology* (New York: Columbia University Press, 1995); Andrew S. Reynolds, *The Third Lens: Metaphor and the Creation of Modern Cell Biology* (Chicago: University of Chicago Press, 2018). Useful jumping-off points on the question are George Lakoff and Mark Johnson, *Metaphors We Live By* (Chicago: University of Chicago Press, 1980); R. C. Paton, "Towards a Metaphorical Biology," *Biology and Philosophy* 7 (1992): 279–294; Theodore L. Brown, *Making Truth: Metaphor in Science* (Chicago: University of Illinois Press, 2003).

26. Three 2019 examples from very different areas of biology highlight the interest of young scientists in the question: Daniel J. Nicholson, "Is the Cell *Really* a Machine?" *Journal of Theoretical Biology* 477 (2019): 108–126; Mark E. Olson, Alfonso Arroyo-Santos, and Francisco Vergara-Silva, "A User's Guide to Metaphors in Ecology and Evolution," *Trends in Ecology and Evolution* 34 (2019): 605–615; Romain Brette, "Is Coding a Relevant Metaphor for the Brain?" *Behavioral and Brain Sciences* 42 (2019): e15.

27. These two examples have been the subject of detailed historical and philosophical exploration. See, for example, Keller, *Refiguring Life;* Lily E. Kay, *Who Wrote the Book of Life? A History of the Genetic Code* (Palo Alto, CA: Stanford University Press, 2000); Matthew Cobb, *Life's Greatest Secret: The Race to Crack the Genetic Code* (London: Profile Books, 2015).

28. Crick wryly pointed this out in his 1988 autobiography, commenting that "genetic code" sounded "a lot more intriguing" than "genetic cipher." Francis

Crick, *What Mad Pursuit: A Personal View of Scientific Discovery* (Cambridge, MA: Basic Books, 1988).

29. Francis H. C. Crick, "On Protein Synthesis," *Symposia of the Society for Experimental Biology* 12 (1958): 138–163.

General Conclusion

1. Nils Roll-Hansen, "The Meaning of Reductionism," in *International Conference on Philosophy of Science* (Vigo, Spain: University of Vigo, 1996), 125–148.

2. François Jacob, *The Logic of Life: A History of Heredity* (Princeton, NJ: Princeton University Press, 1973).

Index